Environmental Chemistry of Arsenic

BOOKS IN SOILS, PLANTS, AND THE ENVIRONMENT

Editorial Board

Soil Analysis: Modern Instrumental Techniques, Second Edition, edited by Keith A. Smith

Soil Analysis: Physical Methods, edited by Keith A. Smith and Chris E. Mullins

Growth and Mineral Nutrition of Field Crops, N. K. Fageria, V. C. Baligar, and Charles Allan Jones

Semiarid Lands and Deserts: Soil Resource and Reclamation, edited by J. Skujiņš

Plant Roots: The Hidden Half, edited by Yoav Waisel, Amram Eshel, and Uzi Kafkafi

Plant Biochemical Regulators, edited by Harold W. Gausman

Maximizing Crop Yields, N. K. Fageria

Transgenic Plants: Fundamentals and Applications, edited by Andrew Hiatt

Soil Microbial Ecology: Applications in Agricultural and Environmental Management, edited by F. Blaine Metting, Jr.

Principles of Soil Chemistry: Second Edition, Kim H. Tan

Water Flow in Soils, edited by Tsuyoshi Miyazaki

Handbook of Plant and Crop Stress, edited by Mohammad Pessarakli

Genetic Improvement of Field Crops, edited by Gustavo A. Slafer

Agricultural Field Experiments: Design and Analysis, Roger G. Petersen

Environmental Soil Science, Kim H. Tan

Mechanisms of Plant Growth and Improved Productivity: Modern Approaches, edited by Amarjit S. Basra

Selenium in the Environment, edited by W. T. Frankenberger, Jr., and Sally Benson

Plant–Environment Interactions, edited by Robert E. Wilkinson

Handbook of Plant and Crop Physiology, edited by Mohammad Pessarakli

Handbook of Phytoalexin Metabolism and Action, edited by M. Daniel and R. P. Purkayastha

Soil–Water Interactions: Mechanisms and Applications, Second Edition, Revised and Expanded, Shingo Iwata, Toshio Tabuchi, and Benno P. Warkentin

Stored-Grain Ecosystems, edited by Digvir S. Jayas, Noel D. G. White, and William E. Muir

Agrochemicals from Natural Products, edited by C. R. A. Godfrey

Seed Development and Germination, edited by Jaime Kigel and Gad Galili

Nitrogen Fertilization in the Environment, edited by Peter Edward Bacon

Phytohormones in Soils: Microbial Production and Function, William T. Frankenberger, Jr., and Muhammad Arshad

Handbook of Weed Management Systems, edited by Albert E. Smith

Soil Sampling, Preparation, and Analysis, Kim H. Tan

Soil Erosion, Conservation, and Rehabilitation, edited by Menachem Agassi

Plant Roots: The Hidden Half, Second Edition, Revised and Expanded, edited by Yoav Waisel, Amram Eshel, and Uzi Kafkafi

Photoassimilate Distribution in Plants and Crops: Source–Sink Relationships, edited by Eli Zamski and Arthur A. Schaffer

Mass Spectrometry of Soils, edited by Thomas W. Boutton and Shinichi Yamasaki

The Rhizosphere: Biochemistry and Organic Substances at the Soil–Plant Interface, Roberto Pinton, Zeno Varanini, and Paolo Nannipieri

Woody Plants and Woody Plant Management: Ecology, Safety, and Environmental Impact, Rodney W. Bovey

Metals in the Environment: Analysis by Biodiversity, M. N. V. Prasad

Plant Pathogen Detection and Disease Diagnosis, Second Edition, Revised and Expanded, P. Narayanasamy

Handbook of Plant and Crop Physiology, Second Edition, Revised and Expanded, edited by Mohammad Pessarakli

Environmental Chemistry of Arsenic, edited by William T. Frankenberger, Jr.

Additional Volumes in Preparation

Enzymes in the Environment: Activity, Ecology, and Applications, edited by Richard G. Burns and Richard Dick

Plant Roots: The Hidden Half, edited by Yoav Waisel, Amram Eshel, and Uzi Kafkafi

Handbook of Postharvest Technology, edited by A. Chakraverty, Arun S. Mujumdar, G. S. V. Raghavan, and H. S. Ramaswamy

Biological Control of Major Crop Plant Diseases, edited by Samuel S. Ganamanickam

Handbook of Plant Growth, edited by Zdenko Rengel

Environmental Chemistry of Arsenic

edited by
William T. Frankenberger, Jr.
University of California
Riverside, California

CRC Press
Taylor & Francis Group
Boca Raton London New York

CRC Press is an imprint of the
Taylor & Francis Group, an **informa** business

CRC Press
Taylor & Francis Group
6000 Broken Sound Parkway NW, Suite 300
Boca Raton, FL 33487-2742

First issued in paperback 2019

© 2002 by Taylor & Francis Group, LLC
CRC Press is an imprint of Taylor & Francis Group, an Informa business

No claim to original U.S. Government works

ISBN-13: 978-0-8247-0676-0 (hbk)
ISBN-13: 978-0-367-39656-5 (pbk)

Visit the Taylor & Francis Web site at
http://www.taylorandfrancis.com

and the CRC Press Web site at
http://www.crcpress.com

Preface

Arsenic-contaminated soils, sediments, and sludges are the major sources of arsenic contamination of the food chain, surface water, ground water, and drinking water. Arsenic is a known carcinogen and mutagen posing serious health risks to humans and animals. Health effects including cancers of the skin and internal organs have been linked to chronic exposure to arsenic in drinking water.

The effect of arsenic on human health is an issue of global concern. A large-scale shift in water resource allocation from surface water to ground water in West Bengal, India, and Bangladesh (tube well water) and the exposure of local populations to ground water containing arsenic at concentrations of several hundred μg/L have resulted in very extreme environmental health effects. Spurred by increasing concern over exposure to low levels of arsenic in drinking water, the U.S. Environmental Protection Agency (EPA) has proposed lowering the maximum contaminant level (MCL) for arsenic from 50 to 5–10 μg/L.

This book contains contributions from world-renowned international scientists on topics that include toxicity of arsenic, analytical methods for determination of arsenic compounds in the environment, health and risk exposure of arsenic, biogeochemical controls of arsenic, treatment of arsenic-contaminated water, and microbial transformations of arsenic that may be useful in bioremediation.

There have been many new exciting discoveries in the environmental chemistry of arsenic. Chapter 1 traces the sordid history of arsenic from ancient times to recent times and discusses its manifold uses in human therapy, daily commodities, pesticides, animal husbandry, and wars. Arsenic is still affecting our daily lives, and millions of people are being chronically exposed to elevated levels of arsenic from food, air, water, and soils with unknown long-term health consequences.

Chapters 2 and 3 focus on the detection of arsenic and arsenic compounds in the environment. New analytical methods and instruments allow determination of trace and ultra-trace concentrations of arsenic and arsenic compounds (Chapter 2). With detection of trace levels, the toxic reputation of arsenic might change from that as a life-threatening element to an essential element. Chapter 3 covers extensive characterization of environmental arsenic compounds including arsenic-containing minerals and specific species found in both terrestrial and marine environments. This chapter focuses on environmental compartments of arsenic compounds including marine life, terrestrial fungi and lichens, plants, and animals.

Chapters 4, 5, and 6 cover human exposure to arsenic ingestion, inhalation, and dermal absorption. Metabolism of arsenic in the body depends on the chemical species of arsenic absorbed. Association of arsenic in human urine is the most suitable biomarker to assess exposure to arsenic. Risk characterization and bioavailability of arsenic upon soil ingestion is critically evaluated. Arsenic occurrence in the United States is compared to that in Taiwan for health implications. Factors that may interfere with arsenic removal during water treatment, as well as significant diurnal and seasonal variations in arsenic concentration in U.S. surface water supplies, are thoroughly discussed.

Chapter 7 covers the biogeochemical control of arsenic occurrence and mobility in water systems. Arsenic concentrations are compared in ground water and surface water supplies. Redox control as well as dissolution, precipitation, adsorption, and desorption are important parameters in the biogeochemistry of arsenic. Arsenic is cycled in soils and natural waters by physical, chemical, and microbiological processes (Chapter 8). A greater understanding of these processes in arsenic cycling has allowed the development of suitable models for predicting arsenic speciation, bioavailability, and subsequent risk exposure. Arsenic may enter the environment naturally from arsenic-containing minerals as well as from natural waste as herbicides or as insecticides. Arsenic in drinking water can be removed by a variety of treatment processes. Chapter 9 focuses on arsenic treatment by metal oxide adsorption, ion exchange and iron III coagulation–microfiltration.

Many microorganisms are known for their ability to reduce, oxidize, and methylate arsenic. Chapter 10 covers the molecular genetics of bacterial arsenic resistance and enzymatic transformations of inorganic arsenic oxyanions. Inorganic arsenic oxyanions, frequently present as environmental pollutants, are very toxic to most microorganisms. Many microbial strains possess genetic determinants that confer resistance by encoding specific efflux pumps able to extrude arsenic from the cell.

Chapter 11 covers the exciting discovery of respiratory reduction of arsenate by prokaryotes, including seven new and highly diverse species of Eubacteria and one new and one previously isolated species of Cremoarchae. Detailed biochemical and genetic characterization of the enzymes involved in these organisms are thoroughly discussed. Chapter 12 reveals a unique mode of arsenate respiration by *Chrysogenes arsenatis* and *Desulfomicrobium* sp. str. Ben-RB. These organisms have specific respiration arsenate reductases involved in energy generation and have potential for removal of arsenic from potable water.

A number of bacteria have the ability to oxidize arsenite including heterotrophs and at least three autotrophic bacteria (Chapters 13–15). Some heterotrophs are able to use arsenite as an auxiliary source of energy, but others seem to oxidize arsenite to arsenate merely as a means of detoxification since arsenite is more toxic than arsenate (Chapter 13). Oxidation of arsenite in freshwater and sewage, soil and the marine environment as well as oxidation of As(III)-containing minerals are extensively reviewed. Chapter 14 provides new information on the discovery of six new chemolithoautotrophic arsenite-oxidizing bacteria isolated from gold mines from different regions of Australia demonstrating that energy for growth can be conserved during the oxidation of arsenite to arsenate. Chapter 15 discusses the oxidation of arsenite by *Alcaligenes faecalis*. The arsenite oxoreductase was purified and characterized and found to effectively catalyze the oxidation of arsenite to arsenate.

Chapter 16 focuses on the methylation and volatilization of arsenic. Biomethylation of arsenic results in formation of mono-, di- and trimethylarsine, which are volatile gases. In developing a bioremediation strategy to clean the environment of arsenic, special attention must be paid to the toxic nature of microbially transformed arsenic compounds. This chapter covers the environmental parameters that promote arsenic volatilization under natural and engineered conditions.

In recent years, we have seen dynamic growth in understanding arsenic as a result of the teamwork of a worldwide community of researchers working on arsenic speciation, transformations, transport kinetics, seasonal cycling, accumulation, geochemistry, and toxicology. New developments in arsenic biological and geochemical behavior will engender better understanding in developing safe levels for consumption of water for humans worldwide. It is now evident that total elemental concentrations of arsenic are not reliable indicators of environmental toxicity. Extensive efforts should be devoted to developing arsenic-specific speciation and soil fractionation techniques. New remediation strategies have been developed that combine biological, chemical, and physical remediation techniques.

This important information is compiled into a single resource for distribution among scientists, regulators, and the general public. This book will have a worldwide impact on the treatment of potable waters contaminated with arsenic.

William T. Frankenberger, Jr.

Joan M. Macy

In Memoriam

This book is dedicated to Professor Joan M. Macy.

Professor Joan Marie Macy's 25 year scientific career was notable primarily for her ability to isolate and characterize unusual bacteria as well as for her gifted teaching.

She obtained her Ph.D. in 1974 from University of California, Davis, under the supervision of Professor W. Hungate, one of the pioneers in the study of anaerobes. An Alexander von Humboldt fellowship made it possible for her to go to Germany, where she worked as a postdoctoral fellow from 1974 to 1976 with Professor G. Gottschalk at the Institut für Mikrobiologie, Georg-August-Universität, Göttingen and continued working with anaerobes. While in Germany she developed an interest in German culture—literature and music in particular. She was awarded a second Fulbright in 1985 and returned to Germany to work with Professor R. Thauer at FB Mikrobiologie, Biologie, Philips Universität, Marburg.

After a 1990–1991 sabbatical at Monash University in Melbourne, in 1995 she accepted the appointment as Chair of Microbiology at La Trobe University, Melbourne, where she remained until her untimely death in 2000.

Her most recent scientific contributions were in the areas of bacterial selenium and arsenic metabolism. Her interest in bioremediation began with selenium-contaminated water found in the San Joaquin Valley in California, from which she isolated the first bacterium able to respire with selenate (reducing it to selenite and then to elemental selenium) using acetate as the electron donor/carbon source. This organism was found to represent a new genus and named *Thauera selenatis*. She studied the organism extensively for the purpose of selenium bioremediation. This involved the design and implementation of lab-scale and then pilot-scale reactors.

More recently, she became interested in the problem of arsenic contamination and isolated the first bacterium able to use arsenate as the terminal electron acceptor and acetate as the electron donor. This organism, a strict anaerobe, reduces arsenate to arsenite. It too was found to represent a new genus and was named *Chrysiogenes arsenatis* (*Chrysiogenes* = sprung from a gold mine; *arsenatis* = of arsenate). More recently, this organism was designated the first representative of a new phylum of the *Bacteria*. A separate group of novel bacteria, which oxidize arsenite to arsenate, were isolated from gold mines in Australia. She had hoped that together these bacteria would prove suitable for use in purifying arsenic-contaminated waste and drinking water. This work led her to present a paper at the International Conference on Arsenic held in Dhaka, Bangladesh, in 1998.

At the time of her death Professor Macy was at the forefront of her field and was just starting to receive the accolades that she well deserved.

Contents

Contributors

Gretchen L. Anderson   Department of Chemistry, Indiana University South Bend, South Bend, Indiana

Muhammad Arshad   University of Agriculture, Faisalabad, Pakistan

Nicholas T. Basta   Department of Plant and Soil Sciences, Oklahoma State University, Stillwater, Oklahoma

Stan W. Casteel   College of Veterinary Medicine, University of Missouri, Columbia, Missouri

Hsiao-wen Chen   Department of Civil, Environmental, and Architectural Engineering, University of Colorado, Boulder, Colorado

Dennis A. Clifford   Department of Civil and Environmental Engineering, University of Houston, Houston, Texas

Marc Edwards   Department of Civil and Environmental Engineering, Virginia Polytechnic Institute and State University, Blacksburg, Virginia

Henry L. Ehrlich   Department of Biology, Rensselaer Polytechnic Institute, Troy, New York

Paul J. Ellis   Department of Structural Molecular Biology, Stanford Synchrotron Radiation Laboratory, Menlo Park, California

Scott Fendorf   Department of Geological and Environmental Sciences, Stanford University, Stanford, California

Kevin A. Francesconi   Institute of Biology, University of Southern Denmark, Odense, Denmark

William T. Frankenberger, Jr.   Department of Soil and Environmental Sciences, University of California, Riverside, California

Ganesh L. Ghurye   Department of Civil and Environmental Engineering, University of Houston, Houston, Texas

Walter Goessler   Institute of Chemistry–Analytical Chemistry, Karl-Franzens University, Graz, Austria

Janet G. Hering   Environmental Engineering Science, California Institute of Technology, Pasadena, California

Russ Hille   Department of Molecular and Cellular Biochemistry, The Ohio State University, Columbus, Ohio

William P. Inskeep   Department of Land Resources and Environmental Sciences, Montana State University, Bozeman, Montana

Brian W. Kail   Department of Biological Sciences, Duquesne University, Pittsburgh, Pennsylvania

Penelope E. Kneebone*   Environmental Engineering Science, California Institute of Technology, Pasadena, California

Doris Kuehnelt   Institute of Chemistry–Analytical Chemistry, Karl-Franzens University, Graz, Austria

Peter Kuhn   Stanford Synchrotron Radiation Laboratory, Stanford University, Menlo Park, California

X. Chris Le   Department of Public Health Sciences, University of Alberta, Edmonton, Alberta, Canada

---

* *Current affiliation*: ENVIRON International Corporation, Emeryville, California

Joan M. Macy† Department of Microbiology, La Trobe University, Melbourne, Victoria, Australia

Timothy R. McDermott Department of Land Resources and Environmental Sciences, Montana State University, Bozeman, Montana

Laurie S. McNeill Utah Water Research Laboratory, Utah State University, Logan, Utah

Dianne K. Newman Division of Geological and Planetary Sciences, California Institute of Technology, Pasadena, California

Jerome O. Nriagu Department of Environmental Health Sciences, School of Public Health, University of Michigan, Ann Arbor, Michigan

Ronald S. Oremland Water Resources Division, U.S. Geological Survey, Menlo Park, California

Le T. Phung Department of Microbiology and Immunology, University of Illinois at Chicago, Chicago, Illinois

Robin R. Rodriguez Stratum Engineering, Inc., Bridgeton, Missouri

Barry P. Rosen Department of Biochemistry and Molecular Biology, School of Medicine, Wayne State University, Detroit, Michigan

Joanne M. Santini Department of Microbiology, La Trobe University, Melbourne, Victoria, Australia

Simon Silver Department of Microbiology and Immunology, University of Illinois at Chicago, Chicago, Illinois

John F. Stolz Department of Biological Sciences, Duquesne University, Pittsburgh, Philadelphia

Rachel N. vanden Hoven Department of Microbiology, La Trobe University, Melbourne, Victoria, Australia

---

† *Deceased.*

1

Arsenic Poisoning Through the Ages

Jerome O. Nriagu
University of Michigan, Ann Arbor, Michigan

I. INTRODUCTION

> Arsenic is mined from deep mines, for it is a material that Nature hides from us, teaching us to leave it alone as harmful, but this does not cause the arrogant miners to leave it (1, p 106)

Arsenic, the king of poisons, has probably influenced human history more than any other element or toxic compound. This enigmatic metal (chemically, arsenic is a metalloid but will be referred to as a metal throughout this chapter) began its long association with human culture by poisoning the god (Hephaestus) who first endeavored to find some beneficial use for it. In subsequent ages, it was used nefariously to kill many aristocratic and noble gentlemen, terrorize others, and engender events that influenced the cultural and social developments in many parts of the world. Until the recent emancipation of women, arsenic was a key instrument employed by women to free themselves from tyrannical husbands and unwanted lovers. Arsenic's misguided benevolence was equally romanticized so that it was able to achieve widespread acclaim, becoming, at the turn of the century, the best agent in pharmacopoeia of the Western world. Millions of people probably received arsenicism rather than cure for trusting their health to arsenical preparations. The long fight with pests was tipped in humanity's favor when arsenic was brought in as pest killer. The pesticide sprays began the practice of contaminating human foods and environment with arsenic, resulting in untold effects on the health of many people and their offspring. By lurking underground, arsenic threatens the drinking water resources in many parts of the world, and the consumption of the tainted water over the years must have resulted in untold

1

suffering by millions of people. Many scientific disciplines, such as toxicology, forensic science, and analytical chemistry, owe their origin and development to the fascination with arsenic and the need to reveal its presence in our air, drinks, and food. This chapter presents a brief overview of arsenic exposure through the ages and makes the case that arsenic has poisoned and killed more people than any toxin known to humankind.

II. ANCIENT TIMES

The role of arsenic in metallurgical development at the beginning of the Bronze Age is well documented (2). In many places, arsenic minerals tend to co-occur with copper minerals, and during primitive smelting operations some of the arsenic would have been alloyed by chance with the copper. Perceptive smiths of those days would have noticed that some copper alloys were more desirable than others and subsequently would have made some conscious effort to select minerals that would yield the preferred copper–arsenic alloys (2,3). The primitive furnaces would have generated copious fumes of toxic, garlic-smelling arsenious oxide (As_2O_3), which could have adversely affected the health and life expectancy of the smiths (4). Exposure to the toxic arsenical fumes conceivably could have resulted in degenerative changes manifested as polyneuritis and muscular atrophy in limbs leading ultimately to lameness. The deformity of Hephaestus (or Hephaistos), the mythical Greek god of fire and technology, his Roman counterpart (Vulcan), and the patron gods of smiths in many other cultures might have been an occupational disease linkable to effects of exposure to toxic fumes of arsenic or lead (5,6). The change in the caricature image of Hephaestus the smith with the passage of time from the buffoonlike achondroplastic walk to the club-footed limp and eventually to normal behavior has been related to improvements in smelting techniques and the switch to widespread use of cassiterite (SnO_2), which minimized the deformatory effects of occupational exposure to arsenic. Regardless as to whether the physical deformities of the patron gods of smiths were related to the unhealthiness of their craft, arsenic poisoning appears to have been among the first occupational diseases to afflict humankind. In all likelihood, arsenic poisoned the first person who managed to obtain a metallic alloy from one of its ores.

The illusory similarity between anthropomorphized Hephaestus and arsenic is striking. Both had questionable fathers—"Hera gave birth to Hephaestus without intercourse with the other sex but according to Homer he was one of her children with Zeus" (7, I, iii, p 5), while the origin of metallic arsenic is shrouded in mystery. Neither had alluring physical qualities—Hephaestus limped about on an ill-matched pair of feeble legs, while metallic arsenic possesses few desirable physical properties and was depicted allegorically as a serpent in alchemical texts.

Both fell from grace—early in his life, Hephaestus was hurled down from heaven, while the use or mention of arsenic was outlawed in many cultures. Both were subsequently rehabilitated—Hephaestus was called back to Olympus where he lived in his palace and worked on his forge, while arsenic achieved fame as the backbone of Western pharmacopoeia until fairly recent times. Both felt unwanted—Hephaestus intoned "I was born mis-shapen . . . would they [his parents] had never begotten me," while no one could find a use for metallic arsenic for a long time. In the end, both helped to forge human history—Hephaestus taught men glorious crafts throughout the world, while human history could never have been the same without arsenic.

Application of arsenic in ancient metallurgy other than as arsenical bronze was limited. It was used by the Egyptians in the third millennium B.C. as well as in ancient China to produce a silvery surface on mirrors and statues (8,9). It was cited in one of the first treatises on glass as one of the fluxing ingredients, its presumed effect being to produce crystallization during the cooling of the glass (10). Arsenic fumes would have been a hazard to the people who made such artifacts. Occupational exposure to arsenic in ancient times would have extended beyond the bronze makers and their families. Many copper, gold, and lead ores contain significant amounts of arsenic. Most of the arsenic went into the smoke, which was inhaled by the artisans and contaminated the surrounding areas. Arsenic poisoning remained a real and persistent hazard to many people involved in the smelting and recovery of copper, gold, and lead in ancient times. Workers so afflicted would have numbered in the millions. Orphiment (As_2S_3, the yellow arsenic sulfide) and realgar (As_2S_2) were also used as depilatories in the leather industry with a high probability of exposure of ancient workers.

The ancients believed that orphiment contained gold, hence the name *auropigmentum*. Pyrite and arsenopyrite, often with high arsenic content, have been known as fool's gold since time immemorial. *Mappae Clavicula*, an eclectic compilation of ancient alchemical methods, listed many recipes for using orphiment to make gold, to make silver from copper, and to cover iron and tin objects with gold color (11). Pliny described the effort of emperor Caligula to extract gold from orphiment and noted that while some gold was obtained, the quantity was so small that despite the low price of orphiment, which was about four denarii per pound, the operation was deemed unprofitable. In A.D. 260, the emperor Diocletian was so disgusted with the failure of Egyptian alchemists to extract gold from orphiment that he collected all the books dealing with transmutation and had them destroyed (12). The alchemists who relied on orphiment to make gold were probably rewarded with arsenicism rather than materially for their effort.

Orphiment was used in ancient times as pigment. It was found in a linen bag in King Tutankhamun's tomb, in wall paintings of the Theban necropolis, and more extensively from the Eighteenth Dynasty onward (13). Since the mineral is not found in Egypt, it must have been imported from Persia, Armenia, or Asia Minor. Orphiment and realgar were occasionally cited in the Akkadian texts as ingredients of ornamental paint and for cosmetic purposes (14). Realgar was identified as the red pigment in a pot excavated at Corinth and orphiment has been identified in yellow pigments found in ancient Greek graves (15). Pliny (16, XXXIII, p 79) noted that in his time there was "a recipe to make gold from orphiment which occurs near the surface of the earth in Syria, and is dug up by painters." The number of ancient painters and artists who were occupationally exposed to arsenic is impossible to estimate at this time. Strabo (17, p 40) referred to a mine of yellow and red sulfides near Pompeiopolis where, because of its poisonous character, only slaves were employed.

Arsenic was featured extensively in the materia medica of ancient cultures. The name arsenic itself is derived from the Greek word *arsenikon*, which means potent (18). Both orphiment and realgar were mentioned in the ancient medical texts of Assyria (19) and some of the so-called malachites, hematites, yellow and red ochres, and iron pyrites mentioned extensively in the Ebers Papyrus of Egypt (about 1550 B.C.) presumably consisted of arsenic sulfides (5). Hippocrates (460–377 B.C.) used orphiment and realgar as escharotics. Dioscorides (20, L, xxx and xxxi) notes that orphiment "is astringent and corrosive, burning the skin with severe pain and produces eschar. It removes fungal flesh and is a depilatory." In describing the medicinal properties of realgar, he notes that when "mixed with resin, it causes hair to grow when destroyed by pellagra, and that with pitch it serves to remove rough and deformed nails. With oil, it is used to destroy lice, to resolve abscesses, to heal ulcers of the nose, mouth and fundament. With wine, it is given for fetid expectoration and its vapor, with those of resin, are inhaled in chronic cough. With honey, it clears the voice and with resin is good for shortness of breath." Pliny furnishes an identical description of the medicinal properties of the arsenic sulfides. Galen (A.D. 130–200) also recommended a paste of orphiment for treatment of ulcers. Other noted ancient medical authorities, including Aristotle, Celsus, Caelius Aurelianus, Aetius, and Soranus, all employed orphiment and realgar for various curative purposes (21). The literary records thus point to the fact that in ancient cultures of the West, medicinal exposure to arsenic was widespread and was a public health hazard.

Arsenic minerals were valued for their medical properties by ancient Indian cultures during the age of Buddha. The Charaka-Samhita medical texts recommend *Ala* or *Haritala* (orphiment) and *Manahsila* (realgar) for external and internal medication (12). The multitude of names for arsenic compounds in Sanskrit (*Sankh*, and *Sabala Kshara*), Hindi (*Sanbul-Khar*, *Sammal Khar*, *Sankhyia Sanbul*, and *Sankyhia*) and Bengali (*Sanka* or *Senko*) suggest general familiarity and extensive use of these compounds in some curative and nefarious ways (12). In the *Rasa* system of Indian medicine, it is claimed that orphiment (*Haritalam*) "cures phlegm, poison, excess air and fear from ghosts. It stops menstrual discharge, is soothing, pungent and produces warm effect on the system. It increases appetite and cures leprosy" (22, vol 2, p 157). By contrast, improperly purified orphiment is said to shorten life, give rise to abnormal excess of air, inflammation, boils, and contraction of the limbs, and hence one is adjoined to purify it very carefully (22). Realgar (*Manas-shila*) is said to be an "improver of complexion, a laxative, producer of heat in the system, destroyer of fat, curer of poison, asthma, bronchitis, impurities of the blood and evil effects of ghosts" (22, vol 2, p 198). Unpurified realgar on the other hand "gives rise to stone disease, stricture, gleet, loss of appetite and constipation" (22, vol 2, p 198). The problem of arsenic poisoning became so common that a number of antidotes for both orphiment and realgar were also provided in the *Ras-Jala Nidhi* medical compendium (22).

Realgar (*Hiung-hwang, Hwang-kin-shih, Ming-hiung* or *T'u-hiung*) was produced from several places in ancient China (23). It is said to be spermatic and masculine, and of the *Yang* principle, just as orphiment is female, and of the germinal or *Yin* principle. It was regarded as the germ of gold, and was used in soldering gold, hence one of its names, *Hwang-kin-shih* (23). Its antifebrile, prophylactic, emetic, expectorant, deobstruent, arthritic, antihelmintic, antidotal, and escharotic properties were noted in *Pen Ts'au* and earlier works (23). Similar medical properties were also ascribed to orphiment. Arsenic was one of the common ingredients of metallic elixirs consumed by the ancient Chinese in their quest for longevity and/or immortality. For example, Sun Ssu-Mo, the great seventh century A.D. alchemist and pharmacist, revealed in his *Thai-Chhing A Cheng-Jen Ta Tan* (The Great Elixir of the Adepts: A Thai-Chhing Scripture) that the secret recipe for gold elixir (*chin tan*) consisted of 8 oz of gold, 8 oz of mercury, 1 lb of realgar, and 1 lb of orphiment (24). Arsenic must therefore bear some of the blame for the metallic poisoning experienced by many Chinese alchemists and their patron emperors. Even today, many traditional Chinese medicinals still contain high levels of arsenic and mercury (25); arsenic remains an important ingredient of many Chinese patent and herbal medicines used in the treatment of psoriasis, asthma, tuberculosis, leukemia, and other diseases (26).

The role of arsenic in ancient warfare was ingenious. It was cited as an ingredient of the devilish incendiary material used by Marcus Graccus to burn the Roman naval fleet (1, p 438). Early Chinese alchemical texts recommended arsenic sulfides for making toxic smoke bombs or "holy smokes" for mass poisoning of soldiers—one of the earliest references to chemical warfare. Death lamps in which oil and wax impregnated with arsenic were burned to poison victims slowly presumably owe their roots to the holy smokes of ancient Chinese (27).

It is debatable whether arsenic was widely used as secret poison in ancient times. Theophrastus spoke of a poison that could be moderated in such a manner as to have effects in two or three months or at the end of a year or two years and remarked that the more lingering the administration, the more miserable the death. Agrippina, being intent on getting rid of Claudius but not daring to dispatch him suddenly and yet wishing not to leave him sufficient time to make new regulations regarding succession to the throne, used such a poison to deprive him of his reason and gradually consume him (28). Later, the nefarious emperor Nero used the same poison to dispatch Britannicus who was in line to succeed Claudius (28). The strategy for administering the ancient poisons was thus symptomatic of the method later used by the celebrated arsenic poisoners of the Middle Ages.

III. MEDIEVAL TIMES AND MIDDLE AGES

Attempts to isolate and study elemental arsenic became as frustrating as the effort to find a universal antidote for its poisonous properties in Medieval times. Geber (Jabir ibn-Haiyan), an Arabian alchemist of the eighth century, discovered white arsenic by heating orphiment (29). Avicena (980–1038) differentiated between white, yellow, and red arsenic (29). Although many people have been cited as the discoverer of metallic arsenic, some people credit Albertus Magnus (1193–1280), a German Dominican scholar and alchemist, with being the first to accomplish the feat (30). A clear distinction between elemental arsenic and arsenic minerals was made by Biringuccio in 1540: "They also say, as I told you before, that orphiment and crystalline arsenic are not at all of the same nature. And I must say, from all that I have seen or that I think I have seen, that their composition is different. For one is clear white and citrine, and the other is of a shining and beautiful golden color" (1, p 106). He then goes on to say that when orphiment is sublimed, it makes realgar, a thing of the same nature and that "in the residue of this sublimation, or when they are roasted in some other way, they leave a regulus [of metallic arsenic] very white like silver but more brittle than glass" (1, p 106). Shroeder, in 1641, described a method for obtaining elemental arsenic by reducing white arsenic with charcoal (29). In 1733, Brand showed that white arsenic was a calx or oxide of elemental arsenic (31). Arsine (AsH_3) was discovered by Scheele, the famous Swedish chemist, in 1775, but little was known of its deadly nature until Gehlen, a professor of chemistry in Munich, died in 1815 by inhaling a minute quantity of the pure gas. By heating a mixture of equal weights of potassium acetate and arsenious oxide, Cadet, in 1760, obtained a heavy brown, strongly fuming liquid of fearful odor (32). Robert Wilhelm Bunsen (1811–1899) was able to isolate cacodyl oxide, $C_4H_{12}As_2O_3$, to which the name cacodyl (from κακωδης, stinking) was given by Berzelius. Although the daring experiments brought Bunsen instant acclaim, he nearly killed himself from arsenic poisoning and he also lost the sight of one eye to an explosion of the compound (32). The discovery and characterization of the various arsenic compounds served to bring the element into more intimate contact with human culture. Poisonings of Gehlen and Bunsen by arsenic were not isolated incidents; many chemists and alchemists who worked with arsenic during the same period probably paid a toll with their health.

A. Uses of Arsenic

Finding metallurgical uses for arsenic continued to elude the craftsmen, except as Biringuccio (1, p 105) noted: "they penetrate very easily into fused metals, and indeed they act in such a way that they corrupt and convert almost into another nature any metal with which they find themselves. . . . With these [orphiment and arsenic] the fraudulent alchemists blanch copper, brass and even lead to the whiteness of silver." *Mappae Clavicula*, an eleventh century compilation of medieval techniques had earlier listed 33 orphiment-containing recipes dealing with metals, pigments, and chemical operations (11). In medieval times especially, orphiment was sought as an ingredient for the philosopher's touchstone, an elusive magical wand for converting base metals into gold.

Until well into the seventeenth century, chemistry was a handmaid of painting, which in turn was closely interwoven with medicine. Colors and other material, when not furnished by monks who retained the ancient habits of the cloister, were provided by the apothecary (33). The most valuable treatises on the arts were composed by physicians, and many famous painters (such as Leonardo de Vinci) were equally noted in the medical sciences. Throughout most of this time, the principal uses for arsenic were in chemistry (homicidal and practical), painting, and medicine.

In its natural state, orphiment has a mica-like sparkle that recalls the luster of metallic gold, and painters then being more practical than the scientists or the alchemists, found the close resemblance to gold to be of immense benefit to their work. Orphiment was a regular ingredient for most of the compounds concocted in the Middle Ages for painting and writing in gold (34,35). Mixtures of orphiment with calcined bone, ground eggshells or oyster shells were used extensively as bone white and inert white where white lead was deemed inappropriate. Greens were sometimes made by mixing yellow orphiment with blue pigments. Realgar, with its beautiful orange-scarlet hue, was not common in medieval paintings although Cennini (36) mentioned it, though not enthusiastically. The poisoning of painters by arsenic in golden orphiment in their palette during the Middle Ages presumably was common. The toll among the painters and the people who made and/or mixed the arsenious pigments must have been high.

Orphiment and realgar used in the Middle Ages were either obtained from mines or prepared artificially. Natural orphiment was obtained in masses from the neighborhood of Naples and in other volcanic countries (35), or imported from Asia Minor (1). The occupational hazard of mining orphiment was well recognized at the time: "On account of its poisonous exhalation, after they have made very deep mines and have found it, they pass through it with mouth and nose closed with a sponge wet with vinegar if they wish to save their lives, and they do not remove the earth around or upon that which they have found it" (1, p 106). Unfortunately, the protective measures employed by miners were generally ignored by people who made the arseniferous products. Many polymetallic ores of copper, zinc, and lead in Europe and elsewhere were said to contain significant amounts of arsenic as an unwanted impurity—"experienced mineralogists say that some of this [white arsenic] is found in the company of almost all the metal ores and that it is this which consumes and carries away in the smelting operations the silver that they contain" (1, p 106). The people might not have lost their silver to arsenic but their health and life expectancy instead. In his *Pharmacologia* published in 1812, John Paris, a British physician, described many cases of cancer among workers involved in the production of tin and copper and in farm animals grazing in the vicinity of copper smelters. He attributed all of these diseases to poisoning by arsenic present in ores of many metals (37). Although Paris's speculation on the arsenic origin for the cancers is difficult to prove, his report drew attention to potential environmental and human health effects of roguish arsenic in many ores.

B.   Homicidal Poisoning and Arsenophobia

Arsenic became practically synonymous with poison during the Middle Ages when the art of secret poisoning became part of the social and political life. Arsenic oxide, reputedly discovered by Geber, was an ideal suicidal and homicidal poison for the time. It is tasteless, odorless, cheap, powdery white and hence can be administered with sugar, does not diminish appetite, is fatal in small doses, and the symptoms of chronic and acute poisoning mimic natural diseases thereby obfuscating the true cause of death. Its effects tend to be cumulative, enabling the poisoner to weaken the victim with small doses over a period of time before administering the fatal dose. One of the earliest accounts (in western Europe) of the use of arsenic for homicidal purposes was in 1314 when Charles the Bad, King of Navarre, tried unsuccessfully to use arsenic trioxide from apothecaries to murder his brother Charles VI, King of France. Karl the Evil, in 1383, noted the ready availability of arsenic all over France: "You are going to Paris . . . you will find arsenic at Pampelune, Bordeaux, at Bayonne, and from every one of the pretty villages where you pass, at the apothecaries' hospitals" (21, p 14). In the late 1600s, the infamous Tophana or Toffana of Palermo and Naples distributed her murderous oil marked *Manna of St. Nicholas of Bari* (a sanctimonious appellation) as charity to wives who wished to have other husbands. It is estimated that over 600 persons perished from *Aqua della Toffana*, which became a generic name for secret poisons sold widely in Europe between 1630 and 1730 (28). Apocryphal vignettes of the *Aqua della Toffana* homicides can still be seen in John Kesselring's (38) play, *Arsenic and Old Lace*, in which old ladies used elderberry wine spiked with arsenic, cyanide, and strychnine to dispatch their gentlemen callers. In the court of France, there were so many notorious poisoners that the name *poudre de succession* (inheritance powder) for white arsenic became a nightmare and a destabilizing influence as the heads of the great families came to regard all relatives and friends with extreme suspicion. The punishment for the treacherous act was harsh: "If a Christian disavows faith or works magic or the mixing of poison and is caught in the act, the person shall be burned on a rack" (21, p 8). For the rich and famous and men with unhappy spouses, nevertheless, arsenic powder remained a hazard to life and property until James Marsh published his method for detecting low levels of arsenic in 1836 (37).

C. Arsenic in Medicine

The Arabian physicians copied and slightly modified the medical uses of arsenic described by Dioscorides. Rhazes (ca. 850–925), for instance, recommended arsenic in asthma, lung diseases, skin diseases, ulcerations, and with unslaked lime and opium for dysentery (21). Avicena (980–1037) used arsenic for skin diseases and lung problems, and with honey and other vehicles to improve the mucous membrane. The introduction of arsenic prophylaxis into the plague literature may be attributed to Nikolaos Myrepsos of Arabia, who alluded to arsenic compounds as "Persian antidote" against plague in his extensive recipe collection (39). The collected works on the materia medica of the ancient and Arabian physicians were propagated through subsequent ages in the various books of secrets or natural magic, which often included liberal interpolations by the author. Typical of these books was *Natural Magik* by John Baptista Porta (40), which listed many complicated remedies containing arsenic that could be used against asthma, skin diseases, and head lice. Because of its high toxicity, white arsenic was first used against external diseases of sheep and horses and was only applied in the fourteenth century in the treatment of human scrofulous ulcers (41). By the middle of the sixteenth century, arsenic compounds were primarily used for external diseases in western European medical practice—as an amulet, cauterizing agent, or as a smoke. Biringuccio was an eyewitness:

> They say that it is a remedy against the plague to carry them [arsenic compounds] in a little bag over the heart, that their fumes are beneficial to those who have asthma, and that they work against chronic cough or bloody sputum. Orphiment mixed with lye and lime shaves every hairy spot without cutting. With these materials, a very powerful corrosive for cauterizing is made (1, p 107).

Paracelsus, who broke the rigid tradition of Galenic medicine, used realgar internally against cancerlike tumors (42) but more importantly gave cogency to the belief that there was a therapeutic window for toxic compounds in which clinical benefits can be had without achieving extreme toxicity. After Paracelsus, the febrifuge qualities of arsenic and its compounds began to be touted throughout Europe in an ever-expanding manner.

IV. SPRINGTIME OF ARSENIC POISONING

The period of 1850 to 1950 can be regarded as the golden age of arsenic poisoning, when human beings were exposed to unprecedented levels of arsenic in their medicine, food, air, water, and at work or accidentally. During this time, worldwide production of arsenic trioxide increased from less than 5000 tons per year and peaked at over 60,000 tons per year around 1940 (Fig. 1). The levels of arsenic in many environmental media also rose, as indicated, for instance, by the increase in the arsenic content of American tobacco from less than 10 ppm in 1917 to over 50 ppm in 1952 (see Fig. 1). It is rather remarkable that arsenic was able to penetrate the economy and human culture pervasively even though its toxic properties were well known (43).

A. Arsenic Became the Panacea of Western Pharmacopoeia

The introduction of *Tasteless Ague Drop* or Fowler's solution (alkaline solution of potassium arsenite) in 1670 began the ascendancy of arsenic in Western pharmacopoeia. By the end of the nineteenth century, every major disease known was

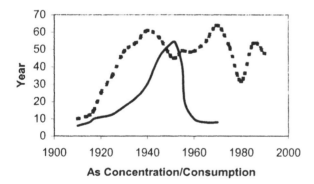

Figure 1 Changes in worldwide consumption of arsenic trioxide (solid line) and concentrations of arsenic in U.S. tobacco crops (broken line) during the last century. (From Refs. 58 and 43.)

being subjected to arsenotherapy. The following synopsis is based on extensive compilations by several authors (44–52).

SKIN DISEASES. Arsenic was used externally and internally to improve the nutrition of the skin and hair. A popular belief was that "arsenic is not only useful but essential and almost unexceptionally successful when properly administered in all cases of chronic idiopathic affections of the skin; provided always and in all cases, that it be administered discretely, perseveringly and with such adjuncts as the case may require" (45, p 834). Skin diseases treated with arsenotherapy included psoriasis, chronic eczema, warts, lupus, epithelioma, recurrent herpes, vesicular or bullous eruptions of children, ulcers, gangrene of the penis, external cancers, inflammation of the skin, pustular and popular forms of acne, leprosy, impetigo, etc.

BLOOD DISEASES. Fowler's solution or arsenous acid was widely used in the treatment of anemia, leukemia, and Hodgkin's disease.

NERVOUS AND RHEUMATIC CONDITIONS. People with chorea, neuralgia, epilepsy, arthritis, even tetanus, and angina pectoris received their daily doses of arsenic.

MALARIA. For years, arsenic preparations were the first line of defense against malaria, and later were used in obstinate cases to reinforce quinine therapy.

OTHER DISEASES. These included diabetes, scarlet fever, diphtheria, influenza, pulmonary tuberculosis, heart diseases (palpitation, edema, intermittency of pulse), disorders of respiration, asthma, hay fever, bronchitis, pneumonia, morning sickness during pregnancy, diarrhea, chronic gastritis, chronic rheumatism, eye inflammation, scrofula, leukemia, breast and other cancers, high blood pressure, bites of snakes and rabid animals, passive dropsies, as an abortifacient, and so on and so on.

Organic arsenicals entered the Western pharmacopoeia toward the end of the nineteenth century. Sodium cacodylate and sodium arsenilate were among the first of these compounds to be used for treatment of malaria, sleeping sickness, and pellagra (53). In his quest for a magic bullet in chemotherapy—a drug that could destroy bacteria circulating in the bloodstream without killing or seriously harming the patient or the patient's organs—Paul Ehrlich was able to synthesize hundreds of organic compounds of arsenic. Preparation "606" (an arsenobenzene compound), which was synthesized in 1907, turned out to be particularly effective in the treatment of syphilis. Arsphenamine, renamed salvarsan afterwards, and neoarsphenamine (neosalvarsan or "909") dominated syphilis therapy until the late 1940s and even later in the Far East where they were also used to treat yaws (53,54). Treatment of syphilis with arsenic was a lengthy and unpleasant business; minimum duration was about 18 months and involved 20 injections of salvarsan and 30–40 injections of bismuth (53). The use of arsenic in fire retardants has been linked to sudden infant death syndrome (SIDS)—the result of the release of poisonous arsine from the infant's mattress in the presence of urine or sweat (moisture and nutrient) and fungal infection (55).

Arsenotherapy was impressive and pervasive, and at no other time in human history has the health of nations depended so much on one element. Of some concern was the fact that the medical profession then was advised to administer the arsenic in progressively increasing doses. The toxic effects of the therapy were sometimes treated with more arsenic, especially if there were a change in the medical provider. The conclusion seems inescapable that more people were hurt or killed than were cured by arsenic—the cure was worse than the disease. The health of the millions of people therapeutically exposed then would have been impaired rather than repaired by the arsenic.

B. Food Contamination with Arsenic Spread

The efficacy of arsenic as an insecticide was discovered serendipitously when an exasperated farmer threw some Paris green (copper arsenate) paint on beetle-infested potato plants and came back a few hours later to find that all the bugs were dead (56). There were, of course, other claims about the first use of Paris green as a pesticide, considering that the lethality of arsenic to many insects had been carefully studied by Orfila (57) and others at the beginning of the nineteenth century. Millingen (45, p 814) summarized the available information in his day: "Arsenic is equally fatal to spiders, flies, worms, leeches, and crustacea such as crawfish, etc; and before killing these creatures it excites in them convulsive movements and a discharge of excretions. Even the salmon and the gudgeon perish with similar symptoms, but more slowly. Birds are still more tardily and incompletely affected, some of them surviving a dose of the poison sufficient to destroy an amphibious animal of equal size." Widespread commercialization of Paris green as an insecticide began around 1867 and in the following year, the editors of *American Agriculturalist* were so alarmed that they put out the following warning against the dangerous practice:

> The following is going the round of the press. "Sure death to potato bugs: take 1 lb Paris green, 2 lbs pulverized lime. Mix together, and sprinkle the vine." We consider this unsafe as there is no intimation of the fact, not generally known, that Paris green is a compound of arsenic and copper and a deadly poison (56, p 13).

Despite the early warnings against its use, Paris green was enthusiastically received by American farmers. By 1872, Paris green had been displaced as the leading pesticide by London purple (a by-product of the aniline industry composed of a mixture of calcium arsenate, arsenite, and organic matter), which was cheaper, easier to apply, adhered better to plants, and had more conspicuous color. The problem of the phytotoxicity of Paris green and London purple was solved with the introduction of lead arsenate (especially for gypsy moth) in 1892, which was gentler to the foliage and its bug-killing properties were more catholic. Ironically, the concern of entomologists then was on the poisoning of plants rather than consumers. With the development of calcium arsenate to treat cotton pests in southern United States in 1906, arsenic became unchallenged as the protector of American crops until the introduction of synthetic organic insecticides during World War II. For nearly three-quarters of a century, farmers in the United States and many European countries sang the "O Spray" ode in praise of arsenic (56, p 47):

Spray, farmer, spray with care,
Spray the apple, peach and pear;
Spray for scab, and spray for blight,
Spray, O spray, and do it right.

Spray the scale that's hiding there,
Give the insects all a share;
Let your fruit be smooth and bright,
Spray, O spray and do it right.

Spray your grapes, spray them well,
Make first class what you've to sell;
The very best is none too good,
You can have it, if you would.

Spray your roses, for the slug,
Spray the fat potato bug;
Spray your cantaloupes, spray them thin,
You must fight if you would win.

The use of lead arsenate in U.S. agriculture rose from about 5.4 million kg in 1919 to over 13 million kg in 1929, averaged about 23 million kg between 1930 and 1940, and peaked at 40 million kg in 1944 (27). The consumption of calcium arsenate likewise increased from about 1.4 million kg in 1919 to an average of 23 million kg between 1930 and 1940, and peaked at 36 million kg in 1944 (58). At the use peak, lead and calcium arsenates were registered for use on 41 and 83 feed and food crops, respectively, in addition to many other nonfood uses on turf and ornamental and shade trees (59). The debate on the public health effects of arsenic and lead residues on foods was long-winded and convoluted (56). Not surprisingly, the "arseno-phobics" lost by default. They were unable to prove that the levels of the residues on foods were a hazard, considering that doctors were regularly prescribing considerably higher doses to restore health. Furthermore, they were unable to provide a safer alternative and were forced to concede that arsenicals were a necessary evil to provide food security to all Americans. American farmers were freed to apply virulent toxins to their food crops however and whenever they wished. The American public was daily "served a steady diet of arsenic and lead" (60). Although a "world tolerance" for arsenic was set at 0.01 gr/lb (about 1.3 ppm As_2O_3), concentrations as high as 11.5 mg per berry were often reported (56). Profligate overspraying of crops with lead arsenate was well documented. A common maxim among the farmers at the time was that "if spraying twice a month is good, four to five times as often would be a great deal better" (56). During the watershed period for arsenical insecticides, the American public served as the *"100,000,000 Guinea Pigs"* (60) in which the measured outcomes were lethal doses and acute arsenic poisoning with little attention paid to long-term chronic arsenicism.

C. Widespread Environmental Contamination with Arsenic Began

Spraying of lead and calcium arsenates marked the beginning of widespread intentional contamination of the environment from dissipative application of a highly toxic material (61).

> We may have, in the free use of Paris green in the garden and field, one explanation of the frequent occurrence of arsenic in the [eco]system. The soil in many localities is doubtless quite highly impregnated with the poison. The question of its absorption by growing vegetables is, in my opinion, one which should be carefully investigated (62, p 454).

Early concerns about a buildup of arsenic and lead in soils (see above) were quickly dismissed by farmers who regarded lead arsenate as the panacea for American agriculture. Over the years, the spraying of over 1 billion lb of arsenical pesticides on American crops has left a legacy of contaminated soils and groundwater that will remain with us for a long time. An orchard soil that was typically sprayed with 30–60 kg/ha of lead arsenate per year would have received 2000–4000 kg/ha during the period of treatment, which could have increased the soil arsenic content by 200–400 ppm if arsenic stayed in the top 15 cm of soil (49). It is not surprising that the levels of arsenic in many wines produced in the United States were for years reported to be well above 50 μg/L, a former limit for arsenic in public drinking waters of most countries (63).

An important reason for the popularity of arsenic as an insecticide is the fact that it is obtained as a cheap by-product during the smelting of copper, silver, gold, nickel, lead, zinc, manganese, and tin ores. Only a small fraction of the arsenic in the ores was commercially recoverable, however. By the end of the nineteenth century, emissions of arsenic from base metal smelters and coal-burning power plants reached unprecedented levels because of low-level technology and little attention devoted to air pollution control. During this time, there were numerous complaints from smelter districts in many parts of the world that the smelter smoke was poisoning the horses and other livestock. In the United

States, attention was drawn to the problem by an investigation done in Anaconda, Montana—the famous copper mining center where the Washoe smelter daily released large quantities of toxic fumes containing 29,270 lb of arsenic trioxide into the surrounding environment (64). From about 1902 onward, farmers began to complain that their cattle, horses, and sheep were dying in alarming numbers. The symptoms displayed by affected livestock (diarrhea, thirst, emaciation, mucus inflammation, and paralysis) pointed to arsenic as the likely cause of the epidemic. The hypothesis was subsequently confirmed by studies that showed that soils within an 8-mile radius of the Washoe smelter were heavily contaminated and that organs of dead animals contained high levels of arsenic (64,65). The Anaconda incidence inspired the study of chronic effects of arsenic by American toxicologists (56) but resulted in no public policy to deal with such an environmental hazard.

Arsenic-containing minerals are common in gold-bearing rocks and gold mining has, for a long time, been associated with release of high levels of arsenic into the environment. A well-known case of arsenic poisoning from this source occurred in Reichenstein, Silesia (western Poland) in 1898 (66). The town's water supply (a spring) was severely contaminated with arsenic from waste tailings of a nearby gold mine. About 60 people were reported to have been poisoned from drinking the contaminated water (with up to 26 ppm arsenic), with many developing skin cancer (later referred to as Reichensteiner's disease) (67). Gold mining in Ghana, which intensified during the Portuguese voyages of exploration, has contaminated many rivers and lakes with arsenic (68), resulting in the poisoning of untold numbers of people. Gold rushes represented a romantic chapter in world history at this time in which dreams were sometimes turned into reality but most often resulted in arsenic poisoning (69). Abandoned and forgotten behind the thundering herd of gold rushers was the arsenic nakedly exposed after the gold had been extracted—still a major environmental hazard in many parts of North America.

Coal contains small amounts of arsenic (typically less than 10 ppm) and burning coal for energy contributed to the arsenic burden in the environment, especially in mining areas. In his *Fumifugion* published in 1661, John Evelyn riled against toxic air pollution:

> New Castle cole, as an expert physician affirms, causeth consumptions,
> phthisicks, and the indisposition of the lungs, not only by the suffocating
> abundance of smoake, but also by its virulency: for all subterrany fuel hath
> a kind of virulent or arsenic vapour arising from it (70, p 121).

John Evelyn might have exaggerated the effects of arsenic in local air pollution but that particular source nevertheless represented a health hazard. An epidemic of arsenic poisoning in China has been linked to burning of arsenic-rich coal in stoves to provide heat (26,71).

D. Arsenic in Animal Husbandry

One of the major uses for arsenic compounds was in animal husbandry—from washing scratches and removing insects on cows and horses to sheep-dip. Small doses of arsenic compounds were also widely believed to have a tonic effect, especially on horses. A common belief was that old and worn-out horses recovered their appetite, activity, and strength under the influence of small doses of arsenic (45). Widespread use of arsenic to improve the appearance of horses and make them more energetic was documented in the following 1851 report:

> It must be noted that the usage of arsenic is very extensive even in Vienna among the grooms and especially government drivers. Either they scatter a strong dose of powder on the hay, or they tie a pea-sized piece of cloth and fasten it to a bit, and when the horse is harnessed, the arsenic is gradually dissolved in the saliva. The glossy, round, beautiful appearance of most fine cart horses, and especially the desired foaming is generally due to feeding arsenic (it is well known that arsenic causes increased secretion of saliva). Quite generally in the mountain regions, when the horse must raise heavy loads steeply, the stable hands sprinkle a dose of arsenic on the last portion of fodder. This has been practiced for years without the least disadvantage (21, p 78).

During the early twentieth century, organic arsenicals were being added to animal (especially chicken, turkey, and pig) feeds to promote growth and control disease. This particular application, still in vogue, was another way devised by human ingenuity to contaminate the meat products and increase the dietary intake of arsenic.

E. Accidental Poisoning Became Pandemic

Widespread accidental poisoning by arsenic was inescapable considering the common use of this element in daily commodities and as over-the-counter remedies for the most common ailments. Pigments containing arsenic were employed in numerous consumer products, including fancy and colored papers in magazines and children's books, sheets for cardboard boxes, labels of all kinds, advertising cards, wrappers for candies, confectionary, and sweetmeats, playing cards, lamp shades, paper hangings for walls and other purposes, artificial leaves and flowers, artificial wreaths, wax ornaments for Christmas trees and other purposes, children's toys, printed or woven fabric intended for use as garments, curtains, furniture coverings, painted India rubber dolls, Venetian and other blinds, leather cloth, printed table baizes, book cloth and fancy bindings, decorative tin plates, oil paintings, carpets, floorcloth linoleum, wallpaper, wall paint (Paris-Scheele's-Vienna-Emerald greens, King's or Naples yellow, magenta, and other aniline-based colors), boxes of watercolors, and surprisingly to give color to confectionery ornaments. In addition, arsenic was used in medicated soaps, embalming solutions, preparation of skins for stuffing, adhesive envelopes, glass, fly-powder, rat poison, and sheep-dip. Practically everybody must have come into contact with these products (29,37,45).

Numerous cases of poisonings and fatalities following contact with arsenic in consumer products appeared frequently in the historical records from about 1820 onward (41–50). An epidemic in Paris in 1828 and 1829, referred to as "acrodynie," in which 4000 people were affected by neuritis, digestive disorders, and skin diseases, was subsequently identified as cases of arsenic poisoning (56). In 1842, 14 children were reported to have been poisoned in Dublin by eating some confectionery ornaments colored with copper arsenite, and several children in Manchester also met the same fate from an identical exposure route (37). In 1857, about 340 children were poisoned after drinking milk diluted with water from a boiler into which 9 lb of arsenic was added under the notion that the alkaline arsenite would cleanse the boiler (37). A Bradford confectioner in 1858 mistakenly adulterated lozenges with arsenic rather than the intended plaster of Paris, and poisoned more than 200 people while killing 17 of those who ate the lozenges (37). Six persons were stricken with arsenicism after consuming Bath buns mistakenly colored with yellow orphiment (37). An epidemic of arsenic poisoning occurred around 1900 in Manchester in which over 6000 beer drinkers were affected and 80 died as a result of arsenic-contaminated glucose used in the brewing (51). Besides these well-known cases of mass poisoning by arsenic, the medical jurisprudence publications contain a rich record of individual fatalities from accidental exposure to arsenic (37,45).

F.   Exposure from Other Uses of Arsenic

During this period in arsenic history, many unusual, sometimes bizarre, uses were found for the metalloid. With the development of the Marsh test in 1836, the use of arsenic was shifted from homicide to suicide. A compilation of the common means of suicide in Prussia in 1913, for instance, showed the following startling results:

Men		Women	
Hanging	51%	Hanging	34%
Drowning	12%	Drowning	31%
Shooting	23%	Shooting	3.7%
Poison	6.9%	Poison	21%
Other	6.5%	Other	11%

The reliance of women on poison as the means of suicide, compared with men, may be attributed to expediency or vanity, but the striking parallel with the female criminal preference for poison as an agent of murder cannot go unnoticed; around the turn of the twentieth century, the ratio of female-to-male poison murderers was estimated to be 5:1 (21).

Arsenic was associated with human reproduction in many guises. Some Indian women are said to take arsenical potions early in their pregnancy to increase their chances of having a son (72). Ingestion of arsenic preparations to increase the male sexual appetite and excite sexual pleasure or as an aphrodisiac was well documented among many Asian and European cultures (21,24). The most common use of arsenic was as an abortifacient, however. It has been shown, for instance, that about 30% of all abortions in Sweden between 1851 and 1880 were committed with arsenic (21). Numerous fatalities from attempted abortions with arsenic reported in the medical records of the nineteenth century (37,45) attest to the fact that the practice was common in many countries.

Arsenophagy or habituated eating of arsenic is deeply rooted in folkloric medicine, religious beliefs, and harmful magic (21). The famed arsenic eaters of Styria (Austrian Alps) take a special place in the history of pharmacotoxicology (73,74). Their antics served to minimize the public concern about the dangers of environmental and occupational exposure to arsenic and perpetuate the myth that arsenic may be good for the human race. Somewhat related to arsenophagy of Styria was the prospective arsenic prophylaxis of the fakirs (snake charmers) said to have been widespread in Persia during the middle of the nineteenth century (75).

G. Arsenic in Wars

The war gas lewisite causes skin lesions that are difficult to heal and was highly effective as a killing agent during World War I (49,76). Other arsenicals have equally desirable properties for chemical warfare but details about their production and use are classified and unavailable to the public (76). Arsenic compounds that are less toxic than lewisite but highly irritating to the skin, eyes, and respiratory tract, and can cause dermal pain, lacrimation, sneezing, and vomiting are still available for use as riot-control agents (49). The possibility of using these compounds to poison community water supplies has been studied (77). Cacodylic acid is registered as a silvicide (forest pesticide) and can defoliate and desiccate a wide range of plant species and was employed extensively in South Vietnam as Orange Blue (49).

The nefarious Harmony policy (''gifts'' of food laced with arsenic) in Australia deserves special mention—it was used by British settlers in the 1840s to wipe out a large percentage of the famished aboriginal population in the Manning River basin (78). The sphere of arsenic killing was greatly extended when arsenic was added to molten lead to increase the sphericity of the lead shot. We will never know whether the outcome of any of the wars of the last three centuries would have been different if a perfect lead shot had never been invented. Dr. Thomas Holmes is credited with inventing or at least popularizing arsenic embalming as a sanitary practice during the American Civil War so that soldiers (killed with arsenical lead bullets) were mercifully preserved with arsenic until given proper burial. Until about 1910, arsenic remained the main ingredient in embalming fluids used widely in the United States (79). This practice further interweaves the history of arsenic with that of both the living and the dead. Arsenic might have also intruded when soldiers celebrated the battle victories—most glass contained arsenic, which could be leached out if such glass were used to drink or store wines and acidic juices.

H. Occupational Poisoning by Arsenic Hidden from Public Scrutiny

Despite the numerous reported cases of arsenic poisoning in the medical record, it does not appear that systematic investigations of occupational exposure to arsenic were ever undertaken. The historical silence about workplace exposure to arsenic was unprecedented. None of the pioneers of occupational medicine (including Paracelsus, George Agricola, Bernardino Ramazzani, Charles Thackrah, T. Oliver, J. T. Arlidge or T. M. Legge) mentioned arsenic in their lists of dangerous trades (3). Although there were concerns about the effects of arsenic emitted by smelters on livestock, exposure of workers and people in surrounding communities was never seriously addressed until about the 1950s (80). It was the poisoning of bees, which threatened the avian industry (56), rather than concern for farm workers who applied the arsenical pesticides, that nearly derailed the commercialization of these dangerous compounds. Ignorance on the part of employers cannot be blamed for poisoning of people at work with arsenic, and there is no evidence to suggest that exposure to arsenic was eliminated in companies that produced the arsenic and its compounds. Occupational poisoning would thus appear to be one of the leading miasmatic crimes of historic proportion that has been committed with arsenic.

V. RECENT TIMES

During the past few decades, the growing public concern about the health effects of arsenic in our food, water, and air has driven down the worldwide consumption of arsenic trioxide to less than 20,000 tons per year, but arsenic still remains a part of our daily lives. Today, we see traces of arsenic everywhere, in our food, water, air, and soil. Arsenic contamination of natural resources has emerged as one of the major public health problems in many countries. Worldwide, a large number of people are exposed to chronic doses of arsenic with health risks that may include vascular diseases, jaundice, hyperkeratosis, and cancers of various organs and tissues. Financial and social implications of any technical solutions to the legacy of extensive and long-term use of the element for human health and welfare are enormous.

A. Organoarsenicals Become Prevalent

Since the early 1900s, over 32,000 organoarsenic compounds have been synthesized and some have become important elements of industry and commerce. The introduction of salvarsan for widespread treatment of syphilis has already been noted. In subsequent years (mostly after World War II) many other efficacious organoarsenical therapies were developed. Some of these still remain the last line of defense against a number of common human and veterinary diseases. Well-known examples include atoxyl and melarsoprol for sleeping sickness, thiacetarsimide as a deworming agent in dogs and cats, and roxarsone and atoxyl to promote growth by eradicating parasites and prevent dysentery in poultry and pigs (53,55). These compounds, often administered by intravenous injection, are toxic and side effects, such as reactive encephalopathy leading to fatalities and degeneration of the optic nerve leading to blindness, have been reported (55).

The commercialization of organoarsenicals for agricultural purposes was a post–World War II development. During the 1970s and 1980s, these compounds, especially monosodium methanoarsonate (MSMA), disodium methanoarsonate (DSMA), dimethylarsenic acid (DMSA), and arsonic acid, became one of the largest herbicides in terms of volume used worldwide (81). In the United States alone, 10–12 million acres were annually treated with 2.1 million kg of MSMA + DSMA and 2.1 million acres were treated with 3.3 million kg of arsonic acid (82). Although the toxicity of these organoarsenic herbicides is generally low, they are ultimately biodegraded into the more toxic inorganic forms, which have been added permanently to the arsenic burden in the human food chain. The number of farm workers poisoned by organoarsenical herbicides will never be known. Large quantities of chromated copper arsenate (patented in 1938) and ammoniacal copper arsenate (patented in 1939) are still being used as wood preservatives (27). There are reported cases of people being poisoned by burning wood so treated as well as children being poisoned by playing in soils contaminated with arsenic leached out of treated lumber.

B. Accidental Poisoning Remains Problematic

In spite of reduced use of arsenic in commercial products, accidental poisoning by arsenic remains a problem in many countries. In 1955, drinking of dry milk made with arsenic-contaminated phosphate used as a stabilizer resulted in chronic and acute poisoning of 13,419 Japanese children, of whom 839 have died (83).

The following year (1956) another 400 Japanese were poisoned by soy sauce contaminated with arsenic (84). Of the 5000 cases of heavy metal ingestion reported to the American Association of Poison Control Centers in 1984, arsenic was found to be the most commonly involved (over 1200 incidents) in the poisonings (85). The medical literature remains replete with reported cases of accidental arsenic poisoning involving individuals and their families.

C. Growing Health Risks from Arsenic in the Environment

Most (over 80%) of all the arsenic ever produced by humankind was used in environmentally dissipative manners—as herbicides, insecticides, desiccants, feed additives, wood treatments, chemical warfare agents, and drugs. Furthermore, a large percentage of the arsenic in the ores mined and smelted was never recovered and was instead released directly to the environment. In many respects, humankind appears to have released the curse of the evil smoke unto itself by ignoring its warnings:

> I am an evil, poisonous smoke
> But when from poison I am freed,
> Through art and sleight of hand,
> Then I can cure both man and beast,
> From dire disease oft-times direct them;
> But prepare me correctly, and take great care
> That you faithfully keep watchful guard over me;
> For else I am poison, and poison remain,
> That pierces the heart of many a one. (Valentini, 1694) (86, p 92)

Contamination of the environment with arsenic from both anthropogenic and natural sources has occurred in many parts of the world and is a global problem (87). In many areas, the levels of arsenic in the environment have exceeded the safe threshold and epidemiological studies have documented various adverse effects on the human population. Examples include areas around base metal smelters, gold mines, power plants that burn arsenic-rich coals or treated lumber, disposal sites for wastes from arsenic processing plants, industrial and municipal dump sites, and soils naturally enriched in arsenic. Many cases of arsenic poisoning have, for instance, been reported in Japan involving pollution around arsenic mines, pollution of groundwater around arsenic-using industries, and pollution of groundwater by industrial waste burial sites (88). In most other areas, arsenic levels are elevated well above background values but the human effects of the exposure dose are not yet clear. Examples include agricultural lands treated with arsenical pesticides, urban areas, war zones defoliated or sprayed with arsenic compounds, etc. Elevated levels of arsenic from natural and industrial sources have been reported in groundwater in Taiwan, China, India,

Bangladesh, Thailand, Chile, Argentina, Mexico, Canada, and the United States. Contamination of groundwater is increasingly being reported in other countries as their water supplies are tested for arsenic; as was the case with lead poisoning, the more one looks, the more one finds arsenic in groundwater. It is estimated that about 50 million people in Bangladesh (89,90) and 6 million people in China (26) are at risk of being poisoned by ingesting water with arsenic levels above 50 µg/L; worldwide, the at-risk population is estimated to be over 150 million.

As a result of wide distribution of arsenic in all environmental media, most people are now daily exposed to a measurable dose of arsenic. Recent studies increasingly find health effects at levels of exposure previously thought to be safe, and the threshold dose for health outcome continues to be progressively lowered. There is therefore a growing worldwide concern that a very large number of people are being chronically exposed to levels of arsenic that may be inimical to health. Chronic exposure to low levels of arsenic can affect the skin, liver, kidney, circulatory system, gastrointestinal tract, nervous system, and heart (91). Recent epidemiological studies have shown that exposure to inorganic arsenic in environmental media (especially in drinking water) increases the risk of skin cancer and other internal tumors of the bladder, liver, kidney, and lung (92). Reproductive effects (such as congenital malformations, low birth weight, and spontaneous abortion), genotoxicity, mutagenicity, and teratogenicity of arsenic are being increasingly reported at environmental levels of exposure (93). The human health implications of exposure to pollutant and natural arsenic in the environment have yet to be fully identified and quantified, and the arsenic problem will likely remain with us for another millennium.

REFERENCES

1. V Biringuccio. The Pirotechnia, 1540. Translated by CS Smith and MT Gnudi, Cambridge, MA: MIT Press, 1941.
2. JA Charles. The coming ages of copper and copper-based alloys and iron: A metallurgical sequence. In: TA Wartime, JD Muhle, eds. The Coming of the Age of Iron. New Haven, CT: Yale University Press, 1980.
3. M Harper. Occupational health aspects of arsenic extractive industry in Britain (1868–1925). Br J Indust Med 45:602–605, 1988.
4. IJ Polmer. Metallurgy of the elements. In: NC Norman, ed. Chemistry of Arsenic, Antimony and Bismuth. London, Chapman & Hall, 1998, pp 39–63.
5. JO Nriagu. Lead and Lead Poisoning in Antiquity. New York: Wiley, 1983.
6. K Alterman. From Horus the child to Hephaestus who limps: A romp through history. Am J Med Genet 83:53–63, 1999.
7. Apollodorus. The Library. Translated by Sir George Frazer. Loeb Classical Library. Cambridge, MA: Harvard University Press, 1956.

8. TK Derry, TI Williams. A Short History of Technology. New York: Oxford University Press, 1961.
9. SC Carapella. Arsenic and arsenic alloys. In: M Grayson, D Eckroth, eds. Kirk-Othmer Encyclopedia of Chemical Technology. Vol. 3. New York: Wiley, 1978, pp 243–250.
10. C Singer, EJ Holmyard, AR Hall, TI Williams. A History of Technology. Clarendon: Oxford University Press, 1957.
11. CS Smith, JG Hawthorne. Mappae Clavicula. Trans Am Philos Soc 64 (part 4):1–128, 1974.
12. KN Bagachi. Poisons and Poisoning—Their History and Romance and Their Detection in Crime. Bengal: University of Calcutta Press, 1969.
13. A Lucas, JR Harris. Ancient Egyptian Materials and Industries. London: Edward Arnolds, 1962.
14. RJ Forbes. Studies in Ancient Technology. Leiden, Netherlands: Brill, 1964.
15. ER Caley. Ancient Greek pigments. J Chem Educ 23:314–316, 1946.
16. Pliny, the Elder. Natural History. Translated by Bostock and Riley. London: Geo, Bell & Sons, 1856.
17. Strabo. The Geography. Translated by HL Jones. Loeb Classical Library. Cambridge, MA: Harvard University Press, 1960–1966.
18. DV Frost. What is in a name? Sci Total Environ 35:1–6, 1984.
19. JR Partington. Origins and Development of Applied Chemistry. London: Longmans, Green & Co, 1935.
20. The Greek Herbal of Dioscorides. Translated by RT Gunther. Clarendon: Oxford University Press, 1934.
21. KH Most. Arsenic as a Poison and Charm in German Folk Medicine with Particular Respect to Styria. Santa Barbara, CA: Scitran, 1939.
22. KB Mookerji. Rasa-Jala Nidsi or Ocean of Indian Chemistry and Alchemy. Ahmedabad, India: Avani Prakashan, 1984.
23. FP Smith. Contributions Towards the Materia Medica and Natural History of China. London: Turner & Co, 1871.
24. J Needham. Clerks and Craftsmen in China and the West. Cambridge: Cambridge University Press, 1970.
25. EO Espinola, MJ Mann, B Bleasdell, S DeKorte, M Cox. Toxic metals in selected traditional Chinese medicinals. J Forensic Sci 41:453–456.
26. C Zhai, B Zheng. Researches on the health effects of arsenic in China. In: JA Centeno, P Collery, G Vernet, RB Finkelman, H Gibb, JC Etienne, eds. Metal Ions in Biology and Medicine. Vol. 6. Rome: John Libbey Eurotext, 1999, pp 32–33.
27. JO Nriagu, JM Azcue. Arsenic: Historical perspectives. In: JO Nriagu, ed. Arsenic in the Environment. Vol. 1. New York: Wiley, 1994, pp 1–15.
28. J Beckman. A History of Inventions, Discoveries and Origins. London: Henry G. Bohn, 1846.
29. JW Mellor. A Comprehensive Treatise on Inorganic and Theoretical Chemistry. Vol. 9. London: Longman, 1952, pp 1–337.
30. Albertus Magnus. Book of Minerals. Translated by D. Wyckoff. Clarendon: Oxford University Press, 1967.
31. L Aitchinson. A History of Metals. London: McDonald & Evans, 1960.

32. D Hunter. The Diseases of Occupations. London: Hodder & Stoughton, 1978.
33. CL Eastlake. Methods and Materials of Painting of the Great Schools and Masters. New York: Dover, 1960.
34. DV Thompson. The Materials and Techniques of Medieval Painting. New York: Dover, 1956.
35. MP Merrifield. Original Treatises on the Arts of Painting. London: John Murray, 1849.
36. CD Cennini. The Craftman's Handbook [1437]. Translated by DV Thompson. New York: Dover, 1954.
37. AS Taylor. On Poisons in Relation to Medical Jurisprudence and Medicine. Philadelphia: Henry C. Lea, 1875.
38. J Kesselring. Arsenic and Old Lace. New York: Random House, 1941.
39. MW van Denburg. A Homeopathic Materia Medica. Port Edward, NY: Published by the author, 1895.
40. JP Porta. Natural Magic [1558]. Reprinted by Basic Books, Norwalk, CT, 1957.
41. RE Griffith. A Dispensatory or Commentary on the Pharmacopoeias of Great Britain and the United States. Philadelphia: Lea & Blanchard, 1848.
42. K. Sudhoff. Bibliographia Paracelsica. Graz: Akademische Druck-u. Verlagsanstalt, 1958.
43. DW Jenkins. The toxic elements in your future and your past. Smithsonian 3:62–69, 1972.
44. N Chapman. Elements of Therapeutics and Materia Medica. Philadelphia: Carey & Lea, 1825.
45. JG Millingen. Curiosities of Medical Experience. Philadelphia: Haswell, Barrington & Haswell, 1838.
46. J Pereira. The Elements of Materia Medica and Therapeutics. Philadelphia: Lea & Blanchard, 1843.
47. JF Royle. Materia Medica and Therapeutics. Philadelphia: Lea & Blanchard, 1847.
48. R Christison. A Treatise on Poisons. Philadelphia: Barrington & Haswell, 1845.
49. National Academy of Sciences. Arsenic. Washington, DC: U.S. National Academy of Sciences Press, 1977.
50. JV Shoemaker. A Practical Treatise on Materia Medica and Therapeutics. Philadelphia: Davis, 1902.
51. AA Stevens. Modern Materia Medica and Therapeutics. Philadelphia: Saunders, 1903.
52. JW Holland. Arsenic. In: F Peterson, WS Haines, eds. A Textbook of Legal Medicine and Toxicology. Philadelphia: Saunders, 1904.
53. DM Jolliffe. A history of the use of arsenicals in man. J Royal Soc Med 86:287–289.
54. AL Thorburn. Paul Ehrlich: Pioneer of chemotherapy and cure by arsenic (1854–1915). Br J Vener Dis 59:404–405.
55. J Reglinski. Environmental and medicinal chemistry of arsenic, antimony and bismuth. In: NC Norman, ed. Chemistry of Arsenic, Antimony and Bismuth. London: Chapman & Hall, 1998, pp 403–445.
56. JC Whorton. Insecticide Residues on Foods as a Public Health Problem: 1865–1938. Dissertation, University of Wisconsin, Madison, Ann Arbor, University Microfilms, 1970.

57. MJB Orfila. A General System of Toxicology. Philadelphia: Carey & Sons, 1817.

58. Minerals Yearbooks. Metals and Minerals. Washington, DC: Bureau of Mines, U.S. Department of the Interior, 1923–1998.

59. JC Aldren. The continuing need for inorganic arsenical pesticides. In: WH Lederer, RJ Fensterheim, eds. Arsenic: Industrial, Biomedical and Environmental Perspectives. New York: Van Nostrand Reinhold, 1983, pp 63–70.

60. A Kallet, FJ Schlink. 100,000,000 Guinea Pigs: Dangers in Everyday Foods, Drugs and Cosmetics. New York: Vanguard, 1933.

61. JO Nriagu, JM Azcue. Food contamination with arsenic in the environment. In: JO Nriagu, MS Simmons, eds. Food Contamination from Environmental Sources. New York: Wiley, 1990, pp 121–143.

62. WB Hills. Chronic arsenic poisoning. Bost Med Surg J 131:453–478, 1894.

63. DV Frost. The two faces of arsenic: Can arsenophobia be cured? In: M Anke, HJ Schneider, C Bruckner, eds. Arsenic. Jena: Friedrich-Schiller University, 1980, pp 17–23.

64. WD Harkins, RE Swain. The chronic arsenical poisoning of herbivorous animals. J Am Chem Soc 30:928–942, 1908.

65. WC Ebaugh. Gases versus solids: An investigation of the injurious ingredients of smelter smoke. J Am Chem Soc 29:951–970.

66. L Geyer. Ueber die chronischen Hautveranderungen biem Arsenicismus und Betrachtungen. Arch Dermatol Syphilol 43:221–289, 1898.

67. GR Peters, RF McCurdy, JT Hindmarsh. Environmental aspects of arsenic toxicity. Critical Rev Clin Lab Sci 33:457–493, 1996.

68. PL Smedley, WM Edmunds, KB Pelig-Ba. Mobility of arsenic in groundwater in the Obuasi gold-mining area of Ghana: Some implications for human health. In: JD Appleton, R Fuge, GJH McCall, eds. Environmental Geochemistry and Health. London: Geological Society Special Publication No. 113, 1996, pp 163–181.

69. JO Nriagu, HKT Wong. Gold rushes and metal pollution. Metal Ions in Biological Systems 34:131–160, 1997.

70. J Lehihan. The Crumbs of Creation. Bristol: Adam Hilger, 1988.

71. Z Aihua, H Xiaoxin, J Xianyao, L Peng, G Yucheng, X Shouzheng. The progress of study on endemic arsenism due to burning arsenic containing coal in Guizhou province. In: JA Centeno, P Collery, G Vernet, RB Finkelman, H Gibb, JC Etienne, eds. Metal Ions in Biology and Medicine. Vol. 6. Rome: John Libbey Eurotext, 2000, pp 53–55.

72. KA Winship. Toxicity of inorganic arsenic salts. Adverse Drug React Acute Poisoning Rev 3:129–160, 1984.

73. C Maclagan. On the arsenic eaters of Styria. Edinburgh Med J 10:200–207, 1864.

74. EW Schwartze. The so-called habituation to "arsenic": Variation in the toxicity of arsenious oxide. J Pharmacol Exp Ther 20:181–203, 1922.

75. L Lewin. Gifte u. Vergiftungen. Berlin: Verlag G. Stille, 1921.

76. JF Bunnett, M Mikolajczk. Arsenic and Old Mustard: Chemical Problems in the Destruction of Old Arsenical and Mustard Munitions. Boston: Kluwer, 1998.

77. CC Ruchhoft, OR Placak, S Schott. The detection and analysis of arsenic in water contaminated with chemical warfare agents. Public Health Rep 58:1761–1771, 1943.

78. N Marr. Aboriginal history of the Great Lakes district, Australia. http:// www.greatlakes.nsw.gov.au/commprof/aborigin.htm, 1995.
79. JL Konefes, MK McGee. Old cemeteries, arsenic and health safety. http:// waterindustry.org/arsenic-3.hml, 2000.
80. PB Mushak, W Gaike, V Hasselblad, L Grant. Health Assessment Document for Arsenic. Research Triangle Park, NC: Environmental Criteria and Assessment Office, Office of Research and Development, U.S. Environmental Protection Agency, 1980.
81. RD Wauchope, LL McDowell. Adsorption of phosphate, methanearsonate and cacodylate by lake and stream sediments: comparison with soils. J Environ Qual 13: 499–504, 1984.
82. JR Abernathy. Role of arsenical chemicals in agriculture. WH Lederer, RJ Fensterheim, eds. Arsenic: Industrial, Biomedical and Environmental Perspectives. New York: Van Nostrand Reinhold, 1983, pp 57–62.
83. A Kanazawa, T Tohyama, Y Baba, et al. 40 years follow-up study on mental sequelae to an accidental mass arsenic poisoning in Japan. In: JA Centeno, P Collery, G Vernet, RB Finkelman, H Gibb, JC Etienne, eds. Metal Ions in Biology and Medicine. Vol. 6. Rome: John Libbey Eurotext, 2000, pp 74–76.
84. Y Tsuda, A Babazono, T Ogawa, H Hamada, et al. Inorganic arsenic: A dangerous enigma for mankind. Appl Organometal Chem 6:309–322, 1992.
85. L Fuortes. Arsenic poisoning: Ongoing diagnostic and social problem. Postgrad Med 83:233–244, 1988.
86. ME Weeks, HM Leicester. Discovery of the Elements. 7th ed. Easton, PA: Journal of Chemical Education, 1968, p 92.
87. I Thornton. Arsenic in the global environment: Looking towards the millennium. In: WR Chappell, CO Abernathy, RL Caldron, eds. Arsenic Exposure and Health Effects. Amsterdam: Elsevier, 1999, pp 1–7.
88. T Tsuda, T Ogawa, A Babazono, H Hamada, S Kanazawa, Y Mino, H Aoyama, E Yamamoto, N Kurumatani. Historical cohort studies in three arsenic poisoning areas in Japan. Appl Organometal Chem 6:333–341, 1992.
89. UK Chowdhury, BK Biswas, RK Dhar, G Samanta, BK Mandal, TR Chowdhury, D Chakraborti, S Kabir, S Roy. Groundwater arsenic contamination and suffering of people in Bangladesh. In: WR Chappell, CO Abernathy, RL Caldron, eds. Arsenic Exposure and Health Effects. Amsterdam: Elsevier, 1999, pp 165–182.
90. AH Smith, ML Biggs, L Moore, R Haque, C Steinmaus, J Chung, A Hernandez, P Lopipero. Cancer risks from arsenic in drinking water: Implications for drinking water standards. In: WR Chappell, CO Abernathy, RL Caldron, eds. Arsenic Exposure and Health Effects. Amsterdam: Elsevier, 1999, pp 191–199.
91. MS Gorby. Arsenic in human medicine. In: JO Nriagu, ed. Arsenic in the Environment. Vol. 2. New York: Wiley, 1994, pp 1–16.
92. CJ Chen, CW Chen, MM Wu, TL Kuo. Cancer potential in liver, lung, bladder, and kidney due to ingested inorganic arsenic in drinking water. Br J Cancer 66:888–892, 1992.
93. Agency for Toxic Substances and Disease Registry. Toxicological Profile for Arsenic. Atlanta, GA: Agency for Toxic Substances and Disease Registry, U.S. Department of Health & Human Services, 1999.

2

Analytical Methods for the Determination of Arsenic and Arsenic Compounds in the Environment

Walter Goessler and Doris Kuehnelt
Karl-Franzens University, Graz, Austria

I. INTRODUCTION

Arsenic has been known for a very long time. The properties of arsenic sulfides (As_4S_4, As_2S_3) and arsenic trioxide (As_2O_3) were already known to physicians and professional poisoners in the fifth century B.C. The dosage of these compounds to achieve toxicosis or death and their beneficial effects were described (1). Albertus Magnus (1193–1280) is credited with the isolation of elemental arsenic by heating auripigment (As_2S_3) with soap.

Arsenic is ubiquitous in the environment. The natural abundance in the earth's crust is 1.8 mg/kg. In regions with abundant volcanic rocks or sulfidic ores, the arsenic concentration is elevated. Also coal, especially brown coal (lignite), usually shows higher arsenic concentrations (up to 1500 mg/kg) (2). Whereas the arsenic concentrations in uncontaminated terrestrial organisms are low (3,4,5), usually <0.5 µg/kg dry mass, the arsenic concentrations in marine organisms may be several hundreds of milligrams of arsenic per kilogram dry mass (6,7).

The beneficial effects of arsenic, or better, its compounds, to human and animal life have been known for a long time. A variety of arsenic-containing formulations were used for curing various diseases. Reduction of fever, prevention of black death, healing of boils, treatment of chronic myelocytic leukemia, and/or healing of psoriasis are only a few positive responses to these formulations (8). These beneficial effects of arsenic were not the driving force to develop methods for its determination, but rather the hundreds of murders that have been committed all over Europe with the help of arsenic trioxide. Its lack of smell, color, and taste made As_2O_3 (white arsenic or huettrach) the ideal poison because it was easily mixed into the food or dissolved in beverages. In 1786, Hahnemann, the German physician, wrote a book entitled *About Arsenic Poisoning* in which he described the determination of arsenic at trace levels with respect to forensic medicine. It took another 40 years until Marsh developed his famous test that allowed the determination of traces of arsenic. The establishment of the Marsh test in forensic laboratories resulted in a steady decrease of murder with As_2O_3 (9).

The detection limits of the "old" methods for the determination of arsenic (10) were too high to determine arsenic in uncontaminated biological samples. With the invention of instrumental techniques, such as flame atomic absorption (emission) spectrometry, graphite furnace atomic absorption spectrometry, neutron activation analysis, inductively coupled plasma atomic emission spectrometry, and inductively coupled plasma mass spectrometry, the ubiquity of arsenic in our environment was proven. The improvement of the analytical techniques has changed the reputation of arsenic from a poisonous substance to an essential trace element at least for warm-blooded animals (11). An arsenic requirement for humans cannot be deduced from these animal experiments. In recent literature, there are certainly more hints that arsenic might be an essential trace element for humans, but there is still a lot of future research work necessary to prove this.

II. ANALYTICAL METHODS FOR THE DETERMINATION OF TOTAL ARSENIC CONCENTRATIONS

The determination of total arsenic concentrations in biological samples requires, in most cases, complete destruction of the organic matrix. During this mineralization, all the organic arsenic compounds should be converted into inorganic arsenic and a loss of the analyte should be prevented. In samples with high halide concentrations, loss of arsenic can occur. Dry ashing or evaporation of an acid digest to dryness might also result in loss of arsenic, as the oxides may sublime. Arsenic trioxide sublimes at 180°C, whereas arsenic pentoxide (As_2O_5) releases oxygen at temperatures above 300°C and decomposes to arsenic trioxide (12). Dry ashing is usually performed with inorganic oxidants such as magnesium nitrate/magnesium oxide mixtures. For wet ashing, acid mixtures of nitric acid and perchloric acid, or mixtures with nitric acid, perchloric acid, and sulfuric acid are commonly employed. Microwave-assisted heating in closed pressurized Teflon® bombs is nowadays widely exploited for the mineralization of samples (13). The closed pressurized bombs for sample preparation have the advantage of lower analytical blanks because no particles from the laboratory atmosphere can reach the sample and lower amounts of acids are necessary for destruction of the sample. Typical temperatures that can be reached in closed Teflon bombs are 260°C. Higher temperatures must be avoided because above 260°C Teflon is destroyed. For higher temperatures, quartz vessels are a good material for trace element analysis. Systematic investigations have shown that arsenobetaine, the dominating arsenic compound in marine animals, is almost entirely converted to trimethylarsine oxide when heated with nitric acid to 220°C in a microwave-assisted autoclave. When inductively coupled plasma mass spectrometry is used for the determination of arsenic, correct results will be obtained, whereas lower recoveries will be attained with the hydride generation technique. An almost quantitative conversion of arsenobetaine to arsenate was only found when the temperature was kept at 300°C for 1 hr (Fig. 1) (14). The mineralization of biological samples is very often conducted without knowledge of the arsenic compounds present. Many of the methods used for the determination of total arsenic are influenced by the arsenic compounds present in solution. The use of improper standard solutions for calibration (very often inorganic salts) will result in false results for the arsenic concentration in the sample. The ideal sample solution, an acidified solution with

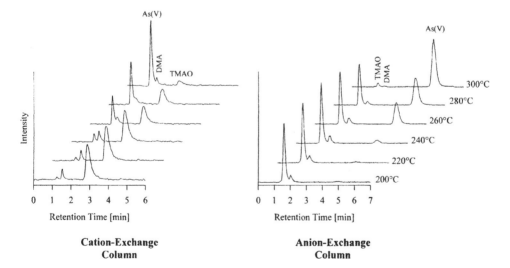

Figure 1 Anion- and cation-exchange chromatograms of nitric acid digests of NRCC DORM 2 heated to different temperatures. An HP4500 inductively coupled plasma mass spectrometer was used as element-specific detector.

no matrix and all analytes present as inorganic ions, is not always easily attainable. It is the job of the analytical chemist to find a compromise among the trueness, accuracy, and costs of the analysis. If the costs are not an important factor (which is very seldom the case) many of the results produced would be more accurate.

In the following sections, several analytical methods are discussed that are easily accessible. Neutron activation analysis is not discussed although it is a reference method for the determination of arsenic.

A. Spectrophotometric Methods

1. Marsh Test

Although the Marsh test is not a spectrophotometric determination for arsenic, it is discussed in this section. In forensic laboratories, the Marsh test has already been replaced by modern analytical instrumentation. Nevertheless, it is worth mention as a very elegant way to prove the presence of arsenic in a sample.

The sample containing arsenite is treated with zinc/hydrochloric acid (Zn/HCl) in a closed vessel. During this reaction, hydrogen and the volatile arsine AsH_3 are formed. When the hydrogen is ignited and the flame is directed to a cold surface, metallic arsenic is formed on this surface. The metallic arsenic is easily redissolved in an alkaline hydrogen peroxide solution.

2. Molybdenum Blue Method

A prerequisite for the molybdenum blue method is that all the arsenic has to be present as arsenate. After digestion with oxidizing acids, such as nitric acid, all the arsenic is converted into arsenate when appropriate heating time and temperatures are applied. The principle of this determination is the reaction of arsenate with ammonium molybdate in acidic medium to form an arsenate containing molybdenum heteropolyacid that can be reduced to molybdenum blue with stannous chloride, hydrazine, or ascorbic acid. Best results are obtained with hydrazine sulfate. The absorption maximum of the blue solution is between 840–860 nm (15). The most severe interferences for this method derive from phosphates and silicates. To remove interfering ions, distillation of arsenic as $AsCl_3$ or $AsBr_3$ is often recommended (12,15).

3. Silver Diethyldithiocarbamate Method

The silver diethyldithiocarbamate method is one of the most popular spectrophotometric methods for the determination of arsenic. The method is based on the generation of arsine (AsH_3) either with zinc and hydrochloric acid or sodium borohydride in acidic solutions. The arsine gas is then flushed through a solution of diethyldithiocarbamate in pyridine or pyridine/chloroform (16). The red-colored complex can be measured at 520 nm. The method is selective for arsenite. Organic arsenic compounds have to be completely converted to arsenate and reduced to arsenite before measurement, as the volatile arsines $Me_x AsH_{3-x}$ (x = 1–3) also form colored complexes but with a different absorption compared with arsine. The silver diethyldithiocarbamate method is mainly used for the determination of arsenic in water samples.

B. Electrochemical Techniques

Generally, several electrochemical methods are available for the determination of arsenic at trace levels. Most of the electrochemical methods suffer from severe matrix interferences. Simple measurements are only possible if the matrix is completely eliminated either by a chromatographic separation or by complete mineralization (17). Direct-current polarography is able to determine arsenite at concentrations above 0.7 mg/L, a concentration certainly too high for the determination of arsenic in uncontaminated environmental samples. The detection limit for arsenite with differential pulse polarography, a very common polarographic technique nowadays, is about 20 µg/L (17). Reasonably low detection limits of 0.3 µg As/L were obtained when potentiometric stripping analysis was used for the determination of arsenic (18). All electrochemical methods that can be exploited for the determination of arsenic have been recently reviewed by Greschonig and Irgolic (19). Electrochemical methods can be used for the determination of arsenic, but the techniques suffer from severe interferences and lack of robustness. Therefore, electrochemical methods are not commonly employed.

C. Atomic Absorption Spectrometry

Since flame atomic absorption spectrometry (FAAS) suffers from interferences of the combustion process and, therefore, high detection limits (~1 mg As/L), this technique was never seriously considered to be practicable for the determination of arsenic in environmental samples. Nevertheless, the combination of FAAS with the hydride generation technique is certainly one of the most widely used techniques for the determination of arsenic in environmental samples (see Sec. II.F). The use of electrothermal atomic absorption spectrometry (ETAAS) or graphite furnace atomic absorption spectrometry (GFAAS) has a long tradition for the determination of arsenic in environmental samples. Without the use of matrix modifiers, arsenic might be lost during the ashing stage (arsenic sublimes at 613°C). It was also shown that different arsenic compounds show large differences in the signal intensities when analyzed without the addition of matrix modifiers (20). The nitrates of nickel (Ni), palladium (Pd), magnesium (Mg), or mixtures of these compounds are successfully utilized as matrix modifiers to prevent volatilization during ashing and to give uniform signal intensities irrespective of the arsenic compounds present in the sample. Recently, the systematic investigation of combinations of magnesium, copper, nickel, and palladium nitrates as matrix modifiers for the determination of arsenic in marine biological tissues was published (21). The authors have shown that mixtures of Ni/Mg and Pd/Mg modifier combinations have similar effects on the response on the arsenic compounds investigated. To correct for spectral interferences, modern instruments are equipped with deuterium or Zeeman background correction. Phosphorus can reduce the signal for arsenic even when matrix modification and Zeeman background correction are used. As a possible solution for this effect, the use of large amounts of modifiers is suggested (21). Slurry sampling for ETAAS has proven to be a popular technique for the analysis of solid samples as no sample mineralization is required. For the preparation of slurries, ultrasonic mixing is frequently employed. Arsenic in baby food was successfully determined by the preparation of slurries in a suspension medium containing 0.1% (w/v) Triton X-100, 20% (v/v) hydrogen peroxide, 1% (v/v) concentrated nitric acid, and 0.3% nickel nitrate. After sonication, the slurry was directly introduced into the furnace and analyzed (22).

Electrothermal atomic absorption spectrometry is certainly a method for the determination of arsenic at trace concentration levels. Nevertheless, it has to be stated that this technique must be used carefully as arsenic might be lost during ashing and matrix interferences may occur. An *ideal* heating program cannot be given without knowledge of the sample composition.

D. Inductively Coupled Plasma Atomic Emission Spectrometry

Inductively coupled plasma atomic emission spectrometry (ICP-AES) involves a plasma, usually argon, at temperatures between 6000 and 8000 K as excitation source. The analyte enters the plasma as an aerosol. The droplets are dried, desolvated, and the matrix is decomposed in the plasma. In the high-temperature region of the plasma, molecular, atomic, and ionic species in various energy states are formed. The emission lines can then be exploited for analytical purposes. Typical detection limits achievable for arsenic with this technique are 30 µg As/L (23). Due to the rather high detection limit, ICP-AES is not frequently used for the determination of arsenic in biological samples. The use of special nebulizers, such as ultrasonic nebulization, increases the sample transport efficiency from 1–2% (conventional pneumatic nebulizer) to 10–20% and, therefore, improves the detection limits for most elements 10-fold. In addition to the fact that the ultrasonic nebulizer is rather expensive, it was reported to be matrix sensitive (24). Inductively coupled plasma atomic emission spectrometry is known to suffer from interferences due to the rather complex emission spectrum consisting of atomic as well as of ionic lines. To reduce the interferences and to increase the detection limits, ICP-AES is frequently coupled to hydride generation systems (12,25).

E. Inductively Coupled Plasma Mass Spectrometry

Among the methods commonly used for arsenic determination, inductively coupled plasma mass spectrometry (ICP-MS) is superior with respect to detection limits, multielement capabilities, and wide linear dynamic range. This technique combines the inductively coupled plasma as ion source with a mass analyzer. Quadrupole mass filters are the most common mass analyzers because they are rather inexpensive. Double-focusing magnetic/electrostatic sector instruments and time-of-flight mass analyzers are also used (26,27). At temperatures of 7500 K, about 50% of the arsenic is ionized to $^{75}As^+$, which can then be detected with the mass analyzer (26). The determination of the monoisotopic arsenic is known to be hampered when samples with a high chloride concentration, such as urine, have to be analyzed, because a molecular interference deriving from $^{40}Ar^{35}Cl^+$ overlaps the arsenic signal if no high-resolution instrument is used. Several strategies have been developed to solve this problem when quadrupole-based instruments are employed. Mathematical corrections, addition of nitrogen to the carrier gas (28,29), addition of ethanol to the samples (29), addition of tertiary amines (30), and chromatographic separation (31) have been successfully applied to overcome this problem. Recently, a 1000-fold reduction of the $^{40}Ar^{35}Cl^+$ interference was reported when a dynamic reaction cell with hydrogen as the reaction gas was used for the determination of arsenic in sodium chloride solutions (32). Sector field ICP-MS instruments operated at the high-resolution mode ($M/\Delta M = 7500$) were successfully applied to overcome the $^{40}Ar^{35}Cl^+$ interference (33,34). When the high-resolution mode is used, the transmission is reduced 100-fold and thus the detection limit is increased. The comparison of ICP-MS operated at the high-resolution mode with hydride generation–ICP-MS (HG-ICP-MS) at the low-resolution mode (34) showed that HG-ICP-MS is advantageous when arsenic in water samples at trace levels has to be determined.

F. Hydride Generation Technique

Hydride generation (HG) is one of the most frequently used methods for the determination of arsenic at trace concentrations. The advantage of this method is that the hydride (arsine) generation is easily connected to various detection systems and improves the detection limits of almost all methods 100-fold. Hydride generation is based on the production of volatile arsines either by zinc/hydrochloric acid or sodium borohydride/acid mixtures. The volatile arsines are transported by an inert gas to the detection system. When the pH of the hydride generation reaction is carefully controlled, differentiation between As(III) and

As(V) is possible. At pH 5, only As(III) compounds can be reduced to volatile arsines. At pH <1, arsenic acid, dimethylarsinic acid, and methylarsonic acid also form volatile arsines. Qualitative detection of the arsines as As^0 is performed in the Marsh test (see Sec. II.A.1). When the arsine reacts with silver nitrate, a yellow compound, $Ag_3As\cdot3AgNO_3$, is formed (Gutzeit test). The mercuric bromide stain method uses $HgBr_2$ instead of $AgNO_3$. A quantification of the formed yellowish brown compounds is performed visually (35). The first coupling of HG with atomic spectroscopic detection was reported in 1969 (36). From this time on, numerous researchers exploited the hydride generation method for arsenic determination. Hydride generation–colorimetry (see Sec. II.A.3), hydride generation–atomic absorption spectrometry (37), hydride generation–atomic fluorescence spectrometry (38), trapping of arsines and subsequent electrothermal atomic absorption quantification (39), HG-ICP-AES (40), HG-ETV-ICP-MS (41), and HG-ICP-MS were successfully employed for the determination of arsenic at trace and ultratrace levels. A prerequisite for accurate results is that all the arsenic is present in one chemical form, generally arsenite, because the arsenic compounds, which are convertible to volatile arsines, give different responses irrespective of the detection system used.

III. ANALYTICAL METHODS FOR THE DETERMINATION OF ARSENIC COMPOUNDS

In 1922, the British scientist Jones reported at the British Pharmaceutical Conference in Nottingham that he examined various seaweed samples and found rather high arsenic concentrations in the range of ~5–94 mg As/kg dry mass (42). He suspected that the arsenic present in the algae is organic arsenic as no poisonous properties were observed upon consumption. Four years later, Chapman (43) examined, in a systematic study, the arsenic concentrations in marine algae and marine fish. In his work, he confirmed the results from Jones and suggested that the arsenic compound present in lobster is a ". . . more or less complex organic substance, or mixture of substances. . . ." Further, he stated that ". . . it is evidently possessed of very slight toxic properties as compared with those of arsenious oxide. . . ." It took over 50 years until this complex organic arsenic compound was identified to be arsenobetaine (44). It took another five years until the major arsenic compounds present in marine algae were identified as dimethylarsinoyl-ribosides (6,7). Currently, about 25 different organic arsenic compounds have been identified. Recently, another arsenic compound frequently detected in marine animals was shown to be trimethylarsoniopropionate (45). The structures of the major arsenic compounds found in the environment are presented in Figure 2.

Even in the early arsenic research, it was recognized that the different arsenic compounds, although not known, have different toxic properties (42,43).

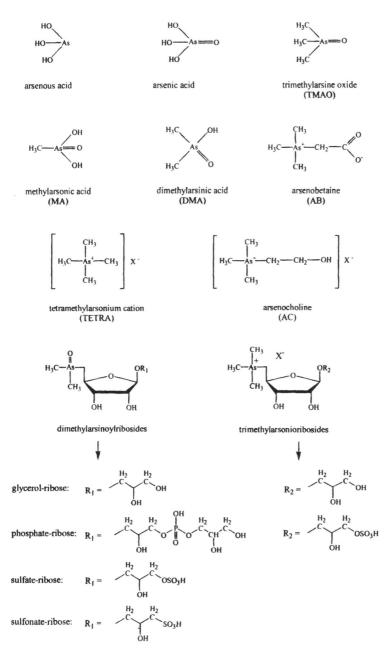

Figure 2 Arsenic compounds commonly detected in the environment (abbreviations used in the text in parentheses).

Nowadays, it is well established that total arsenic concentrations provide no information about possible risks. The determination of individual molecular species of arsenic is absolutely necessary for risk assessment. Acute toxicities of the arsenic compounds are given in Table 1. Moreover, knowledge about the chemical form of an element provides useful information on its bioavailability, transport, and metabolism.

A. Hyphenated Techniques

For the determination of arsenic compounds, three major steps must be taken into consideration. The arsenic species must be extracted from the sample (unfortunately in situ determination of arsenic compounds is not possible at environmental concentrations). During the extraction step, the arsenic compounds must not change or decompose chemically. Therefore, the extraction step should be as mild as possible and almost all the arsenic present in a sample must be extracted (51). A combination of various extractants is often necessary to reach all the arsenic. Polar organic solvents or water are commonly used for these purposes.

After the analytes (arsenic species) have been brought into the analyte solution, a separation step has to be employed to separate the different compounds. Due to the different chemical properties of the arsenic compounds—they might be anionic, neutral, or cationic—a reliable separation within one single run is not possible. A combination of various separation procedures must be employed.

Table 1   Acute Toxicities of Various Arsenic Compounds to Mice and Rats

Arsenic compound	LD_{50} [g/kg body mass]	Animal	Mode of administration	Reference
As_2O_3	0.035	Mouse	oral	47
As_2O_3	0.0045	Rat	intraperitoneal	46
H_3AsO_4	0.014–0.018	Rat	intraperitoneal	46
MA	1.8	Mouse	oral	48
DMA	1.2	Mouse	oral	48
TMAO	10.6	Mouse	oral	48, 49
AB	>10	Mouse	oral	47
AC	6.5	Mouse	oral	49, 50
AC	0.19	Mouse	intravenous	50
TETRA Iodide	0.9	Mouse	oral	49

Hydride generation, liquid chromatography, gas chromatography, or capillary electrophoresis are commonly utilized.

After the compounds have been separated, they must be detected. All the methods discussed in Sections II.C to II.E have been utilized more or less successfully in the detection of the arsenic compounds. Figure 3 summarizes the hyphenated techniques that are used for the identification and determination of arsenic compounds.

B. Chromatographic Methods

Volatile arsines are produced from several inorganic and organic arsenicals using the hydride generation technique (see Sec. II.F). After cryotrapping of the volatile arsines AsH_3 (bp. $-55°C$), $MeAsH_2$ (bp. $2°C$), Me_2AsH (bp. $36°C$), and Me_3As (bp. $70°C$), these compounds can be detected after fractionated distillation (25). An improvement for the determination of all arsenic compounds is only possible when a chromatographic separation precedes the detection of the arsenic compounds.

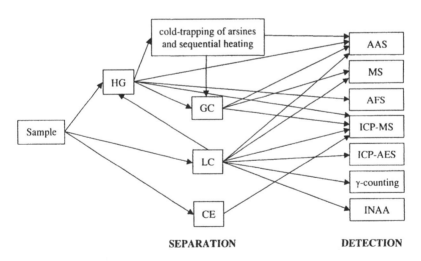

SEPARATION   **DETECTION**

Figure 3 Instrumental methods for the determination of arsenic compounds (Abbreviations: AAS, atomic absorption spectrometry; AFS, atomic fluorescence spectrometry; CE, capillary electrophoresis; GC, gas chromatography; HG, hydride generation; ICP-AES, inductively coupled plasma–atomic emission spectrometry; ICP-MS, inductively coupled plasma-mass spectrometry; INAA, instrumental neutron activation analysis; LC, liquid chromatography; MS, mass spectrometry).

1. Liquid Chromatography

Due to the fact that not all the arsenic compounds can be easily reduced to volatile hydrides, various chromatographic separations have been utilized to examine all the arsenic compounds present in a sample. Most of the arsenic compounds can be present in solution as neutral, anionic, or cationic species, depending on pH. Only arsenocholine (AC) and the tetramethylarsonium cation (TETRA) carry a positive charge irrespective of pH. The "anionic" arsenic compounds, arsenous acid, arsenic acid, methylarsonic acid (MA), and dimethylarsinic acid (DMA), are commonly separated on anion-exchange columns (all the "cationic" arsenic compounds elute with or close to the void volume). Arsenobetaine (AB), arseno-choline (AC), trimethylarsine oxide (TMAO), and the tetramethylarsonium cation (TETRA) are separated on cation-exchange columns (all the "anionic" arsenic compounds elute with or close to the void volume). For the separation of the arsenoriboses, anion- and cation-exchange chromatography as well as ion-pairing reversed-phase chromatography are utilized.

Anion-exchange chromatography with a 30 mM sodium phosphate buffer solution at pH 6 was used for the separation of arsenous acid, DMA, MA, and arsenic acid on the polymer-based Hamilton PRP-X100 anion-exchange column (52). The addition of 30% methanol and a change of the pH to 5 allowed the additional separation of methylarsonous acid and dimethylarsinous acid, potential arsenic metabolites (53). The separation of arsenous acid, arsenic acid, MA, DMA, and four arsenosugars on the same column has been recently demonstrated using a 20 mM ammoniumdihydrogen phosphate solution at pH 5.6 (54). Under these conditions, the gycerol-ribose (see Fig. 2) coeluted with arsenous acid. With anion-exchange chromatography, AB and arsenous acid are difficult to separate when buffers in the pH range of 5–9 are employed. At higher pH values, arsenous acid is deprotonated and is therefore retained on an anion-exchange column. At these high pH values, it is difficult to elute arsenic acid from the column within a reasonable analysis time. Gradient elution has been employed to overcome this problem. The "cationic" arsenic compounds TMAO, AB, AC, and TETRA are best separated on cation-exchange columns. As mobile phases, pyridine-formate buffer solutions are frequently exploited (55,56). In a recent review, chromato-graphic separations for the determination of arsenic compounds were discussed (57). A set of guidelines for the separation of arsenic compounds using ion-exchange chromatography and ICP-MS as element-specific detector has been published (58).

The simultaneous separation of the eight common arsenic compounds (arsenous acid, arsenic acid, MA, DMA, AB, TMAO, AC, TETRA) within one chromatographic run was achieved with a Dionex Ion Pac AS 7 high-capacity anion-exchange column with a nitric acid gradient and the addition of 1,2-benzenedisulfonic acid as ion-pairing reagent (59). This separation was only successful when matrix-free samples were analyzed. Due to the fact that most organoarsenic compounds are not retained on reversed-phase columns, the addition of anion-pairing reagents, such as tetraethylammonium salts (60), or cation-pairing reagents, such as hexanesulfonates (61), is required. The application of several chromatographic separation methods applied to real life samples, such as food products and urine samples, has been summarized by Benramdane et al. (62).

Ion-exclusion chromatography has been successfully applied for the determination of seven arsenic compounds in biological samples with a column packed with a carboxylated methacrylate resin and a 0.35 mM sodium sulfate solution at pH 3.8 as mobile phase (63). However, the time required for one chromatographic run (60 min) makes this method not useful for routine analysis. Size-exclusion chromatography was proposed for the fractionation of the arsenic compounds and matrix removal prior to another chromatographic separation (64).

Due to the fact that no chromatographic method allows the simultaneous separation of all the currently known arsenic compounds, at least two independent chromatographic methods must be employed for positive identification of arsenic compounds in unknown samples based on the retention time.

2. Gas Chromatography

Gas chromatography is not commonly used for the determination of arsenic compounds. The reason for this is that not all arsenic compounds are easily volatilized. A rapid method for the determination of arsine, methylarsine, dimethylarsine, and trimethylarsine in air based on gas chromatography–mass spectrometry (GC-MS) was recently published (65). In another study, DMA and MA present in urine were derivatized with thioglycol methylate and extracted with a 100 μm solid-phase microextraction fiber in 40 minutes. Thereafter, the two arsenic compounds were determined with GC-MS (66). The combination of purge and trap gas chromatography with atomic fluorescence spectrometry was used for the determination of arsenous acid, arsenic acid, MA, and DMA in a mushroom sample (67). Low-temperature gas chromatography coupled to ICP-MS was used to determine the volatile arsenic compounds in intraoral air (68). This method is also applicable to the determination of volatile arsenic compounds in landfill gases.

The fact that only the volatile arsenic compounds are easily separated by gas chromatographic separation is the reason that this technique is nowadays rarely used for the determination of arsenic compounds in biological samples.

3. Capillary Zone Electrophoresis

Capillary zone electrophoresis (CZE) offers high chromatographic resolution and fast analysis time. The amount of sample that is usually injected onto the capillary is in the nanoliter range. Ultraviolet (UV) detection, which is not element specific, was often utilized in the early studies. Conductivity detection was employed for the determination of arsenous acid, arsenic acid, and DMA in water samples from a tailing of tin mine processes. Conductivity detection was found to be superior over UV detection because the analyte ion must not necessarily be UV active. With this method, detection limits of ~0.05 mg As/L were achievable (69). Proton-induced x-ray emission was also investigated as a possible element-specific detection method for the determination of arsenic compounds. The arsenic concentrations required for this detector (~1 mg As/L) are certainly too high for a possible application to uncontaminated real life samples (70). To improve the detection limits for the determination of arsenic acid, arsenous acid, and DMA with UV-detection, the possibility of large volume stacking was investigated. A 10- to 20-fold improvement in the detection limits could be obtained (71). Organic solvent modification was investigated with the purpose to reduce sample conductivity for the application of field amplified samples stacking capillary electrophoresis with direct UV detection. The optimized method allowed the determination of arsenous acid, arsenic acid, and DMA in the range from 0.5–20 mg As/L. For two phenylarsonic acids, the working range was much better due to the higher absorption coefficients of these compounds (72). The separation of arsenous acid, arsenic acid, DMA, and three phenylarsonic acids on a fused silica capillary with UV detection has been shown recently. A 15 mM phosphate solution containing 10 mM sodium dodecylsulfonate at pH 6.5 served as background electrolyte. The detection limits obtained were too high for the determination of arsenic compounds in unexposed human urine samples (73).

The difficulty in connecting CZE with other than on-line UV detectors is that most of the other detection systems require a certain flow. This is generally done with a sheath flow between the outlet of the capillary and the nebulizer of the detector. Concentric nebulizers used in ICP-MS are self-aspirating and they induce a suction flow in the CE capillary, which causes dispersion. The employed sheath flow rates should not be too high because they result in dilution of the analyte ions.

Arsenous acid, arsenic acid, DMA, and MA were separated on a fused silica capillary using a 50 mM phosphate buffer. For the detection of the arsenic compounds, hydride generation-ICP-AES was employed. The first connection of CZE with ICP-MS was published by Olesik et al. in 1995 (74). In this work, they showed the determination of arsenite and arsenate with this technique. The concentration required for the determination of these two arsenic species was in the mg/L range. Since then, several attempts have been made to connect CZE with ICP-MS. To improve the detection limits, different ultrasonic nebulizers were investigated for ICP-MS detection. The detection limits obtained with electrokinetic injection were 0.2 µg/L for As(V) and 0.5 µg/L for DMA (75). Arsenous acid, arsenic acid, DMA, and MA were separated on a fused silica capillary with a 20 mM potassium hydrogen phthalate–20 mM boric acid buffer at pH 9.03. Excellent detection limits (<0.06 µg As/L) were obtained when HG-ICP-MS was employed as element-specific detector (76). The four anionic arsenic compounds, As(III), As(V), DMA, and MA, and the two cationic arsenic compounds, AB and AC, were successfully separated on a fused silica capillary with a 20 mM borate buffer at pH 9.4. The capillary was connected via a stainless steel-T-piece to a microconcentric nebulizer. Under the optimized condition, the detection limits were between 1–2 µg As/L. The method was successfully applied to the determination of arsenic in a mineral water but was not useful for the determination of arsenic compounds in human urine (77). The coupling of CZE to electrospray ionization mass spectrometry was demonstrated for the determination of arsenous acid, arsenic acid, DMA, MA, AB, and AC. The method was applied for the determination of these arsenic species in standard mixtures and urine samples. The detection limits obtained are certainly too high to determine arsenic compounds in unexposed urine. The excellent chromatographic resolution and the molecular information obtainable with the electrospray mass spectrometer make this technique a potentially valuable tool for the identification of unknown arsenic compounds at higher concentrations (78).

There is still a lot of work to do to make CZE a robust method for the determination of arsenic compounds in environmental samples. The proposed methods are too matrix sensitive and the difficulties of connecting CZE to element-specific detectors, such as ICP-MS, are not yet solved. At the moment, liquid chromatography in connection with element-specific detectors is certainly favored over CZE.

C. Detectors

The identification and quantification of arsenic compounds is easily achieved with element-specific detectors: ETV-AAS, HG-FAAS, AFS, ICP-AES, and ICP-MS are commonly used for the detection of arsenic compounds in the effluents of the chromatographic systems (HPLC, CZE). These detection systems have the advantage that only the arsenic-containing compounds have to be separated. The chromatographic separation is therefore much easier as not all the other species present in a natural sample have to be separated from the analytes of interest. When nonspecific detection, such as UV-visible or conductivity detection is employed, the analytes have to be separated from the sample matrix. Nevertheless, it is worth mentioning that one should not forget to think about matrix constituents coeluting with the analyte of interest. This might result in peak splitting and misinterpretation of the results. For the quantification of arsenic compounds in a real matrix after a chromatographic separation, possible matrix effects such as signal suppression must be taken into consideration and analytical strategies must be developed to overcome this problem (56).

The ETV-AAS method has certainly excellent detection limits but it is not easily connected on-line to a chromatographic system. The major reason is the time necessary for one single measurement ($\sim$60 sec). This requires that the flow of the chromatographic system has to be reduced providing that no signal of interest gets lost (79). The application of ETV-AAS as element-specific detector has certainly lost its attractiveness with the invention of ICP-MS.

The HG-F(QF)AAS method is a widespread detector for the determination of arsenic compounds because it is inexpensive. The disadvantage of this detection is that it is restricted to hydride-forming arsenic compounds. There are numerous articles published using postcolumn UV photo-oxidation (67,80,81) or microwave-assisted oxidation (81,82) to convert the non–hydride-active arsenic compounds, such as AB, AC, and TETRA, to the hydride-active compounds.

Atomic fluorescence spectrometry (AFS) has turned out to be an attractive detection system for the determination of arsenic compounds after their separation. Atomic fluorescence spectrometry is rather simple and inexpensive. In an early work, it was found that AFS detection after chromatographic separation of arsenous acid, arsenic acid, DMA, and MA and ultrasonic nebulization of the effluent had high detection limits and was prone to matrix interferences. In this work, it was already recognized that AFS might be a promising detector for hydride generation (83). Le et al. (84) used ion-pair separation of seven arsenic compounds followed by microwave digestion hydride generation AFS for the determination of arsenic compounds in urine samples after consumption of seaweed. Atomic fluorescence spectrometry was used as element-specific detector after purge and trap gas chromatographic separation of the volatile arsines for the determination of arsenic compounds in a mushroom sample (67). The cationic arsenic compounds in seafood were determined with AFS detection after their separation on a polymer-based cation-exchange column and thermo-oxidation and generation of the hydrides in the effluent of the column (85). The performance of AFS detection was recently compared with ICP-MS for the determination of arsenic compounds in environmental samples. The detection limits of both methods were comparable after hydride generation on-line UV photo-oxidation (86).

Atomic fluorescence spectrometry is certainly an excellent detector for the determination of arsenic compounds. Necessary low detection limits can be obtained only after hydride generation. This is certainly a drawback as the non–hydride-forming arsenic compounds have to be converted into the volatile hydride-forming ones.

Among the detectors discussed thus far, ICP-MS is certainly not the cheapest one. The advantage of ICP-MS lies in its multielement capabilities, excellent detection limits, and wide linear range. Moreover, low detection limits are not restricted to the hydride-forming arsenic compounds. The application of ICP-MS as an element-specific detector changed the knowledge about arsenic compounds in the terrestrial environment significantly when, in 1995, AB was detected for the first time in terrestrial mushroom samples (87). From this time on almost all arsenic compounds formerly attributed to the marine environment, such as AB, AC, TETRA, and arsenoriboses, could be detected in the terrestrial environment with ICP-MS as the element-specific detector (88,89,90). The usefulness of ICP-MS as element-specific detector for liquid chromatography was recently reviewed (91). Advantages and limitations of ICP-MS, when used as element-specific detector, were discussed by Szpunar et al. (92).

The identification of arsenic compounds using element-specific detectors in general is based on matching the retention time with known standards. This works perfectly as long as known standards are available. When unknown signals are obtained in a chromatogram, ICP-MS cannot provide any structural information and other detectors must be applied.

1. Detectors that Provide Direct Structural Information

Nuclear magnetic resonance (NMR) is very powerful for structural analysis but has a rather high detection limit and cannot be used as an on-line detector after a chromatographic separation. Mass spectrometers with soft ionization can certainly provide important information about unknown arsenic signals obtained after chromatographic separation. Ion-spray mass spectrometry was successfully used for the characterization of arsenic compounds in extracts of plaice, oysters, and mussels (93). Depending on the settings of the ionization source, elemental as well as molecular information can be obtained. Arsenosugars were characterized and identified in a partially purified algal extract with fast atom bombardment tandem mass spectrometry (94). Liquid chromatography coupled to an electrospray mass spectrometer (ES-MS) was used for simultaneous recording of elemental and molecular mass spectra of seven organoarsenic compounds. The developed method allowed the determination of the major arsenosugars present in brown algal extracts. Arsenous acid and arsenic acid could not be determined with this method (95). Good agreement was obtained when ES-MS was compared with ICP-MS as element-specific detector for the determination of arsenosugars in an algal extract (96). A direct identification of unknown arsenic compounds at low concentrations in an extract of a biological sample without partial purification and preconcentration is to date not possible because the detection limit of ES-MS is approximately two orders of magnitude higher compared with ICP-MS. The possibilities of ES-MS in the case of organoarsenic compounds are hampered due to the fact that arsenic is monoisotopic and, therefore, a clear attribution of signals in the mass spectrum is not always possible. However, the potential of ES-MS was recently demonstrated with the identification of trimethylarsoniopropionate (45) in a fish muscle extract. This compound was frequently reported as unknown in marine animals (56).

Mass spectrometry with various ionization sources is certainly not the ultimate technique for the unequivocal identification of unknown arsenic compounds, but it provides important information about the structure of the compounds and is a big step toward a clear identification of an unknown arsenic compound.

IV. CONCLUSIONS AND FUTURE PROSPECTS

Nowadays, most of the arsenic-orientated analytical work is dealing with the determination of total arsenic concentrations in biological samples. The methods developed for this purpose allow the determination of trace and ultratrace concentrations simultaneously with other elements. The huge differences in the acute toxicities make it evident that total arsenic concentrations are not sufficient for risk assessment. Nowadays, methods for the determination of arsenic compounds are becoming more robust and are (or will be soon) routine methods. Scientists who investigate toxic and beneficial effects of arsenic must concentrate on the determination of arsenic compounds.

There is still a large need for the preparation of reference materials certified for arsenic compounds in various matrices, such as food, urine, and water. Currently, only standard reference materials prepared from fish are available. These materials will help to prepare quality-controlled data on arsenic compounds and, of course, establish speciation analysis in routine laboratories.

The analytical chemists must augment the possibilities of their methods with respect to robustness, reliability, detection limits, and multielement capabilities. Methods for the extraction of arsenic compounds from environmental samples without changing the species must be improved, too. Moreover, techniques that allow the direct determination of arsenic compounds (without prior extraction) should be developed to get a clearer picture of arsenic in our environment.

As soon as more reliable data are available, the reputation of arsenic might change from a toxic, life-threatening element to an essential trace element.

REFERENCES

1. KH Most. Arsen als Gift und Zaubermittel in der deutschen Volksmedizin mit besonderer Beruecksichtigung der Steiermark. Ph.D. dissertation, Karl-Franzens University, Graz, Austria, 1939.
2. PJ Peterson, CA Griling, LM Benson, R Zieve. Metalloids. In: NW Lep, ed. Effect of Heavy Metal Pollution of Plants. London: Applied Science, 1981, pp 299–322.
3. WR Cullen, KJ Reimer. Arsenic speciation in the environment. Chem Rev 89:713–764, 1989.

4. D Kuehnelt, J Lintschinger, W Goessler. Arsenic compounds in terrestrial organisms IV: Green plants and lichens from an old arsenic smelter site in Austria. Appl Organomet Chem 14:411–420, 2000.
5. KJ Irgolic, W Goessler, D Kuehnelt. Arsenic compounds in terrestrial biota. In: WR Chappell, CO Abernathy, RL Calderon, eds. Arsenic Exposure and Health Effects. Amsterdam: Elsevier, 1999, pp 61–68.
6. KA Francesconi, JS Edmonds. Arsenic in the sea. Marine Biol Annu Rev 31:111–151, 1993.
7. KA Francesconi, JS Edmonds. Arsenic and marine organisms. Adv Inorg Chem 44: 147–189, 1997.
8. RM Allesch. Arsenik. Klagenfurt: Ferd. Kleinmayr, 1959, pp 239–250.
9. RM Allesch. Arsenik. Klagenfurt: Ferd. Kleinmayr, 1959, pp 276–278.
10. WD Treadwell. Tabellen und Vorschriften zur Quantitativen Analyse. Wien: Franz Deuticke, 1947, pp 182–186.
11. MA Anke, A Henning, M Gruen, B Groppel, H Luedke. Arsen: Ein neues essentielles Spurenelement. Arch Tierernaehr 26:742–743, 1976.
12. KJ Irgolic. Arsenic. In: M Stoeppler, ed. Hazardous Metals in the Environment. Amsterdam: Elsevier, 1992, pp 287–350.
13. PJ Walter, S Chalk, HM (Skip) Kingston. Overview of microwave assisted sample preparation. In: HM (Skip) Kingston, SJ Haswell, eds. Microwave-Enhanced Chemistry, Fundamentals, Sample Preparation and Applications. Washington, DC: American Chemical Society, 1997, pp 55–400.
14. M Pavkov, W Goessler. Stability of arsenic compounds in nitric acid after microwave assisted digestion. J Anal At Spectrom, in preparation.
15. B Lange, ZJ Vejdelek. Photometrische Analyse. 1. Auflage. Weinheim: Verlag Chemie, 1980, pp 41–44.
16. Deutsche Einheitsverfahren zur Wasser-, Abwasser- und Schlammuntersuchung. Weinheim: Wiley, 2000, D12.
17. H Greschonig, KJ Irgolic. Electrochemical methods for the determination of total arsenic and arsenic compounds. Appl Organomet Chem 6:565–577, 1992.
18. D Jagner, L Renman, SH Stefansdottir. Determination of arsenic by stripping potentiometry on gold electrodes using partial least-squares (PLS) regression calibration. Electroanalysis 6:201–208, 1994.
19. H Greschonig, KJ Irgolic. Arsenic and arsenic compounds, electrochemical determination. In: RA Meyers, ed. Encyclopedia of Environmental Analysis and Remediation. New York: Wiley, 1998, pp 402–413.
20. EH Larsen. Graphite furnace atomic absorption spectrometry of inorganic and organic arsenic species using conventional and fast furnace programmes. J Anal At Spectrom 6:375–377, 1991.
21. M Deaker, W Maher. Determination of arsenic in arsenic compounds and marine biological tissues using low volume microwave digestion and electrothermal atomic absorption spectrometry. J Anal At Spectrom 14:1193–1207, 1999.
22. P Vinas, M Pardo-Martinez, M Hernandez-Cordoba. Slurry atomization for the determination of arsenic in baby foods using electrothermal atomic absorption spectrometry and deuterium background correction. J Anal At Spectrom 14:1215–1219, 1999.

23. S Greenfield, M Foulkes. Introduction. In: SJ Hill, ed. Inductively Coupled Plasma Spectrometry and Its Applications. Sheffield: Sheffield Academic Press, 1999, pp 1–34.

24. A Fisher, SJ Hill. Instrumentation for ICP-AES. In: SJ Hill, ed. Inductively Coupled Plasma Spectrometry and Its Applications. Sheffield: Sheffield Academic Press, 1999, pp 71–97.

25. KA Francesconi, JS Edmonds, M Morita. Determination of arsenic and arsenic species in marine environmental samples. In: JO Nriagu, ed. Arsenic in the Environment. Part I: Cycling and Characterization. New York: Wiley, 1994, pp 189–219.

26. KE Jarvis, AL Gray, RS Houk. Handbook of Inductively Coupled Plasma Mass Spectrometry. Glasgow: Blackie, 1992.

27. G O'Connor, EH Evans. Fundamental aspects of ICP-MS. In: SJ Hill, ed. Inductively Coupled Plasma Spectrometry and Its Applications. Sheffield: Sheffield Academic Press, 1999, pp 119–144.

28. S Branch, L Ebdon, M Ford, M Foulkes, P O'Neill. Determination of arsenic in samples with high chloride content by inductively coupled plasma mass spectrometry. J Anal At Spectrom 6:151–154, 1991.

29. CJ Amarasiriwardena, N Lupoli, V Potula, S Korrick, H Hu. Determination of total arsenic concentrations in human urine by inductively coupled plasma mass spectrometry: A comparison of the accuracy of three analytical methods. Analyst 123: 441–445, 1998.

30. A Krushevska, M Kotrbai, A Lasztity, R Barnes, D Amarasiriwardena. Application of tertiary amines for arsenic and selenium signal enhancement and polyatomic interference reduction in ICP-MS analysis of biological samples. Fresenius J Anal Chem 355:793–800, 1996.

31. W Goessler, D Kuehnelt, KJ Irgolic. Determination of arsenic compounds in human urine by HPLC-ICP-MS. In: CO Abernathy, RL Calderon, WR Chappell, eds. Arsenic Exposure and Health Effects. London: Chapman & Hall, 1997, pp 33–44.

32. SD Tanner, VI Baranov, U Vollkopf. A dynamic reaction cell for inductively coupled plasma mass spectrometry (ICP-DRC-MS) Part III. J Anal At Spectrom 15: 1261–1269, 2000.

33. AT Townsend. The determination of arsenic and selenium in standard reference materials using sector field ICP-MS in high resolution mode. Fresenius J Anal Chem 364:521–526, 1999.

34. B Klaue, JD Blum. Trace analyses of arsenic in drinking water by inductively coupled plasma mass spectrometry: High resolution versus hydride generation. Anal Chem 71:1408–1414, 1999.

35. H Greschonig, KJ Irgolic. The mercuric-bromide-stain method and the Natelson method for the determination of arsines: Implications for assessment of risks from exposure to arsenic in Taiwan. In: CO Abernathy, RL Calderon, WR Chappell, eds. Arsenic Exposure and Health Effects. London: Chapman & Hall, 1997, pp 17–32.

36. W Holak. Gas-sampling technique for arsenic determination by atomic absorption spectrometry. Anal Chem 41:1712–1713, 1969.

37. DL Tsalev. Hyphenated vapour generation atomic absorption spectrometric techniques. J Anal At Spectrom 14:147–162, 1999.

38. Y Cai. Speciation and analysis of mercury, arsenic, and selenium by atomic fluorescence spectrometry. Trends Anal Chem 19:62–66, 2000.
39. H Matusiewicz, RE Sturgeon. Atomic spectrometric determination of hydride forming elements following in situ trapping within a graphite furnace. Spectrochim Acta B 51:377–397, 1996.
40. J Dedina, DL Tsalev. Arsenic. In: JD Winefordner, ed. Hydride Generation Atomic Absorption Spectrometry. Chichester: Wiley, 1995, pp 182–245.
41. H Uggerud, W Lund. Determination of arsenic by inductively coupled plasma mass spectrometry: Comparison of sample introduction techniques. Fresenius J Anal Chem 368:162–165, 2000.
42. AJ Jones. The arsenic content of some of the marine algae. In: CH Hampshire, ed. Yearbook of Pharmacy, Transactions of the British Pharmaceutical Conference. London: Churchill, 1922, pp 388–395.
43. AC Chapman. On the presence of compounds of arsenic in marine crustaceans and shell fish. Analyst 51:548–563, 1926.
44. JS Edmonds, KA Francesconi, JR Canon, CL Raston, BW Skelton, AH White. Isolation, crystal structure and synthesis of arsenobetaine, the arsenical constituent of the western rock lobster *Panulirus longipes cygnus* George. Tetrahedron Lett 18:1543–1546, 1977.
45. KA Francesconi, S Khokiattiwong, W Goessler, SN Pedersen, M Pavkov. A new arsenobetaine from marine organisms identified by liquid chromatography–mass spectrometry. Chem Commun 2000:1083–1084, 2000.
46. KW Franke, AL Moxon. A comparison of the minimal fatal doses of selenium, tellurium, arsenic, and vanadium. J Pharmacol Exp Ther 58:454–459, 1936.
47. T Kaise, S Watanabe, K Itoh. The acute toxicity of arsenobetaine. Chemosphere 14: 1327–1332, 1985.
48. T Kaise, H Yamauchi, Y Horiguchi, T Tani, S Watanabe, T Hirayama, S Fukui. A comparative study on acute toxicity of methylarsonic acid, dimethylarsinic acid and trimethylarsine oxide in mice. Appl Organomet Chem 3:273–277, 1989.
49. T Kaise, S Fukui. The chemical form and acute toxicity of arsenic compounds in marine organisms. Appl Organomet Chem 6:155–160, 1992.
50. T Kaise, Y Horiguchi, S Fukui, K Shiomi, M Chino, T Kikuchi. Acute toxicity and metabolism of arsenocholine in mice. Appl Organomet Chem 6:369–373, 1992.
51. D Kuehnelt, KJ Irgolic, W Goessler. Comparison of three methods for the extraction of arsenic compounds from the NRCC standard reference material DORM-2 and the brown alga *Hijiki fuziforme*. Appl Organomet Chem 15:445–456, 2001.
52. J Gailer, KJ Irgolic. The ion-chromatographic behavior of arsenite, arsenate, methylarsonic acid, and dimethylarsinic acid on the Hamilton PRP-X100 anion-exchange column. Appl Organomet Chem 8:129–140, 1994.
53. J Gailer, S Madden, WR Cullen, M Bonner Denton. The separation of dimethylarsinic acid, methylarsonous acid, methylarsonic acid, arsenate and dimethylarsinous acid on the Hamilton PRP-X100 anion exchange column. Appl Organomet Chem 13:837–843, 1999.
54. G Raber, KA Francesconi, KJ Irgolic, W Goessler. Determination of 'arsenosugars' in algae with anion-exchange chromatography and an inductively coupled plasma

mass spectrometer as element-specific detector. Fresenius J Anal Chem 367:181–188, 2000.

55. SH Hansen, EH Larsen, G Pritzl, C Cornett. Separation of seven arsenic compounds by high-performance liquid chromatography with on-line detection by hydrogen-argon flame atomic absorption spectrometry and inductively coupled plasma mass spectrometry. J Anal At Spectrom 7:629–634, 1992.

56. W Goessler, D Kuehnelt, C Schlagenhaufen, Z Slejkovec, KJ Irgolic. Arsenobetaine and other arsenic compounds in the NRCC SRM DORM 1 and DORM 2. J Anal At Spectrom 13:183–187, 1998.

57. T Guerin, A Astruc, M Astruc. Speciation of arsenic and selenium compounds by HPLC hyphenated to specific detectors: A review of the main separation techniques. Talanta 50:1–24, 1999.

58. EH Larsen. Method optimization and quality assurance in speciation analysis using high performance liquid chromatography with detection by inductively coupled plasma mass spectrometry. Spectrochim Acta B 53:253–265, 1998.

59. S Londesborough, J Mattusch, R Wennrich. Separation of arsenic species by HPLC-ICP-MS. Fresenius J Anal Chem 363:577–581, 1999.

60. J Yoshinaga, Y Shibata, T Horiguchi, M Morita. New reference materials: NIES certified reference materials for arsenic speciation. Accred Qual Assur 2:154–156, 1997.

61. X-C Le, X-F Li, V Lai, M Ma, S Yalcin, J Feldmann. Simultaneous speciation of selenium and arsenic using elevated temperature liquid chromatography with inductively coupled plasma mass spectrometry detection. Spectrochim Acta B 53:899–909, 1998.

62. L Benramdane, F Bressolle, JJ Vallon. Arsenic speciation in humans and food products: A review. J Chromatogr Sci 37:330–344, 1999.

63. T Nakazato, T Taniguchi, H Tao, M Tominaga, A Miyazaki. Ion-exclusion chromatography combined with ICP-MS and hydride generation-ICP-MS for the determination of arsenic species in biological matrices. J Anal At Spectrom 15:1546–1553, 2000.

64. S McSheehy, J Szpunar. Speciation of arsenic compounds in edible algae by bi-dimensional size-exclusion anion-exchange HPLC with dual ICP-MS and electro-spray MS/MS detection. J Anal At Spectrom 15:79–87, 2000.

65. M Pantsar-Kallio, A Korpela. Analysis of gaseous arsenic species and stability studies of arsine and trimethylarsine by gas chromatography–mass spectrometry. Anal Chim Acta 410:65–70, 2000.

66. Z Mester, J Pawliszyn. Speciation of dimethylarsinic acid and monomethylarsonic acid by solid-phase microextraction–gas chromatography–ion trap mass spectrometry. J Chromatogr A 873:129–135, 2000.

67. Z Slejkovec, JT van Elteren, AR Byrne. A dual arsenic speciation system combining liquid chromatography and purge and trap–gas chromatographic separation with atomic fluorescense spectrometric detection. Anal Chim Acta 358:51–60, 1998.

68. J Feldmann, T Riechmann, AV Hirner. Determination of organometallics in intra-oral air by LT-GC/ICP-MS. Fresenius J Anal Chem 354:620–623, 1996.

69. D Schlegel, J Mattusch, R Wennrich. Speciation analysis of arsenic and selenium compounds by capillary electrophoresis. Fresenius J Anal Chem 354:535–539, 1996.

70. C Vogt, J Vogt, H Wittrisch. Element-sensitive detection for capillary electrophoresis. J Chromatogr A 727:301–310, 1996.
71. M Albert, L Debusschere, C Demesmay, JL Rocca. Large volume stacking for quantitative analysis of anions in capillary electrophoresis. I: Large volume stacking with polarity switching. J Chromatogr A 757:281–289, 1997.
72. EP Gil, P Ostapczuk, H Emons. Determination of arsenic species by field amplified injection capillary electrophoresis after modification of the sample solution with methanol. Anal Chim Acta 389:9–19, 1999.
73. H Greschonig, MG Schmid, G Gübitz. Capillary electrophoretic separation of inorganic and organic arsenic compounds. Fresenius J Anal Chem 362:218–223, 1998.
74. JW Olesik, JA Kinzer, SV Olesik. Capillary electrophoresis inductively coupled plasma mass spectrometry for rapid elemental speciation. Anal Chem 67:1–12, 1995.
75. PW Kirlew, MTM Castillano, JA Caruso. An evaluation of ultrasonic nebulizers as interfaces for capillary electrophoresis of inorganic anions and cations with inductively coupled plasma mass spectrometric detection. Spectrochim Acta B 53:221–237, 1998.
76. ML Magnuson, JT Creed, CA Brockhoff. Speciation of arsenic compounds in drinking water by capillary electrophoresis with hydrodynamically modified electroosmotic flow detected through hydride generation inductively coupled plasma mass spectrometry with a membrane gas-liquid separator. J Anal At Spectrom 12:689–695, 1997.
77. M Van Holderbeke, Y Zhao, F Vanhaecke, L Moens, R Dams, P Sandra. Speciation of six arsenic compounds using capillary electrophoresis–inductively coupled plasma mass spectrometry. J Anal At Spectrom 14:229–234, 1999.
78. O Schramel, B Michalke, A Kettrup. Application of capillary electrophoresis–electrospray ionization mass spectrometry to arsenic speciation. J Anal At Spectrom 14:1339–1342, 1999.
79. M Slekovec, W Goessler, KJ Irgolic. Inorganic and organic arsenic compounds in Slovenian mushrooms: Comparison of arsenic specific detectors for liquid chromatography. Chem Spec Bioavailab 11:115–123, 1999.
80. X Zhang, R Cornelis, J De Kimpe, L Mees. Arsenic speciation in serum of uremic patients based on liquid chromatography with hydride generation atomic absorption spectrometry and on-line UV photooxidation digestion. Anal Chim Acta 319:177–185, 1996.
81. R Sur, J Begerow, L Dunemann. Determination of arsenic species in human urine using HPLC with on-line photooxidation or microwave-assisted oxidation combined with flow-injection HG-AAS. Fresenius J Anal Chem 363:526–530, 1999.
82. KJ Lamble, SJ Hill. Arsenic speciation in biological samples by on-line high performance liquid chromatography–microwave digestion–hydride generation–atomic absorption spectrometry. Anal Chim Acta 344:261–270, 1996.
83. A Woller, Z Mester, P Fodor. Determination of arsenic species by high-performance liquid chromatography–ultrasonic nebulization atomic fluorescence spectrometry. J Anal At Spectrom 10:609–613, 1997.
84. X-C Le, M Ma, NA Wong. Speciation of arsenic compounds using high-performance

liquid chromatography at elevated temperature and selective hydride generation atomic fluorescence detection. Anal Chem 68:4501–4506, 1996.

85. MA Suner, V Devesa, I Rivas, D Velez, R Montoro. Speciation of arsenic compounds in seafood by coupling liquid chromatography with hydride generation atomic fluorescence detection. J Anal At Spectrom 15:1501–1507, 2000.

86. JL Gomez-Ariza, D Sanchez-Rodas, I Giraldez, E Morales. A comparison between ICP-MS and AFS detection for arsenic speciation in environmental samples. Talanta 51:257–268, 2000.

87. AR Byrne, Z Slejkovec, T Stijve, L Fay, W Goessler, J Gailer, KJ Irgolic. Arsenobetaine and other arsenic species in mushrooms. Appl Organomet Chem 9:305–313, 1995.

88. D Kuehnelt, W Goessler, KJ Irgolic. Arsenic compounds in terrestrial organisms II: Arsenocholine in the mushroom *Amanita muscaria*. Appl Organomet Chem 11:459–470, 1997.

89. A Geiszinger, W Goessler, D Kuehnelt, KA Francesconi, W Kosmus. Determination of arsenic compounds in earthworms. Environ Sci Technol 32:2238–2243, 1998.

90. I Koch, J Feldmann, L Wang, P Andrewes, KJ Reimer, WR Cullen. Arsenic in the Meager Creek hot springs environment, British Columbia, Canada. Sci Total Environ 236:101–117, 1999.

91. KL Sutton, JA Caruso. Liquid chromatography–inductively coupled plasma mass spectrometry. J Chromatogr A 856:243–258, 1999.

92. J Szpunar, S McSheehy, K Ploec, V Vacchina, S Mounicou, I Rodriguez, R Lobinski. Gas and liquid chromatography with inductively coupled plasma mass spectrometry detection for environmental speciation analysis: Advances and limitations. Spectrochim Acta B 55:779–793, 2000.

93. JJ Corr, EH Larsen. Arsenic speciation by liquid chromatography coupled with ionspray tandem mass spectrometry. J Anal At Spectrom 11:1215–1224, 1996.

94. SA Pergantis, KA Francesconi, W Goessler, JE Thomas-Oates. Characterization of arsenosugars of biological origin using fast atom bombardment tandem mass spectrometry. Anal Chem 69:4931–4937, 1997.

95. SN Pedersen, KA Francesconi. Liquid chromatography electrospray mass spectrometry with variable fragmentor voltages gives simultaneous elemental and molecular detection of arsenic compounds. Rapid Comm Mass Spectrom 14:641–645, 2000.

96. AD Madsen, W Goessler, SN Pedersen, KA Francesconi. Characterization of an algal extract by HPLC-ICP-MS and LC-electrospray MS for use in arsenosugar speciation studies. J Anal At Spectrom 15:657–662, 2000.

3
Arsenic Compounds in the Environment

Kevin A. Francesconi
University of Southern Denmark, Odense, Denmark

Doris Kuehnelt
Karl-Franzens University, Graz, Austria

I. INTRODUCTION

There have been several reviews of environmental arsenic compounds published in the recent past. The 1989 review by Cullen and Reimer (1) was the first to fully cover the large number of naturally occurring arsenic compounds that had been identified in the 1970s and 1980s. That paper discussed not only the types of arsenic compounds but also their basic chemistry, analysis, biotransformations, and fluxes in the various environmental compartments. Subsequent reviews focused on aquatic systems, usually marine (2–4) although freshwater systems were also covered (5). The emphasis initially was on marine systems because arsenic compounds occur widely, and in high concentrations, in marine organisms. Later work has shown that terrestrial organisms contain many of the arsenicals found in marine samples, although the concentrations are generally lower. The data on organoarsenic compounds present in terrestrial samples has recently been comprehensively reviewed (6).

Thus, the literature is already well served with reviews of environmental arsenic compounds, so the chapter presented here will not endeavour to emulate them. Rather, we present a summary of the arsenic compounds relevant to discussions on arsenic's environmental chemistry, and, in so doing, hope to provide information that will complement the detailed material presented in the accompanying chapters dealing with specific aspects of this topic.

Arsenic compounds found in the environment will be presented in two ways. First, they will be briefly discussed individually or under a classification

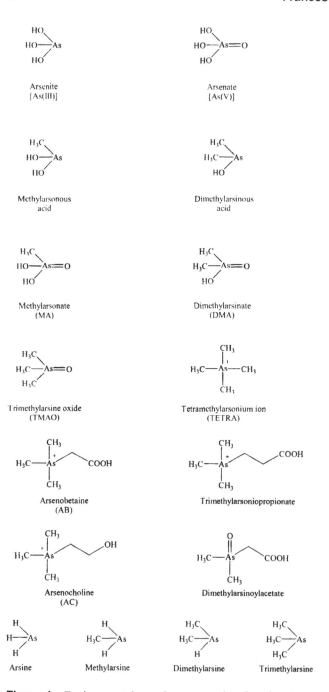

Figure 1   Environmental arsenic compounds referred to in text and tables by name or acronym. The compounds are depicted in their fully protonated form.

of compound type (e.g., arsenosugars). This should quickly convey information on the general distribution of the compounds and their relative importance. Second, the various environmental compartments (e.g., marine animals) will be discussed in terms of the arsenic compounds that they contain. This information is presented primarily in tables, and text is used only to highlight major points and trends. The emphasis is on biological samples because they contain a much greater variety of arsenicals than do abiotic samples. In addition, we focus on

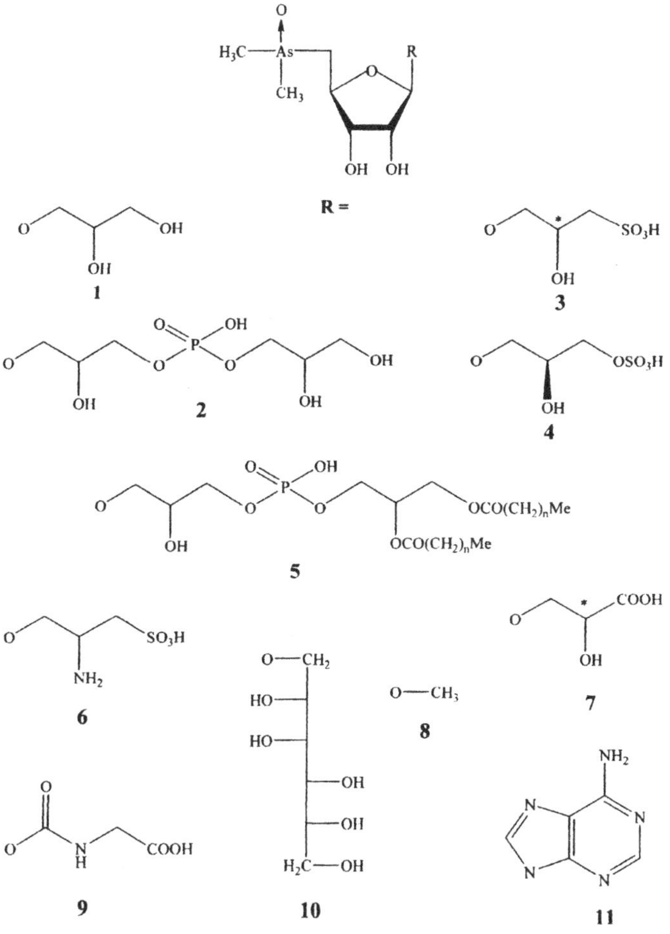

Figure 2 Environmental arsenic compounds referred to in text and tables by structure number. Carbons marked * represent epimeric centers where both diastereoisomers have been identified as natural products.

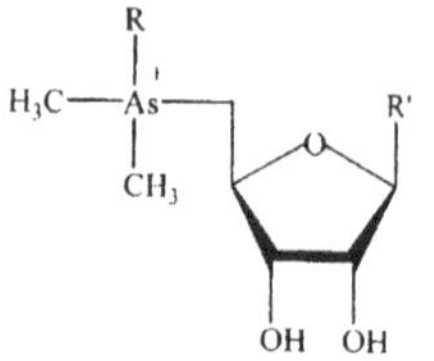

12   **R**= CH_3 **R'**= (structure)

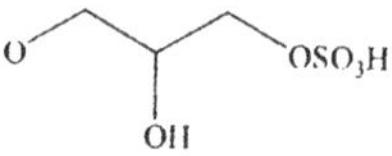

13   **R**= CH_3 **R'**= (structure)

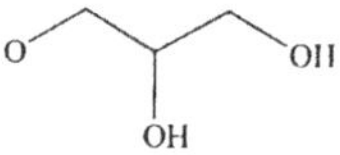

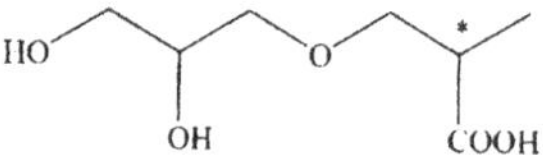

14   **R**= (structure)   **R'**= (structure)

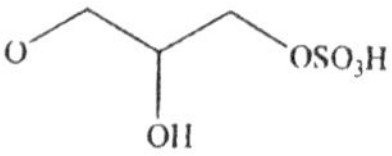

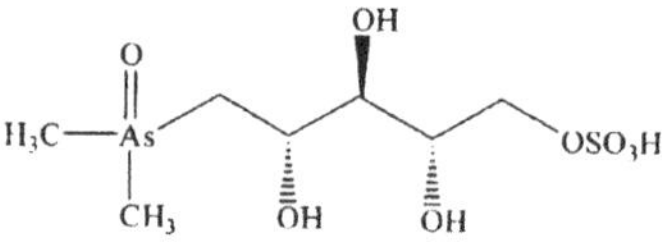

15

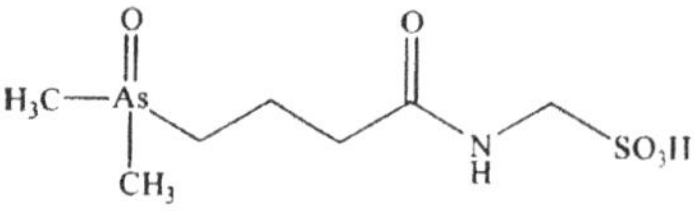

16

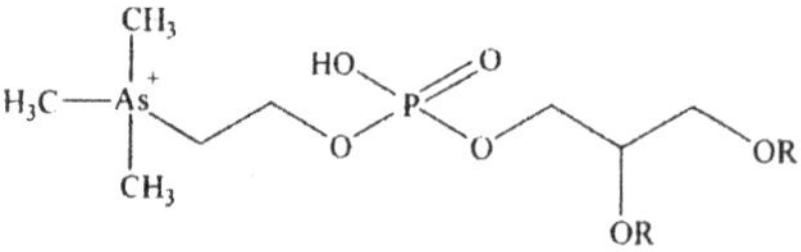

17   R= H

18   R= $CO(CH_2)_nCH_3$

Figure 2   Continued

natural unperturbed systems; only few data from "contaminated" sites are presented, and compounds produced in laboratory experiments (e.g., products from studies on arsenic metabolism) are mentioned only where they may have direct relevance to natural environmental samples.

To simplify the reporting of the various arsenic species, we will not distinguish between degree of protonation. For example, As(III) present as the oxyanion is referred to as arsenite even though it would be present in many environmental samples as the fully protonated species $As(OH)_3$. Figures 1 and 2 give the structures of the arsenic species discussed in this review; the simpler compounds (Fig. 1) are designated by name and generally accepted acronyms (e.g., AB for arsenobetaine), whereas more complicated compounds (e.g., arsenosugars) are designated by structure number (bold Arabic numerals in Fig. 2). Although some of the arsenicals may be depicted as ions in the figures, the term arsenic compound is used synonymously with arsenic species throughout this chapter. This is the generally accepted practice among researchers in the field; it simplifies the discussion and avoids possible ambiguity, or at least discordance, with the biological use of the term species.

Finally, we wish to point out the limitations of the following data set. Analytical techniques for determining organoarsenic species have improved steadily since the first of such methods was reported in the 1970s. Nevertheless, most of the methods used today involve separations (or derivatizations) in aqueous media. Consequently, these methods are only capable of determining water-soluble arsenicals, and arsenic compounds that are not soluble in water remain unidentified. The presence of such compounds has often been ascertained by the difference between concentrations of total arsenic and water-soluble arsenic. Future work should profit from techniques capable of determining water-insoluble arsenic species.

II. ENVIRONMENTAL ARSENIC COMPOUNDS

A. Arsenic-Containing Minerals

Arsenic is the 20th most abundant element in crustal rock, with an average concentration of 2–3 mg/kg (1,7). There is, however, a large variability depending on rock type, with sedimentary rock generally containing higher levels than igneous rock. It has been estimated that of the total arsenic contained in the various natural reservoirs (rocks, oceans, soils, biota, atmosphere), more than 99% is associated with rocks and minerals (8). Arsenopyrite (FeAsS) is the most abundant arsenic-containing mineral, and other important minerals include arsenolite (As_2O_3), olivenite (Cu_2OHAsO_4), cobaltite (CoAsS), and proustite (Ag_3AsS_3) (9).

B. Arsenite and Arsenate

The two arsenic oxyanions, arsenite and arsenate, are widely distributed in the environment. They are readily interconverted and hence are usually found together, with arsenate being the thermodynamically favored form under normal environmental oxygen levels (10). Aqueous extracts of soils usually contain arsenate with smaller amounts of arsenite (8). However, such mild extraction procedures remove only a small proportion of the total arsenic present in the samples (11). Arsenate is the major arsenical found in freshwater, including drinking water supplies, and arsenite also occurs, sometimes as a significant constituent (12). The arsenate and arsenite concentrations in groundwater can often be very high, and widespread toxic effects in humans have resulted from such water being used as drinking water (12). Arsenate is the dominant arsenical in seawater; arsenite is also found in seawater samples at levels correlated with biological activity (1).

Most marine algae contain only small quantities of arsenate, although in some species of brown algae it can be the major arsenic form (2). The presence of high concentrations of arsenate in *Hizikia fusiforme* (marketed as a dried food product called hijiki) is of particular interest because this alga is widely consumed, especially in Japan. Arsenite is less commonly reported in marine algae (4).

Both arsenite and arsenate have been reported in marine animals, but usually as minor constituents. Early analytical methods did not distinguish between these two forms and reported values for their sum (termed "inorganic arsenic"). These data demonstrated that inorganic arsenic generally constituted less than 5% of the total arsenic in marine animals (13).

Arsenate and arsenite are major forms of arsenic in freshwater and terrestrial plants.

C. Arsine

Arsine has been detected in landfill gases (14), in headspace samples collected from a hot spring environment (15), and in gases from anaerobic wastewater treatment facilities (16). Possibly, difficulties in collecting and analyzing this arsenical (it boils at $-55°C$) may have led to underestimates of its distribution in the environment, and future work might show it to be more widespread.

D. Arsenobetaine

Arsenobetaine is by far the major form of arsenic in marine animals, often representing >80% of the extractable arsenic (4). Its concentration varies widely, from 1–300 µg As/g dry mass. Although arsenobetaine is readily accumulated from food by marine animals, there is no clear correlation between concentration and trophic level. This might suggest that marine animals have the ability to biosynthesize arsenobetaine de novo from arsenate in seawater, but no experimental data have been produced so far to support this hypothesis.

Despite the ubiquity of arsenobetaine in marine animals, it has not been found in seawater or marine algae. Similarly, arsenobetaine has not been detected in freshwater samples or freshwater algae and plants, yet it is found in freshwater animals (fish and probably mussels), albeit at low concentrations (6). Arsenobetaine also occurs in a variety of terrestrial organisms, such as fungi, plants, earthworms, and ants (6). Although the origin of arsenobetaine is still unknown, its widespread occurrence in diverse organisms suggests a common nonspecific biogenetic pathway. In this regard, the pathway proposed by Edmonds based on amino acid synthesis appears well founded (17).

It is still not clear why arsenobetaine should occur at such high levels in marine animals relative to freshwater and terrestrial animals. The structural similarity of arsenobetaine to the important osmolyte glycine betaine suggests that it may be utilized in some osmotic role within marine animals (18).

E. Tetramethylarsonium Ion

Tetramethylarsonium ion is a common arsenic species in marine animals. It is present usually as a minor constituent, although exceptions do occur: for example, it is a significant arsenical in several species of bivalve mollusks (1). Traces of tetramethylarsonium ion have been found in some freshwater animals (6). It has also been reported in several fungal species, usually at low levels, and as a trace constituent of some terrestrial plants as well (6). There have been no reports of tetramethylarsonium ion in seawater or freshwater.

F. Arsenocholine

Arsenocholine also occurs commonly in marine animals, usually as a trace constituent. One exception is the turtle *Dermochelys coriacea* where arsenocholine was a significant arsenic constituent in the liver (19). Arsenocholine has been found in terrestrial organisms and can occur as the major arsenical in some species. For example, it is the major arsenical in the fungus *Sparassis crispa*, although the concentration of total arsenic in *S. crispa* is quite low (~1 µg As/g dry mass) (20).

Arsenocholine may serve as a precursor to arsenobetaine; laboratory studies have shown that it is rapidly biotransformed to this compound (21).

G. Methylarsonate and Dimethylarsinate

Methylarsonate and dimethylarsinate are common arsenic metabolites found in most environmental compartments. They often occur together, with dimethylarsinate being the more abundant in most samples. Their joint occurrence might be expected since they are metabolites of the same biogenetic pathway involving reduction and methylation of arsenate (1,22). They are usually minor constituents, although notable exceptions have been reported in fungi. For example, *Sarcosphaera coronari* (up to 2120 µg As/g dry mass) contains methylarsonate as its major arsenical (23), and *Laccaria amethystina* (up to 200 µg As/g dry mass) has dimethylarsinate as the major arsenic constituent (24).

H. Trimethylarsine Oxide

Trimethylarsine oxide has been reported in several marine animals, where it is almost always a trace constituent. The one exception is the fish *Kyphosus sydneyanus*, which has trimethylarsine oxide as the major arsenical (25). That trimethylarsine oxide is not more widespread is perhaps surprising since it is likely to be a metabolite of the same pathway producing methylarsonate and dimethylarsinate, both of which are more commonly found. Trimethylarsine oxide chromatographs rather poorly on cation-exchange columns often used for determining arsenic species, and the resultant poor detection limits for this compound may partly explain the data indicating its apparent absence in many samples. Trimethylarsine oxide is usually only rarely reported in terrestrial organisms, but more recent work (with better detection limits) has shown it to be present in various terrestrial plants and two lichen samples (26).

I. Methylarsines

Trimethylarsine has been found as a trace constituent in six prawn species and two lobster species (27). Possibly, it results from microbial breakdown of arsenobetaine via trimethylarsine oxide (18). Trimethylarsine has also been detected in landfill and sewage gases (14,16,28), and in headspace gases from hot springs environments (15,29). Methylarsine and dimethylarsine have also been detected in these gas samples.

J. Arsenosugars

Arsenosugars are the major arsenic compounds in marine algae. They chiefly comprise water-soluble dimethylarsinoyl compounds (see Fig. 2, compounds **1–4** and **6–11**), although lipid-soluble derivatives (see Fig. 2, compound **5**) and quaternary arsonio analogues (see Fig. 2, compounds **12–14**) are also found (4). Although many arsenosugars have been identified in algae, there are only four (see Fig. 2, compounds **1–4**) that are commonly found.

Arsenosugars are also present in many marine animals, but their origin in most cases is clearly the algae on which the food chain is based (4). Arsenosugars have also been reported recently in freshwater algae where they can occur as major arsenicals (15). They have also been reported at low levels in vascular freshwater plants (30) and in terrestrial organisms, including earthworms, fungi, and plants (6).

Arsenosugars appear to be end metabolites in a biogenetic pathway that begins with arsenate and involves processes of reduction, methylation, and, finally, adenosylation. This pathway, originally proposed for algae (31), was based on the methylation pathway put forward by Challenger (22) to explain the production of volatile methylated arsenicals by molds. The pathway is yet to be proven for algae, although recent work has provided experimental support (32).

K. Other Compounds

An unusual dimethylarsinoyl compound containing a taurine moiety (see Fig. 2, compound **16**) was found in trace amounts in clam kidney, although its origin is almost certainly algae associated with the clam (33). A possible biogenetic pathway for this compound has been discussed recently (17). Glycerophospho-arsenocholine (see Fig. 2, compound **17**) was identified in lobster hepatopancreas following hydrolysis of the lipid fraction (34). Presumably, it resulted from deacylation of a phosphatidylarsenocholine (see Fig. 2, compound **18**) originally present in the lipid fraction. Experiments with fish had indicated that both these compounds (and arsenobetaine) were metabolites of arsenocholine (35).

L. New Arsenic Compounds

Included here are novel arsenic compounds reported in environmental samples over the last five years. Dimethylarsinoylacetate was identified as a naturally occurring arsenical in marine reference materials of mussel, oyster, and lobster hepatopancreas (36). This compound had been proposed as a possible intermediate in the formation of arsenobetaine (31). More recently, however, arsenobetaine was found to degrade to dimethylarsinoylacetate under aerobic microbial conditions (37), and such a biotransformation suggests an alternative hypothesis for the presence of dimethylarsinoylacetate in marine samples.

The dimethylarsinoyl derivative of sulfated ribitol (see Fig. 2, compound **15**) was isolated from the red alga *Chondria crassicaulis* (38). It had been observed as a major arsenical in *C. crassicaulis* by high performance liquid chromatography–inductively coupled plasma mass spectrometry (HPLC-ICPMS), and was initially reported as an unknown because it did not match any available standard (39). Subsequently, the compound was isolated and a chemical structure was proposed chiefly on nuclear magnetic resonance (NMR) data; chemical synthesis of authentic material confirmed the proposed structure (38). This compound, which appears to be distributed widely in red algae, may be derived from arsenosugars (38).

A trimethylarsoniosugar (see Fig. 2, compound **13**) was identified in three species of gastropods by HPLC-ICPMS (40). This compound is a trimethylated analogue of one of the most common arsenosugars found in marine samples. A trimethylated analogue (see Fig. 2, compound **12**) of another common arseno-sugar was earlier reported in algae (41).

Two quaternary arsoniosugars (diastereoisomers), previously isolated as an unresolved mixture (see Fig. 2, structure **14**, epimeric at the carbon bonded to the carboxyl group) from the brown alga *Sargassum lacerifolium*, were separated by HPLC and their structures proposed on the basis of NMR spectral data (17). Electrospray mass spectrometric analysis of the unresolved mixture supports the proposed structures by demonstrating a $[M + H]^+$ molecular species with m/z 569 (Pedersen and Francesconi, unpublished results). The presence of these two compounds in *S. lacerifolium* formed the basis of a more general scheme for the formation of naturally occurring arsenic compounds (17).

An arsenobetaine analogue, trimethylarsoniopropionate, was reported in a coral reef fish *Abudefduf vaigiensis*, and in the fish reference material DORM-2 (42). The identification was proposed on the basis of LC electrospray MS data, and supported by chromatographic comparison with synthetic material.

III. ENVIRONMENTAL COMPARTMENTS AND THEIR ARSENIC COMPOUNDS

The major environmental compartments are considered here with respect to the types of arsenic compounds that have been found there. In most cases, the text merely summarizes the contents of the tables. Although the tables do not cover all available data, we have attempted to provide representative data wherever possible. In order to present the data in this manner, certain assumptions, approximations, and interpretations have been made. The quality of the analytical data from the many different studies varies considerably both in terms of quantification and, perhaps more importantly, the rigor applied when identifying the various arsenic compounds. We offer no judgment on these matters; the reader may wish to refer to the original paper before making an assessment.

A. Arsenic Compounds in Seawater and Freshwater (Table 1)

The world's oceans contain about 0.5–2.0 µg As/L, although higher concentrations can occur in estuaries, particularly those receiving anthropogenic discharges

Table 1 Arsenic Compounds in Seawater and Freshwater[a]

Sample	Site	As concn. (µg/L)	As(V)	As(III)	MA	DMA	TMAO	Other cpds. and unknowns	Ref.
Seawater	Causeway, Tampa Bay, Florida	1.77	major	minor	nd	sig	nr	nd	43
Seawater	Tidal flat, Tampa Bay, Florida	2.28	major	sig	minor	sig	nr	nd	43
Seawater	McKay Bay, Florida	1.48	sig	minor	minor	major	nr	nd	43
Seawater	Southern California Bight	~1–2	major	nd/minor	nd/tr	minor/sig	nd	nr	44
Seawater	Northeast Pacific and California Shelf	~1–2	major	tr/sig	tr	tr	nd	nr	45
Surface seawater	East Indian Ocean	0.87	major	sig	minor	sig	nr	nr	46
Surface seawater	Antarctic Ocean	1.1	major	tr	tr	minor	nr	nr	46
Surface seawater	North Indian Ocean	0.85	major	sig	minor	sig	nr	nr	46
Surface seawater	China Sea	0.64	major	nd	minor	sig	nr	nr	46
Seawater	Northwest coast of Spain	4.3	major	minor	nd	minor	nr	nd	47
Seawater	Northwest coast of Spain	2.9–5.0	sig	minor	minor	major	nr	nd	47
Seawater	Northwest coast of Spain	2.9	sig	nd	minor	sig	nr	nd	47
Seawater	Northwest coast of Spain	1.0–8.0	sig	sig	nd	sig	nr	nd	47
Seawater	Northwest coast of Spain	2.9, 3.5	major	minor	nd	sig	nr	nd	47
Seawater	Northwest coast of Spain	2.2, 6.2	sig	major	nd	sig	nr	nd	47
Seawater	Northwest coast of Spain	2.9	major	nd	nd	major	nr	nd	47
Seawater	Northwest coast of Spain	2.0	major	nd	nd	minor	nr	nd	47
Seawater	Northwest coast of Spain	6.2	sig	sig	minor	sig	nr	nd	47
Seawater	Northwest coast of Spain	12.5	sig	sig	sig	sig	nr	nd	47
Estuarine water	Southampton water, United Kingdom	0.76–1.00	[As(V) + As(III)] major		sig	sig	nd	sig	48
Estuarine water	Tagus estuary, Portugal	24.3	[As(V) + As(III)] major		(MA + DMA) minor		nd	sig	49
Estuarine water	Beaulieu estuary, England	~1	[As(V) + As(III)] major		sig	minor/sig	nd	nr	50
Estuarine water	Schelde estuary, Belgium	1.5–6	major	minor/sig	nr	nr	nr	nr	51

Table 1 Continued

Sample	Site	As concn. (μg/L)	As(V)	As(III)	MA	DMA	TMAO	Other cpds. and unknowns	Ref.
Estuary seawater	Estuary of Tinto and Odiel Rivers, Spain	31	major	nd	nd	nd	nd	nd	52
Estuary freshwater	Estuary of Tinto and Odiel Rivers, Spain	8.7	sig	sig	sig	sig	nd	nd	52
River water	Hayakawa River, Japan	30	major [As(V) + As(III)]		nd	nd	minor	nd	53
Well water	Near Whitlacoochee River, Florida	0.68	sig	nd	sig	sig	nr	nd	43
Well water	Northern Alberta, Canada	0.8–21.7	sig	major	nd	nd	nr	nd	54
Well water	Northern Alberta, Canada	4.9	major	sig	nd	nd	nr	nd	54
Well water	Taiwan	543	major [As(V) + As(III)]		tr	minor	tr	nd	55
Well water	Taiwan	74.8, 75.4	major [As(V) + As(III)]		minor	minor	minor	nd	55
Well water	Taiwan	27.4, 32.5	major [As(V) + As(III)]		tr	minor	sig	nd	55
Well water	Taiwan	570/730	major [As(V) + As(III)]		tr	tr	tr	nd	55
River water	Hillsborough River, Florida	0.25	major	nd	nd	nd	nr	nd	43
River water	Whitlacoochee River, Florida	0.42	sig	nd	sig	sig	nr	nd	43
River water	Colorado River, Arizona	2.2	major	sig	minor	minor	nr	nr	44
River water	Alamo River, California	7.5	major	tr	minor	nd	nr	nr	44
River water	Owens River, California	43	major	tr	tr	tr	nr	nr	44
River water	River Loa and Sloman Reservoir, Chile	970–2960	major	nd	nr	nr	nr	nd	56
River water	River Loa and Sloman Reservoir, Chile	1160–15270	major	minor	nr	nr	nr	nd	56
River water near realgar mining site	Presa River, Corsica	1860–2300	major	nd	nd	nd	nr	nd	57
River water near realgar mining site	Bravona River, Corsica	0.96 to 101	major	nd	nd	nd	nr	nd	57

Water type	Location	Concentration							Ref.
River water near old gold and arsenic mine	Orbiel River, France	35 to 7600	major	nd	nd	nd	nr	nd	57
Pond water	Remote pond, Whitlacoochee Forest, Florida	1.06	sig	nd	sig	major	nr	nd	43
Pond water	University Research Pond, USF, Florida	1.95	sig	sig	minor	minor	nr	nd	43
Lake water	Lake Echols, Tampa, Florida	3.58	sig	major	minor	minor	nr	nd	43
Lake water	Lake Magdalene, Tampa, Florida	1.75	sig	major	sig	minor	nr	nd	43
Lake water	Donner Lake, California	0.14	major	sig	nd	tr	nr	nr	44
Lake water	Squaw Lake, California	3.7	major	sig	tr	minor	nr	nr	44
Lake water	Biwa Lake, Japan	0.5–2	major	major [As(V) + As(III)]	nd/tr	minor/sig	nr	nr	58
Underground water	Sun Yat-Sen University, Taiwan	nr	major	tr	minor	minor	nr	minor	59
Tap water	Sun Yat-Sen University, Taiwan	nr	major	minor	minor	minor	nr	minor	59
Spring water	Kaohsiung, Taiwan	0.29	major	nd	nd	nd	nr	nd	59
Thermal water	nr	5300	major	tr	tr	tr	nr	nd	57
Seepage water of tin ore tailing	Altenberg, Germany	nr	major	minor	nr	nd	nd	tr	60
Surface freshwater	Ron Phibun, Thailand, 24 sites	4.8–583	major	minor	nr	nr	nr	nr	61
Groundwater (shallow)	Ron Phibun, Thailand, 23 sites	1–5114	major	minor	nr	nr	nr	nr	61
Groundwater (deep)	Ron Phibun, Thailand, 13 sites	1–133	major	sig	nr	nr	nr	nr	61

a Major = major arsenic constituent, >50% of total arsenic; sig = significant arsenic constituent, >10–50% of total arsenic; minor = minor arsenic constituent, 1–10% of total arsenic; tr = trace arsenic constituent, <1% of total arsenic; nr = not recorded; nd = not detected.

(1). Arsenate is the major arsenic form in most seawater samples, and arsenite can occur at significant levels as a consequence of reduction by marine phytoplankton and bacteria (4). Methylarsonate and dimethylarsinate also occur in seawater, and they can be significant species in surface waters where primary productivity is high (44,45). Their presence there is thought to be the result of biotransformation processes involving reduction and methylation within phytoplankton. Methylarsonate and dimethylarsinate are not particularly stable in seawater, however, and beyond the photic zone arsenate levels rise sharply with a concomitant decrease in the methylated species (45). As might be expected, seasonal factors and associated biological activity also influence the amounts of methylarsonate and dimethylarsinate in waters (46,62–64).

There have been reports of unknown arsenic compounds in seawater (48,49,65,66) and freshwater (65,67). These compounds are often referred to as "hidden" arsenic because they do not produce volatile arsine analytes necessary for their detection. The presence of hidden arsenic is inferred by the increase in volatile analytes following a sample degradation step.

Trivalent monomethylated and dimethylated arsenic species have also been reported in lake water (58,68,69). These arsenicals are probably methylarsonous acid and dimethylarsinous acid, although their precise chemical structures in natural waters have not been demonstrated. Most analytical methods for determining arsenic species in water samples convert the original arsenic species into volatile hydrides, which then serve as the analytes. Since the trivalent methylated arsenicals generate the same analyte as their respective pentavalent analogues, they must be separated before the hydride generation step so that they can be determined independently. Solvent extraction has been used to effect this separation (58). Possibly, the presence of these trivalent methylated arsenicals has been underestimated because few studies include a solvent separation step. However, in one study at least, dimethylarsenic in estuarine and coastal waters, as determined by hydride generation techniques, was shown to be present solely as the pentavalent dimethylarsinate species in three out of the four samples tested (50).

B. Arsenic Compounds in Marine Algae and Plants (Table 2)

Arsenosugars are by far the major arsenic constituents of marine algae. Although 16 arsenosugars have been identified from marine algal sources, there are four compounds (see Fig. 2, compounds 1–4) that are commonly found in large amounts. There are clear patterns between the various algal divisions, and also between orders (2). Thus, whereas arsenosugars 1 and 2 are common constituents of green and red algae, arsenosugars 3 and 4 are generally not found there. In brown algae of the class Fucales, however, all four arsenosugars are found and compounds **1** and **2** are present as minor constituents. In brown algae of the class Laminariales, compound **3** is dominant, **1** and **2** occur as minor compounds, and **4** is generally not found at all. The concentrations of arsenosugars in brown algae are much higher than those in both red and green algae.

Proposed intermediates in the biogenesis of arsenosugars (31) are arsenate, arsenite, methylarsonate, and dimethylarsinate, all of which occur in marine algae. Arsenate occurs generally as a minor compound, although some species (e.g., *Hizikia fusiforme*) can contain approximately half their total arsenic burden as inorganic (77). Dimethylarsinate is commonly found at low levels in marine algae. Both arsenite and methylarsonate, however, appear to be trace constituents only, and they are often not detected at all.

Marine algae also contain a considerable amount (up to 50%) of lipid-soluble arsenic (39). This arsenic has been rigorously identified on only one occasion, and shown to be a phospholipid derivative (see Fig. 2, compound **5**) of an arsenosugar (82,83).

Arsenic compounds in marine plants have not been well studied so far. They appear not to contain arsenosugars (39).

C. Arsenic Compounds in Marine Animals (Table 3)

Although marine animals contain many arsenic compounds, most species contain arsenobetaine as the major arsenical. Fish tend to have a simple pattern of arsenic compounds dominated by arsenobetaine. The silver drummer *Kyphosus sydneyanus*, however, contains trimethylarsine oxide as its major arsenical (25). Crustaceans also generally contain arsenobetaine as a high percentage of their total arsenic. It should be noted, however, that most work on fish and crustaceans has examined the muscle tissue, and the pattern of compounds may be more complex in other tissues (88).

Gastropods (mollusks) often contain very high arsenic concentrations (e.g., 339 µg/g) (40), most of which is usually arsenobetaine. The high concentrations of arsenobetaine in gastropods are not readily explainable—they do not appear to be diet related but may result from a particular ability of gastropods to bioaccumulate this arsenical. Bivalve mollusks, in addition to arsenobetaine, can contain large quantities of teramethylarsonium ion, particularly in the gill (94). High levels of arsenosugars have been reported in some scallop species, and, interestingly, these compounds were concentrated in the scallops' gonads (96). Presumably, the origin of the arsenosugars in scallops is their diet of unicellular algae, and useful information on the scallops' feeding preference might be revealed when further data on the arsenic species in phytoplankton become available. The arsenosugars found in the kidney of the clam *Tridacna maxima* reflect the presence of these compounds in the symbiotic unicellular algae residing in the clams' mantles (33,93,109).

Table 2 Arsenic Compounds in Marine Algae and Plants[a]

Species	Location	As concn. (µg/g)	As-sugar 1	As-sugar 2	As-sugar 3	As-sugar 4	Other As-sugars	As(V)	DMA	Other cpds. and unknowns	Ref.
Green algae											
Bryopsis maxima	Japan	19.4 dry	minor	sig	nd	nd	nd	nd	nd	sig	39
Caulerpa brachypus	Japan	11.6 dry	nd	nd	nd	nd	nd	nd	nd	major	39
Codium fragile	Japan	0.6 wet	major	sig	nd	nd	nd	nd	minor	minor	39
Codium fragile	Japan	18.2 dry	sig	sig	nd	nd	nd	nd	nd	nd	39
Ulva araksakii	Japan	15.5 dry	major	sig	nd	nd	nd	nd	nd	sig	39
Ulva pertusa	Japan	17.1 dry	sig	minor	nd	nd	nd	nd	nd	sig	39
Red algae											
Ahnfeltia paradoxa	Japan	11.7 dry	minor	sig	nd	nd	nd	nd	nd	sig	39
Campylaephora crassa	Japan	11.5 dry	major	sig	nd	sig	nd	nd	nd	sig	39
Carpopeltis flabellata	Japan	14.9 dry	sig	major	nd	nd	nd	nd	nd	sig	39
Carpopeltis crispata	Japan	13.4 dry	sig	major	nd	nd	nd	nd	nd	sig	39
Centroceras clavulatum	Japan	8.8 dry	sig	major	nd	nd	nd	sig	nd	nd	39
Chondria crassicaulis	Japan	22.5 dry	minor	minor	nd	minor	nd	nd	nd	major[h]	39
Chondrus ocellatus	Japan	13.2 dry	minor	sig	nd	nd	nd	nd	nd	nd	39
Chondrus verrucosus	Japan	17.9 dry	minor	sig	nd	nd	nd	nd	nd	sig	39
Coeloseira pacifica	Japan	23.1 dry	minor	major	nd	nd	nd	nd	nd	sig	39
Corallina pilulifera	Japan	21.6 dry	sig	sig	nd	nd	nd	nd	nd	sig	39
Cyrtymenia sparsa	Japan	44.8 dry	minor	sig	nd	nd	nd	nd	nd	major	39
Gelidium divaricatum	Japan	33.1 dry	sig	sig	nd	nd	nd	nd	nd	nd	39
Gigartina intermedia	Japan	19.8 dry	minor	sig	nd	nd	nd	nd	nd	sig	39
Gloiopeltis furcata	Japan	25.0 dry	sig	major	nd	nd	nd	nd	nd	minor	39
Grateloupia okaurai	Japan	16.1 dry	sig	sig	nd	nd	nd	nd	nd	nd	39
Grateloupia ramosissima	Japan	16.9 dry	sig	sig	nd	nd	nd	nd	nd	sig	39
Grateloupia turuturu	Japan	7.1 dry	sig	sig	nd	nd	nd	nd	nd	sig	39
Hypnea charoides	Japan	16.7 dry	sig	major	nd	nd	nd	nd	nd	sig	39
Hypnea japonica	Japan	9.9 dry	nd	major	nd	nd	nd	nd	nd	sig	39
Hypnea variabilis	Japan	6.0 dry	sig	major	nd	nd	nd	nd	nd	sig	39
Laurencia okamurai	Japan	19.2 dry	minor	sig	nd	nd	sig	nd	nd	minor	39
Lomentaria catenata	Japan	6.6 dry	sig	sig	nd	nd	nd	nd	nd	nd	39
Palmaria palmata	Japan	7.6 dry	major	minor	nd	nd	nd	nd	minor	nd	70
Porphyra tenera	Japan	7.6 dry	major	sig	nd	nd	nd	nd	nd	nd	70

Species	Location	Content									Ref.
Porphyra tenera	China	16 dry	major	sig	nd	nd	nd	nd	nd	nd	70
Porphyra tenera	Taiwan	21 dry	major	sig	nd	nd	nd	nd	nd	minor	70
Psilothallia dantata	Japan	22.4 dry	sig	sig	nd	nd	nd	nd	nd	sig	39
Schizymenia dubyi	Japan	12.0 dry	minor	major	nd	nd	nd	nd	nd	minor	39
Brown algae											
Dictyopteris prolifera	Japan	8.1 dry	minor	minor	sig	nr	nr	nr	nd	nd	39
Ecklonia radiata	Australia	10 wet	sig	nr	major	nr	tr	nr	nr	nr	71
Ecklonia radiata	Australia	nr	nr	sig	major	nr	nd	nr	nr	nr	72
Eisenia bicyclis	Japan	15 dry	sig	minor	major	nd	nd	nd	sig	nd	70
Fucus gardneri	Canada	9–17 dry	sig	nd	major	sig	nd	nr	minor	nd	73
Fucus serratus	Denmark	7 wet	minor	minor	major	sig	nd	tr	tr	tr	74
Fucus spiralis	Denmark	31.5 dry	nr	sig	sig	major	nd	nd	tr	minor, prob. As sugar 1	75
Fucus vesiculosis	England	140 dry	nr	minor	major	sig	nr	nr	nr	nr	76
Halidrys siliquosa	Denmark	21.3 dry	nr	minor	sig	major	nd	minor	nd	sig, prob. As sugar 1	75
Heterochordaria abietina	Japan	56.8 dry	sig	sig	sig	nd	nd	nd	nd	nd	39
Hizikia fusiforme	Japan	10 wet	nr	tr	minor	major	tr	major	nr	nr	77
Laminaria digitata	Denmark	43 dry	nr	sig	major	nd	nd	nr	nr	nr	76
Laminaria digitata	Scotland	72.1 dry	minor	sig	major	nd	nd	nd	nd	nd	78
Laminaria groenlandica	Canada	18.4 dry	sig	sig	major	nd	nd	nd	minor	nr	70
Laminaria japonica	Japan	4 wet	minor	minor	major	nr	nr	nr	nr	nr	79
Myelophycus caespitosus	Japan	33.3 dry	sig	sig	sig	nd	nd	nd	nd	minor	39
Nereocystis leutkeana	Canada	39 dry	sig	sig	major	nd	nd	tr	tr	nr	70
Pachydictyon coriaceum	Japan	16.7 dry	minor	minor	sig	nd	nd	nd	nd	nd	39
Padina arborescens	Japan	17.6 dry	minor	minor	sig	nd	nd	nd	nd	nd	39
Sargassum lacerifolium	Australia	40 wet	sig	minor	major	major	tr (4 cpds)	nr	tr	tr	80
Sargassum thunbergii	Japan	4 wet	nr	nr	sig	sig	tr	major	nr	nr	41
Sphaerotrichia divaricata	Japan	2 wet	major	sig	nr	nr	nr	nr	nr	nr	81
Spathoglossum pacificum	Japan	16.3 dry	minor	minor	minor	minor	nd	nd	nd	nd	39
Undaria pinnatifida	Japan	33.8 dry	minor	sig	sig	nd	nd	nd	nd	nd	39
Undaria pinnatifida	Japan	2.8 wet	sig	sig	sig	major	sig, cpd 5	nr	nr	nr	82,83
Seagrass											
Phyllospadix japonica	Japan	4.5 dry	nd	nd	nd	nd	nd	nd	nd	major	39

a Major = major arsenic constituent, >50% of extractable arsenic; sig = significant arsenic constituent, >10–50% of extractable arsenic; minor = minor arsenic constituent, 1–10% of extractable arsenic; tr = trace arsenic constituent, <1% of extractable arsenic; nr = not recorded; nd = not detected.

b This unknown was subsequently isolated and identified as compound **15** (38).

Table 3 Arsenic Compounds in Marine Animals[a]

Type/Species	As concn. (μg/g)	AB	AC	Tetra	TMAO	DMA	MA	As(III)	As(V)	As sugars	Other cpds. and unknowns	Ref.
Fish												
Mackerel	3.8 dry	sig	nr	nr	nr	minor	nd	nr	sig	nr	minor	84
Mixed (4 species)	13.7–30.7 dry	major	nr	nr	nr	nd	nd	nr	nd	nr	nr	84
Mixed (3 species)	62–196 dry	major	nr	nr	nr	nd	nd	nr	nd	nr	nr	84
Plaice *Pleuronectes platessa*	41.9 dry	major	tr	tr	nr	nr	nd	nr	nd	nr	tr	85
Tuna	3.2 dry	major	tr	tr	tr	nr	nd	nr	nd	nr	tr	85
Silver drummer *Kyphosus sydneyanus*	1 wet	nd	nd	minor	major	nd	nd	nr	nd	minor	nd	25
Estuary catfish *Cnidoglanus macrocephalus*	~0.5 wet	major	nr	nr	sig	nr	nr	nr	nr	nr	nr	86
Flounder *Pleuronectes platessa*	36.9 dry	major	nd	nr	nr	nd	nd	nd	nd	nr	nd	87
Sole *Solea solea*	24.8 dry	major	nd	nr	nr	nd	nd	nd	nd	nr	nd	87
Coral reef fish *Abudefduf vaigiensis*	nr	major	nr	nr	nr	nr	nr	nr	nr	nr	minor[b]	42
Sea mullet *Mugil cephalus* blood	1.5 dry	sig	nd	nd	nd	major	nd	nr	minor	nd	nd	88
M. cephalus intestine	7.8 dry	sig	minor	nd	sig	sig	nd	nr	minor	nd	nd	88
M. cephalus stomach	5.0 dry	major	minor	nd	minor	sig	nd	nr	minor	nd	nd	88
M. cephalus liver	19.2 dry	major	minor	nd	nd	sig	nd	nr	nd	nd	nd	88
M. cephalus heart	5.4 dry	major	minor	nd	nd	sig	nd	nr	nd	nd	nd	88
M. cephalus gill	6.9 dry	major	minor	nd	nd	sig	nd	nr	nd	nd	nd	88
M. cephalus kidney	8.3 dry	major	minor	nd	minor	sig	nd	nr	nd	nd	nd	88
M. cephalus muscle	4.7 dry	major	nd	nd	nd	nd	nd	nr	minor	nd	nd	88
M. cephalus gonad	7.9 dry	major	minor	nd	nd	minor	nd	nr	minor	nd	nd	88
Cod	10.1 dry	major	nd	tr	nr	minor	nd	nd	tr	nr	nd	89
Tuna	5.7 dry	major	nd	nr	nr	minor	nd	nd	tr	nr	nd	89
Ray	96.0 dry	major	nd	nr	nr	nd	nd	nd	nd	nr	nd	90
Place	54.0 dry	major	nd	nr	nr	nd	nd	nd	nd	nr	nd	90
Sole	25.5 dry	major	nd	nr	nr	nd	nd	nd	nd	nr	nd	90
Crustaceans												
Lobster *Panulirus cygnus* digestive gland	nr	major	nr	nr	nr	nr	nr	nr	nr	minor 2	tr 5.18	34
Shrimp *Rimicaris exoculata*	21.8 dry	major	nr	nr	nr	tr	nd	minor	tr	nr	nd	91

Species	Conc.											Ref.
Shrimp *Rimicaris exoculata*	4.4 dry	major	nr	nr	nr	nd	nd	nd	nd	nd	nd	91
Shrimp (3 species)	15–44 dry	major	tr/nd	minor	nd	nr	nd	tr/nr	tr/nd	nr	tr	85
Crab *Cancer pagurus*	118 dry	major	nd	tr	tr	tr	nd	tr	tr	nr	tr	85
Amphipods *Allochestes compressa*	nr	major	nd	nd	nd	nd	nd	nd	nd	sig	nd	25
Krill *Euphausia superba*	nr	major	nd	nd	nd	sig	nd	nd	nd	minor	nd	25
Shrimp *Crangon crangon* muscle	9.2 dry	major	nd	tr	nr	nd	nd	nd	nd	tr	minor	92
Shrimp	3.8 dry	major	nd	nr	nr	tr	nd	nr	nd	nr	tr	90
Molluscs												
Clam kidney *Tridacna maxima*	200 wet	nr	nr	nr	nr	nr	nr	nr	nr	major 4; sig 1; minor 3,7,9; tr 11,12	tr 16	33,93
Clam *Meretrix lusoria* muscle	3.6 wet	major	nd	sig	nr	nr	nr	nr	nr	nr	sig	94
M. lusoria midgut gland	6.6 wet	sig	nd	minor	nr	nr	nr	nr	nr	nr	sig	94
M. lusoria gill	23.8 wet	minor	nd	major	nr	nr	nr	nr	nr	nr	minor	94
Clams (5 species)	1.2–2.2 wet	sig	nr	sig	nr	nd/minor	nd	nr	nr	nr	sig	95
Mussel *Bathymodiolus puteoserpentis* adductor/mantle	11.4 dry	sig	nr	nr	nr	nd	nd	minor	tr	major 2; minor 1	minor	91
B. puteoserpentis gill	70.6 dry	minor	nr	minor	nr	nd	nd	minor	tr	major 2; sig 1	minor	91
Scallop *Placopecten magellanicus* cultured pre-spawn (muscle)	3.9–7.2 dry	major	nd	sig	nd	nd	nd	nd	nd	nd	nd	96
P. magellanicus cultured pre-spawn (gonad)	44.1 dry	minor	nd	nd	nd	nd	nd	nd	nd	major 4; minor 1,2,3	minor	96
P. magellanicus wild pre-spawn (muscle)	7.9–9.4 dry	major	nd	sig	nd	nd	nd	nd	nd	sig 4	nd	96
P. magellanicus wild pre-spawn (gonad)	15.2 dry	sig	nd	nd	nd	nd	nd	nd	nd	major 4; minor 1,2,3	minor	96
Scallop *Chlamys islandica* wild (muscle)	2.3/2.4 dry	major	nd	nd	nd	nd	nd	nd	nd	nd	sig	96
C. islandica wild (gonad)	7.3/10.7 dry	sig/minor	nd	nd	nd	nd	nd	nd	nd	major 4; minor 1,2,3	nd	96
Mussel *Mytilus edulis*	8.5 dry	major	nd	nr	nr	sig	nd	nd	nd	nr	minor	87
Cockle *Cardium edule*	5.7 dry	major	nd	nr	nr	sig	nd	nd	nd	nr	nd	87
Mussels	9.3 dry	major	nd	nr	nr	minor	nd	nd	nd	nr	nd	89
Clams	36.6 dry	major	nd	nr	nr	tr	nd	nd	nd	nr	nd	90

Table 3 Continued

Type/Species	As concn. (μg/g)	AB	AC	Tetra	TMAO	DMA	MA	As(III)	As(V)	As sugars	Other cpds. and unknowns	Ref.
Clams	25.0 dry	major	nd	nr	nr	tr	nd	nd	nd	nr	tr	90
Mussel	2.7 dry	minor	nr	nr	nr	minor	nd	minor	minor	nr	major	97
Cockle	2.8 dry	minor	nr	nr	nr	nd	nd	minor	sig	nr	major	97
Oyster	3.2 dry	minor	nr	nr	nr	minor	nd	tr	minor	nr	major	97
Abalone *Haliotis roeii* muscle	1.0 wet	major	nd	minor	nd	nd	nd	nd	nd	minor	nd	25
H. roeii midgut gland	nr	sig	nd	minor	nd	nd	nr	nr	nr	major	nd	25
Gastropod *Tectus pyramis* muscle	4.2 wet	major	nd	minor	nr	nr	nr	nr	nr	nr	sig	98
T. pyramis midgut gland	7.5 wet	sig	nd	sig	nr	nr	nr	nr	nr	nr	sig	98
Gastropod *Thais distinguenda*	153 dry	major	minor	minor	nd	nd	nd	nd	nd	tr 13	tr	40
Gastropod *Thais bitubercularis*	131–150 dry	major	minor	tr/minor	nd	nd	nd	nd	nd	tr 13	tr	40
Gastropod *Morula musiva*	112–339 dry	major	minor	tr	nd	nd	nd	nd	nd	tr 13	tr	40
Gastropod *Morula marginalba*	233 dry	major	minor	tr	nd	nd	nd	tr	tr	minor 1	tr	99
Gastropod *Austrocochlea constricta*	74.4 dry	major	tr	minor	nd	nd	nd	tr	tr	minor 1	minor	99
Gastropod "sea snail"	48.5 dry	major	nd	nr	nr	nd	nd	nr	nr	nr	nd	90
Other												
Pilot whale *Globicephalus melas*	0.27–1.27 wet (2 specimens)	major	tr/minor	nd	nd	nd/tr	nd/tr	nr	nd	nr	minor	100
Ringed seal *Phoca hispida*	0.57–2.4 wet (10 specimens)	major	tr/minor	minor	nd	tr/minor	nd/tr	nr	nd	nr	minor	100
Beluga whale *Delphinapterus leucas*	0.17 wet	major	minor	nd	nd	sig	minor	nr	nd	nr	minor	100
Bearded seal *Erignatus barbatus*	0.48 wet	major	minor	minor	nd	minor	minor	nr	nd	nr	minor	100
Turtle *Dermochelys coriacea* muscle	4.4 wet	major	nd	nd	nd	nd	nd	nd	tr	nd	nd	19
D. coriacea heart	0.7 wet	major	minor	nd	nd	nd	nd	nd	minor	nd	nd	19
D. coriacea liver	1.2 wet	major	sig	nd	nd	nd	nd	nd	sig	nd	nd	19
Jellyfish *Aurelia aurita*	0.039 wet	major	tr	minor	nd	nd	nd	nd	nd	nd	tr	101
Jellyfish *Carybdea rastonii*	0.135 wet	major	nd	minor	nd	nd	nd	nd	nd	nd	tr	101

Marine reference materials

Material												Ref
NRCC DORM-1 dogfish muscle	17.7 dry	major	tr	minor	nd	minor	nd	nd	nd	nr	minor	102
NRCC DORM-1 dogfish muscle	17.7 dry	major	nd	minor	tr	minor	tr	nd	nd	nd	minor	103
NRCC DORM-2 dogfish muscle	18.0 dry	major	tr	minor	nd	minor	nd	nd	nd	nr	tr	102
NRCC DORM-2 dogfish muscle	18.0 dry	major (not resolved)		nr	nr	minor	tr	nd	minor	nr	tr	104
NRCC DOLT-1 dogfish liver	10.1 dry	major	tr	tr	tr	sig	nd	nd	minor	nr	minor	85
NRCC DOLT-1 dogfish liver	10.1 dry	major	nd	nd	minor	sig	tr	nd	tr	nd	minor	103
NRCC TORT-1 lobster midgut gland	24.6 dry	major	tr	tr	tr	minor	nd	nd	nd	minor 1,2	minor	36
NRCC TORT-1 lobster midgut gland	24.6 dry	major	nd	tr	tr	minor	tr	nd	minor	tr 2	tr	103
NIES No. 6 mussel	9.2 dry	sig	tr	tr	tr	major	nd	nd	minor	sig 1,2	minor	36
NIES No. 15 scallop	3.29 dry	major	nd	tr	nd	nd	nd	nd	nd	1, not quantified	tr	103
NIST SRM 1566a oyster	13.4 dry	sig	tr	tr	tr	major	tr	nd	minor	sig 1,2	sig	36
NIST SRM 1566a oyster	13.4 dry	sig	nr	nr	nr	sig	nr	nr	nr	sig 1,2	nr	70
CRM 278 mussel tissue	5.9 dry	major	nd	nr	nr	sig	nd	nr	nd	nr	sig	105
CRM 278 mussel tissue	5.9 dry	major	nd	nr	nr	sig	nd	nd	nd	nr	minor	89
CRM 422 cod muscle	21.1 dry	major	nd	nr	nr	nd	nr	nr	nr	nr	nd	105
CRM 627 tuna fish	4.8 dry	major	nr	nr	nr	minor	nr	nr	nr	nr	nr	106
CRM 627 tuna fish	4.8 dry	major	nr	nr	nr	minor	nd	nr	nr	nr	nd	97
NFA-Plaice	43.2 dry	major	tr	tr	tr	nr	nr	nr	nr	nr	nr	107
NFA-Plaice	43.2 dry	major	tr	tr	nd	nr	tr	nr	nr	nr	nr	108
NFA-Shrimp	42.2 dry	major	nd	tr	nd	nr	nd	nd	tr	nr	tr	85
NFA-Shrimp	42.2 dry	major	nd	tr	nd	nr	nd	nr	nr	nr	nr	108

[a] Major = major arsenic constituent, >50% of extractable arsenic; sig = significant arsenic constituent, >10–50% of extractable arsenic; minor = minor arsenic constituent, 1–10% of extractable arsenic; tr = trace arsenic constituent, <1% of extractable arsenic; nr = not recorded; nd = not detected.

[b] Trimethylarsoniopropionate (see Fig. 1).

Marine mammals examined so far have contained only low levels of arsenic, but in all cases, arsenobetaine was the dominant arsenical present (100). The liver of the turtle *Dermochelys coriacea* is unusual among marine animals because it contains arsenocholine at significant levels (19).

Lipid arsenic compounds also occur in marine animals (34,110). The compounds originally present in the lipid fraction were subjected to base and/or acid hydrolysis, and the water-soluble products identified by HPLC-ICPMS. In this way, phosphatidylarsenocholine (see Fig. 2, compound **18**) and a phosphatidyl-arsenosugar (see Fig. 2, compound **5**) were identified in the digestive gland of lobster (34), and evidence was presented for the presence of lipids containing arsenocholine and dimethylated arsenic moieties in shark tissues (110).

D. Arsenic Compounds in Freshwater Algae and Plants (Table 4)

Freshwater algae have been little studied compared with their marine counterparts, but their pattern of arsenic compounds appears to be similar (15,70). Arsenosugars are present as major or significant compounds (only compounds **1** and **2** in Fig. 2 have been reported so far). Arsenate can also be a major arsenical in freshwater algae, and small amounts of dimethylarsinate and methylarsonate have also been reported. Arsenite has not been detected.

Freshwater plants present a different picture. Here, arsenite and/or arsenate are found as significant or major arsenicals, and arsenosugars are only occasionally detected (30). Methylarsonate and dimethylarsinate are also present in most samples at low levels. Interestingly, tetramethylarsonium ion was also reported in several freshwater plant species, generally as a trace constituent (30).

A considerable quantity of unspecified lipid arsenic appears to be associated with freshwater algae and plants (5).

E. Arsenic Compounds in Freshwater Animals (Table 5)

In contrast to the large amount of data available for marine animals, there are very few results for freshwater species. The total arsenic concentrations are low, but it appears that arsenobetaine is generally the major arsenical.

F. Arsenic Compounds in Terrestrial Fungi and Lichens (Table 6)

Fungi cover a wide range of arsenic concentrations ($<1->2000$ µg/g), and they contain a most interesting array of arsenic compounds. They are the only organisms containing methylarsonate at appreciable levels, and some species (e.g., *Sarcosphaera coronaria*) contain $>50\%$ of the total extractable arsenic as this

Table 4 Arsenic Compounds in Freshwater Algae and Plants[a]

Species	As concn. (µg/g) dry mass	As(III)	As(V)	MA	DMA	TMAO	AB	AC	Tetra	As sugars	Other cpds. and unknowns	Ref.
Algae												
Nostoc flagelliforme commercially available powder	2.70	nr	nr	nr	minor	nr	nd	nr	nr	major 1	nd	70
deep green algae	249	nd	major	nd	nd	nd	nd	nd	nd	sig 1 / minor 2	nd	15
green algae	56	nd	major	minor	minor	nd	nd	nd	nd	sig 1 / minor 2	nd	15
Plants												
Equisetum fluviatile	48	sig	major	nd	nd	nd	nd	nd	nd	nd	nd	30
E. fluviatile	260	major	sig	nd	nd	nd	nd	nd	nd	nd	nd	30
E. fluviatile	30	sig	sig	nd	nd	nd	nd	nd	nd	nd	nd	30
Typha latifolia	3.8/5.0	sig	major	tr	tr	nd	nd	nd	nd	nd	nd	30
T. latifolia	0.52	sig	major	tr	tr	nd	nd	nd	tr	nd	nd	30
Sparganium augustifolium	2.5	sig	major	nd	nd	nd	nd	nd	nd	nd	nd	30
Potomogetan richardsonii	20	sig	major	nd	tr	nd	nd	nd	tr	minor 1	nd	30
Myriophyllum sp.	39	major	sig	minor	nd	nd	nd	nd	tr	nd	nd	30
Myriophyllum sp.	17.4	sig	major	minor	minor	nd	nd	nd	tr	tr 1 / tr 2	nd	30
Myriophyllum sp.	78	major	sig	minor	tr	nd	nd	nd	tr	nd	nd	30
Lemna minor	28	sig	major	minor	minor	nd	nd	nd	minor	minor 1 / minor 2	nd	30
Carex aquatilis	23.9	sig	major	tr	nd	nd	nd	nd	tr	nd	nd	30
C. aquatilis	8.14	sig	major	tr	tr	nd	nd	nd	tr	nd	nd	30
C. aquatilis	8.54	sig	major	minor	tr	nd	nd	nd	tr	nd	nd	30
C. aquatilis	23.4	minor	major	minor	tr	nd	nd	nd	minor	nd	nd	30
C. aquatilis	31.3	minor	major	minor	minor	nd	nd	nd	nd	nd	nd	30
C. aquatilis	9.44	sig	major	tr	minor	nd	nd	nd	nd	nd	nd	30
C. aquatilis	25.2	sig	major	nd	nd	nd	nd	nd	nd	nd	nd	30
C. aquatilis	45.9	minor	major	nd	tr	nd	nd	nd	nd	nd	nd	30

[a] major = major arsenic constituent, >50% of extractable arsenic; sig = significant arsenic constituent, >10–50% of extractable arsenic; minor = minor arsenic constituent, 1–10% of extractable arsenic; tr = trace arsenic constituent, <1% of extractable arsenic; nr = not recorded; nd = not detected.

Table 5 Arsenic Compounds in Freshwater Animals[a]

Species	As concn. (μg/g)	As(III)	As(V)	MA	DMA	TMAO	AB	AC	Tetra	Other cpds. and unknowns	Ref.
Freshwater fish											
Rainbow trout *Salmo gairdneri* muscle	1.46 wet	nr	nd	nd	nd	nd	major	nd	nd	nd	111
Baked rainbow trout	0.56 wet	nd	nd	nd	major	nr	nr	nr	nr	nd	112
Japanese smelt *Hypomesus nipponensis* muscle	1.08 wet	nr	nd	nd	nd	nd	major	nd	nd	nd	111
Bream muscle	0.67 dry	sig (not resolved)		minor	sig	sig (not resolved)		nr	minor	nd	113
Bream kidney	0.76 dry	sig (not resolved)		minor	sig	sig (not resolved)		nr	tr	nd	113
Freshwater mussels											
Dreisena polymorpha	3.70 dry	sig (not resolvled)		minor	minor	major (not resolved)		nr	tr	nd	113

[a] major = major arsenic constituent, >50% of extractable arsenic; sig = significant arsenic constituent, >10–50% of extractable arsenic; minor = minor arsenic constituent, 1–10% of extractable arsenic; tr = trace arsenic constituent, <1% of extractable arsenic; nr = not recorded; nd = not detected.

compound (23). Almost all species examined contain dimethylarsinate, often (e.g., *Laccaria amethystina*) as the major arsenic constituent (23,24,117). Arsenite and arsenate are also commonly found in fungi.

Of particular interest, however, is the presence of arsenobetaine as a major arsenical in many species of fungi (20). In addition, several fungal species contain arsenocholine and/or tetramethylarsonium ion, and *Sparassis crispa* contains arsenocholine as the major arsenical (20). These three arsenic compounds had traditionally been considered metabolites of marine animals before their discovery in fungi.

Trimethylarsine oxide is not a common constituent of fungi, being confirmed in only one species so far (117). There have been only two reports of an arsenosugar in fungi (15,115). Arsenosugars are more widespread in lichens (26,115), not surprising in view of the algal component in these composite organisms (fungus/alga). Although arsenosugars and other organoarsenic compounds are common in lichens, inorganic arsenic in the form of arsenite and arsenate are the dominant species.

G. Arsenic Compounds in Terrestrial Plants (Table 7)

The arsenic chemistry of terrestrial plants is dominated by inorganic arsenic. With few exceptions, all plant species examined so far contain either arsenite or, more usually, arsenate as their major arsenical. Methylarsonate and dimethylarsinate are also commonly found.

An interesting aspect of the work on terrestrial plants is the widespread occurrence of trimethylarsine oxide. It occurs in many of the plants examined so far, and is the major compound in aqueous methanol extracts of *Alnus incana* (26). Arsenobetaine and tetramethylarsonium ion have been detected in several plant species, but the levels are low. Arsenocholine may have been present in some species but this is not proved because it was unresolved from trimethylarsine oxide (120). Arsenosugars (compounds 1 and 2 only in Fig. 2) have been reported in some plants at low levels (15,26).

Some plants are able to accumulate exceedingly high concentrations of arsenic (in the order of 1% dry mass). Such arsenic hyperaccumulators were first reported in 1975 (126), and there have been several subsequent reports (127,128,118,119). Recent work has shown that arsenic-hyperaccumulating ferns store their high arsenic burden primarily in the fronds as arsenite (118,119). Arsenite is generally thought to be the most toxic of the arsenic species; the physiological processes at play in the fern are of considerable interest.

H. Arsenic Compounds in Terrestrial Animals

Concentrations of total arsenic in terrestrial animals, including those used as human food, are low and the dominant compounds are arsenite and arsenate (9,129).

Table 6 Arsenic Compounds in Fungi and Lichen[a]

Type/species	As concn. (μg/g)	As(III)	As(V)	MA	DMA	TMAO	AB	AC	Tetra	As sugars	Other cpds. and unknowns	Ref.
Ascomycetes: Pezizales												
Sarcosphaera coronaria	360/332 dry	nd	nd	major	nd	nd	nd	nr	nd	nr	nd	23
Sarcosphaera coronaria	2120 dry	tr (not resolved)		major		tr		nr	nd	nr	nd	23
Sarcosphaera coronaria	15 wet	tr (not resolved)		major		minor (not resolved)	nd	nr	nd	nr	nd	23
Turzetta cupularis	nr	tr	major	nd	sig	nd	nd	nd	nd	tr. 1	nd	15
Basidiomycetes: Agaricales												
Macrolepiota procera	0.42 dry	tr	tr	tr	tr	nd	major	nd	nd	nr	nd	20
Leucocoprinus badhamii	2.9 dry	tr	minor	tr	sig	nd	sig	tr	tr	nr	minor	20
Agaricus abruptibulbus	3.49 dry	minor	minor	minor	minor	nd	major	nd	nd	nr	nd	20
Agaricus bisporus	1.00 dry	nd	sig	minor	sig	nd	major	nd	nd	nr	nd	20
Agaricus campester	1.32 dry	tr	nd	tr	tr	nd	major	nd	minor	nr	nd	20
Agaricus elvensis	2.43 dry	minor	minor	minor	minor	nd	major	nd	nd	nr	minor	20
Agaricus fuscofibrillosus	2.68 dry	tr	minor	nd	minor	nd	major	nd	nd	nr	nd	20
Agaricus lilaceps	1.78 dry	minor	minor	nd	sig	nd	major	nd	nd	nr	minor	20
Agaricus macrosporus	3.32 dry	nd	minor	minor	minor	nd	major	nd	nd	nr	nd	20
Agaricus silvicola	6.2 dry	minor	minor	nd	minor	nd	major	nd	nd	nr	minor	20
Agaricus subrutilescens	10.8 dry	tr	minor	minor	minor	nd	major	nd	nd	nr	nd	20
Agaricus haemorrhoidarius	8.2/9.3 dry	nd	nd	nd	minor	nd	major	nr	nd	nr	nd	23
Agaricus haemorrhoidarius	8.2/9.3 dry	nd	nd	tr	nd	nd	major	nr	nd	nr	nd	23
Agaricus placomyces	8.1/9.2 dry	nd	nd	minor	nd	nd	major	nr	nd	nr	nd	23
Amanita caesarea	0.5 dry	sig	sig	nd	sig	nd	nd	nd	sig	nr	nd	20
Amanita magniverrucata	0.50 dry	tr	tr	tr	tr	nd	nd	nd	nd	nr	nd	20
Amanita phalloides	0.55 dry	tr	tr	tr	tr	nd	tr	nd	nd	nr	nd	20
Amanita rubescens	0.1 dry	tr	nd	nd	tr	nd	nd	nd	nd	nr	nd	20
Amanita muscaria	3.1 dry	sig	minor	minor	minor	nd	major	minor	minor	nr	minor	20

Species	Amount											Ref.
Amanita muscaria	21.9 dry	minor	minor	nd	minor	nd	sig	sig	minor	nr	major	114
Psathyrella candolleana	13.6 dry	sig	major	minor	minor	nd	minor	nd	minor	nd	nd	115
Coprinus comatus	410 dry	minor	minor	nd	minor	nd	major	tr	nd	nd	tr	115
Entoloma rhodopolium	0.55 dry	sig	major	nd	sig	nd	nd	nd	nd	nr	nd	20
Pluteus cervinus	nr	sig	major	nd	tr	nd	nd	nd	nd	nr	nd	15
Volvariella volvacea	0.82 dry	tr	minor	minor	major	nd	minor	nd	nd	nr	nd	20
Volvariella volvacea	1.05 dry	tr	minor	tr	major	nr	tr	nr	nd	nr	nd	20
Collybia maculata	30.0 dry	tr	nd	nd	tr	nr	major	nr	nr	nr	nd	116
Collybia butyracea	10.9 dry	nd	minor	nd	sig	nd	major	nr	nr	nr	minor	116
Laccaria laccata	0.66 dry	tr	sig	tr	major	nd	tr	nd	nd	nr	nd	20
Laccaria laccata	4.26 dry	minor	minor	tr	major	nd	minor	nr	nr	nr	nd	20
Laccaria amethystina	109 to 200 dry	nd	minor	minor	major	nd	nr	nr	nr	nr	nd	24
Laccaria amethystina	3.4 wet	tr	tr	tr	major	minor	tr	nr	nd	nr	nd	23
Laccaria amethystina	40.5 wet	nd	nd	nd	major	(not resolved)	(not resolved)	nr	nd	nr	nd	23
Laccaria amethystina	23 dry	nd	minor	minor	major	tr	nd	nd	nd	nd	nd	117
Laccaria amethystina	77 dry	nd	minor	minor	major	minor	nd	nd	nd	nd	nd	117
Laccaria amethystina	1420 dry	nd	nd	tr	major	minor	tr	nd	nd	tr	tr	117
Lyophyllum conglobatum	0.63 dry	nd	tr	nd	sig	nd	major	nd	nd	nr	nd	20
Tricholoma inamoenum	0.38 dry	tr	nd	tr	major	nd	nd	nd	nr	nr	nd	20
Tricholoma pardinum	0.63 dry	sig	minor	nd	sig	nd	major	nd	nd	nr	nd	20
Tricholoma sulphureum	0.26 dry	major	tr	nd	sig	nd	nd	nd	nd	nr	nd	20
Basidiomycetes: Boletales												
Paxillus involutus	36 dry	minor	minor	nd	major	nd	nd	nd	tr	minor 2	sig	115
Leccinum scabrum	8.3 dry	minor	minor	nd	major	nd	nd	nd	nd	nd	nd	115
Basidiomycetes: Cantharellales												
Sparassis crispa	1.03 dry	nd	minor	nd	nd	nd	sig	major	nd	nr	nd	20
Sparassis crispa	0.57 dry	tr	tr	nd	tr	nd	tr	sig	tr	nr	major	20
Basidiomycetes: Gomphales												
Gomphus clavatus	4.47 dry	nd	tr	tr	minor	nd	major	minor	nd	nr	minor	20
Ramaria pallida	3.7 dry	minor (not resolved)	minor	minor	minor	nd	major	sig (not resolved)	nd	nr	nd	20

Table 6 Continued

Type/species	As concn. (µg/g)	As(III)	As(V)	MA	DMA	TMAO	AB	AC	Tetra	As sugars	Other cpds. and unknowns	Ref.
Basidiomycetes: Lycoperdales												
Geastrum sp.	3.12 dry	minor	minor	tr	minor	nd	major	nd	nd	nr	nd	20
Calvatia excipuliformis	0.72 dry	minor	minor	nd	sig	nd	major	nd	nd	nr	nd	20
Calvatia utriformis	0.79 dry	tr	minor	nd	minor	nd	major	nd	nd	nr	nd	20
Lycoperdon echinatum	1.23 dry	tr	minor	nd	sig	nd	major	nd	nd	nr	nd	20
Lycoperdon perlatum	2.81 dry	tr	tr	minor	minor	nd	major	nd	nd	nr	nd	20
Lycoperdon perlatum	0.23 wet	minor (not resolved)		minor	minor	major (not resolved)		nr	nr	nr	nd	23
Lycoperdon pyriforme	0.46 dry	minor	sig	tr	tr	nd	major	nd	nd	nr	nd	20
Lycoperdon pyriforme	1010 dry	minor	sig	minor	sig	nd	sig	tr	tr	nd	nd	115
Lycoperdon pyriforme	nr	sig	minor	nd	minor	nd	major	tr	tr	nd	tr	115
Basidiomycetes: Poriales												
Albatrellus cristatus	7.7 dry	nd	nd	nd	minor	nd	major	minor	nd	nr	minor	20
Albatrellus ovinus	0.26 dry	sig	sig	tr	tr	nd	major	tr	tr	nr	tr	20
Albatrellus pes-caprae	0.77 dry	tr	minor	tr	nd	nd	major	nd	nd	nr	nd	20
Basidiomycetes: Thelephorales												
Thelephora terrestris	15.9 dry	sig	sig	minor	nd	nd	nd	nd	nd	nr	nd	20
Sarcodon imbricatum	24 dry	minor (not resolved)		sig	sig	major (not resolved)		nr	minor	nr	nd	23
Sarcodon imbricatum	0.9 dry	nd	nd	nd	sig	sig (not resolved)		nr	sig	nr	nd	23

Lichens												
Alectoria ochroleuca methanol/water extraction	4.29 dry	sig	nd	tr	nd	minor	major	nd	minor	minor 1	minor	26
A. ochroleuca water extraction	4.29 dry	nd	sig	tr	nd	minor	sig	nd	nd	sig 1	minor	26
Alectoria sp.	nr	tr	tr	nd	nd	nd	nd	nd	nd	tr 2	nd	15
Alectoria sp.	nr	nd	major	nd	nd	nd	nd	nd	nd	nd	nd	15
Alectoria sp.	0.55 dry	sig	sig	nd	nd	nd	nd	nd	nd	sig 2	nd	15
Bryoria sp.	0.30 dry	tr	major	nd	nd	nd	nd	nd	nd	tr 2	nd	15
Bryoria sp.	0.30 dry	nd	major	nd	nd	nd	nd	nd	nd	nd	nd	15
Cladina sp.	38 dry	major	sig	minor	minor	minor	minor	nd	nd	nd	minor	115
Cladonia sp.	nr	sig	major	tr	tr	nd	nd	nd	nd	tr 1	nd	15
Cladonia sp.	14.3 dry	sig	sig	tr	tr	nd	sig	nd	nd	nd	nd	115
Cladonia sp.	29 dry	sig	sig	minor	minor	sig	sig	nd	tr	minor 1	minor	115
Cladonia sp.	520 dry	major	sig	minor	minor	nd	minor	nd	nd	nd	nd	115
Cladonia sp.	49 dry	major	sig	minor	nd	minor	minor	nd	nd	nd	minor	115
Cladonia sp.	55 dry	sig	sig	minor	minor	minor	minor	nd	nd	minor 1	minor	115
Usnea articulata methanol/water extraction	1.12 dry	sig	nd	nd	nd	sig	sig	nd	nd	sig 1	minor	26
Usnea articulata water extraction	1.12 dry	nd	minor	minor	nd	sig	sig	nd	nd	sig 1	minor	26
Unidentified	2300 dry	minor	major	minor	minor	nd	tr	nd	nd	nd	nd	115

[a] Major = major arsenic constituent, >50% of extractable arsenic; sig = significant arsenic constituent, >10–50% of extractable arsenic; minor = minor arsenic constituent, 1–10% of extractable arsenic; tr = trace arsenic constituent, <1% of extractable arsenic; nr = not recorded; nd = not detected.

Source: Ref. 6 (mushroom data).

Table 7 Arsenic Compounds in Terrestrial Plants[a]

Species	As concn. (μg/g)	As(III)	As(V)	MA	DMA	TMAO	AB	AC	Tetra	As sugars	Other cpds. and unknowns	Ref.
Bryophyta (Mosses)												
Drepanocladus sp.	490/1220 dry	sig	major	nd	nd	nd	nd	nd	minor	nd	nd	30
Drepanocladus sp.	880 dry	sig	major	tr	nd	nd	nd	nd	nd	nd	nd	30
Drepanocladus sp.	770 dry	sig	sig	nd	minor	nd	nd	nd	nd	nd	nd	30
Fumaria hygrometrica	237 dry	nd	major	nd	nd	nd	nd	nd	nd	sig 1	nd	15
Fumaria hygrometrica	350 dry	nd	major	nd	sig	nd	nd	nd	nd	minor 1 tr 2	nd	15
Fumaria hygrometrica	91 dry	sig	sig	minor	minor	nd	nd	nd	nd	sig 1	nd	15
Pteridophyta: Ferns												
Asplenium viride leaves. methanol/water extraction	7.1 dry	sig	sig	sig	sig	minor	nd	nd	minor	tr 1	nd	26
Asplenium viride leaves. water extraction	7.1 dry	major	sig	minor	nd	tr	nd	nd	tr	tr 1	nd	26
Dryopteris dilata leaves. methanol/water extraction	2.0 dry	sig	major	nd	nd	minor	nd	nd	minor	minor 1	nd	26
Dryopteris dilata leaves. water extraction	2.0 dry	major	sig	nd	tr	minor	nd	nd	tr	tr 1	nd	26
Pteris vittata fronds	3.280–4.980 wet?	major	sig	nr	nr	nr	nr	nr	nr	nr	nr	118
Pteris vittata roots	~300 wet?	minor	major	nr	nr	nr	nr	nr	nr	nr	nr	118
Pityrogramma calomelanos roots	88/180 dry	minor	major	nd	nd	nd	nd	nd	nd	nd	nd	119
Pityrogramma calomelanos stalk	150/230 dry	sig	major	tr	nd	nd	nd	nd	nd	nd	nd	119
Pityrogramma calomelanos young frond	5130/5210 dry	major	sig	nd	nd	nd	nd	nd	nd	nd	nd	119
Pityrogramma calomelanos old frond	600 dry	minor	major	tr	tr	nd	nd	nd	nd	nd	nd	119
Pteridophyta: Horse-tails												
Equisetum pratense leaves. methanol/water extraction	8.3 dry	sig	sig	minor	nd	sig	nd	nd	nd	tr 1	nd	26

Equisetum pratense leaves, water extraction	8.3 dry	major	major	minor	nd	minor	nd	nd	nd	tr 1	nd	26
Gymnospermae												
Larix decidua needles, methanol/water extraction	3.7 dry	sig	sig	nd	nd	sig	nd	nd	nd	nd	nd	26
Larix decidua needles, water extraction	3.7 dry	major	sig	nd	nd	minor	nd	nd	nd	nd	nd	26
Picea abies needles, methanol/water extraction	0.91 dry	sig	major	nd	minor	minor	nd	nd	nd	minor 1	nd	26
Picea abies needles, water extraction	0.91 dry	major	sig	tr	tr	minor	nd	nd	nd	tr 1	nd	26
Thuja plicata	0.96 dry	major	sig	minor	minor	nd	nd	nd	nd	nd	nd	15
Angiospermae: Monocotyledonea												
Agrostis capillaris	nr	major	sig	nd	tr	nd	tr	nd	nd	nr	nd	120
Agrostis scabra Willd.	53.1 dry	sig	major	nd	tr	nd	nd	nd	nd	nd	nd	30
Agrostis scabra Willd.	71.9 dry	sig	major	tr	nd	nd	nd	nd	nd	nd	nd	30
Calamagrostis epigejos	nr	major	sig	nd	tr	tr (sum of TMAO and AC)	tr	tr (sum of TMAO and AC)	nd	nr	nd	120
Carex leporina	nr	minor	major	tr	tr	tr (sum of TMAO and AC)	nd	tr (sum of TMAO and AC)	nd	nr	nd	120
Carex sp.	35.3 dry	sig	major	tr	nd	nd	nd	nd	nd	nd	nd	30
Carex sp.	17.0 dry	sig	major	nd	nd	nd	nd	nd	tr	nd	nd	30
Carex sp.	22.3/28.4 dry	sig	major	tr	nd	nd	nd	nd	nd	nd	nd	30
Carex sp.	57.5 dry	sig	major	minor	tr	nd	nd	nd	tr	nd	nd	30
Carex sp.	8.21 dry	minor	major	nd	sig	nd	nd	nd	nd	nd	nd	30
Carex sp.	3.87/136 dry	sig	major	nd	tr	nd	nd	nd	tr	nd	nd	30
Carex sp.	40.3 dry	sig	major	minor	minor	nd	nd	nd	nd	nd	nd	30
Carex sp.	30.9 dry	sig	major	minor	minor	nd	nd	nd	nd	nd	nd	30
Dactylis glomerata	1.6 dry	sig	major	nd	minor	minor	minor	nd	tr	nd	nd	121
Deschampsia cespitosa, methanol/water extraction	2.0 dry	minor	major	minor	minor	minor	tr	nd	nd	minor 1	nd	26

Table 7 Continued

Species	As concn. (mg/kg)	As(III)	As(V)	MA	DMA	TMAO	AB	AC	Tetra	As sugars	Other cpds. and unknowns	Ref.
Deschampsia cespitosa, water extraction	2.0 dry	sig	major	minor	minor	minor	tr	nd	nd	tr 1	nd	26
grass (unidentified) stalks and leaves	nr	major	minor	nd	minor	sig (sum of TMAO and AC)	minor	sig (sum of TMAO and AC)	nd	nr	nd	120
Holcus lanatus blooms	nr	major	sig	nd	sig	minor (sum of TMAO and AC)	minor	minor (sum of TMAO and AC)	nd	nr	nd	120
Holcus lanatus stalks and leaves	nr	major	minor	nd	minor	sig (sum of TMAO and AC)	minor	sig (sum of TMAO and AC)	nd	nr	nd	120
Holcus lanatus leaves	9.90 dry	sig	major	nd	minor	nd	nd	nd	nd	nr	nd	122
Holcus lanatus leaves	2.67 dry	major	sig	nd	minor	nd	nd	nd	nd	nr	nd	122
Hordeum jubatum	74.1 dry	sig	major	minor	nd	nd	nd	nd	nd	nd	nd	30
Hordeum jubatum	3.61 dry	sig	sig	tr	tr	nd	nd	nd	nd	nd	nd	30
Hordeum jubatum	8.01 dry	sig	major	tr	tr	nd	nd	nd	nd	nd	nd	30
Juncus effusus stalks and leaves	nr	major	minor	nd	nd	minor (sum of TMAO and AC)	minor	minor (sum of TMAO and AC)	nd	nr	nd	120
Phleum pratense	nr	minor	major	nd	tr	tr (sum of TMAO and AC)	nd	tr (sum of TMAO and AC)	nd	nr	nd	120
Rice grain	0.76 dry	major (not resolved)		minor	minor	nr	nr	nr	nr	nr	nd	123
Rice polished	0.063 to 0.22 dry	major (not resolved)		sig	sig	nr	nr		nr	nr	nd	123
Rice polished	0.14/0.19 dry	major (not resolved)		minor	sig	nr	nr		nr	nr	nd	123
Rice polished	0.15 dry	major (not resolved)		sig	minor	nr	nr	nr	nr	nr	nd	123

	Concentration										Ref
Rice polished	0.23 dry	major (not resolved)	minor	minor	nr	nr	nr	nr	nr	nd	123
Scirpus sp.	7.1 dry	sig	minor	nd	nd	nd	nd	nd	nd	nd	15
Scirpus sp.	4.5 dry	sig	nd	minor	nd	nd	nd	nd	nd	nd	15
Yams	0.005 to 0.52 dry	major (not resolved)	nd	nd	nr	nr	nr	nr	nr	nd	123
Yams	0.028 dry	major (not resolved)	sig	nd	nr	nr	nr	nr	nr	nd	123
Yams	0.40 dry	sig (not resolved)	major	nd	nr	nr	nr	nr	nr	nd	123
Yams	0.055 dry	major (not resolved)	sig	sig	nr	nr	nr	nr	nr	nd	123
Angiospermae: Dicotyledoneae											
Achillea millefolium leaves, methanol/water extraction	2.1 dry	sig	sig	sig	minor	nd	nd	nd	minor 1	nd	26
Achillea millefolium leaves, water extraction	2.1 dry	sig	minor	minor	minor	nd	nd	nd	nd	nd	26
Alnus incana leaves, methanol/water extraction	0.42 dry	nd	sig	tr	major	nd	nd	nd	tr 1	nd	26
Alnus incana leaves, water extraction	0.42 dry	sig	sig	tr	sig	nd	nd	nd	nd	nd	26
Apples	0.081 dry	sig	minor	sig	nr	nr	nr	nr	nr	nd	124
Apples	0.008 dry	sig	nd	sig	nr	nr	nr	nr	nr	nd	124
Apples	0.021 dry	major	nd	minor	nr	nr	nr	tr	nr	nd	124
Bidens cernua	100 dry	minor	minor	tr	nd	nd	nd	tr	nd	nd	30
Daucus carota	0.112 to 0.246 dry	major	nd	nd	nr	nr	nr	nr	nr	nd	125
Daucus carota	0.387 to 1.85 dry	sig	major	nd	nr	nr	nr	nr	nr	nd	125
Erigeron sp.	14 dry	major	nd	sig	nd	nd	nd	nd	nd	nd	15
Erigeron sp.	3.9 dry	sig	nd	sig	nd	nd	nd	nd	nd	nd	15
Fragaria vesca leaves, methanol/water extraction	3.8 dry	nd	minor	sig	sig	minor	nd	minor	minor 1	nd	26
Fragaria vesca leaves, water extraction	3.8 dry	major	minor	tr	minor	tr	nd	tr	tr 1	nd	26
Mimulus sp.	8.7 dry	major	nd	nd	nd	tr	nd	tr	nd	nd	15

Table 7 Continued

Species	As concn. (mg/kg)	As(III)	As(V)	MA	DMA	TMAO	AB	AC	Tetra	As sugars	Other cpds. and unknowns	Ref.
Plantago lanceolata	5.9 dry	sig	sig	nd	sig	tr	tr	tr	tr	nd	nd	121
Polygonum persicaria	nr	sig	major	nd	tr	nd	tr	nd	nd	nr	nd	120
Potentilla fruticosa L.	11.9 dry	sig	major	nd	nd	nd	nd	nd	nd	nd	nd	30
Rubus idaeus leaves. methanol/water extraction	2.6 dry	sig	major	nd	tr	sig	nd	nd	nd	tr I	nd	26
Rubus idaeus leaves. water extraction	2.6 dry	major	sig	nd	nd	minor	nd	nd	nd	nd	nd	26
Rubus idaeus leaves. water extraction	83.6 dry	sig	major	nd	nd	nd	nd	nd	nd	nd	nd	30
Silene vulgaris	nr	major	sig	nd	tr	tr (sum of TMAO and AC)	nd	tr (sum of TMAO and AC)	nd	nr	nd	120
Silene vulgaris	nr	major	sig	nd	tr	tr (sum of TMAO and AC)	tr	tr (sum of TMAO and AC)	nd	nr	nd	120
Trifolium pratense	3.2 dry	sig	sig	sig	sig	minor	tr	nd	tr	nd	nd	121
Vaccinium myrtilis leaves. methanol/water extraction	1.4 dry	major	sig	minor	tr	minor	nd	nd	nd	tr I	nd	26
Vaccinium myrtilis leaves. water extraction	1.4 dry	major	sig	tr	nd	minor	nd	nd	nd	nd	nd	26
Vaccinium vitis idaea leaves. methanol/water extraction	0.29 dry	nd	sig	sig	nd	minor	nd	nd	minor	tr I	nd	26
Vaccinium vitis idaea leaves. water extraction	0.29 dry	major	sig	sig	nd	minor	nd	nd	minor	nd	nd	26

ª Major = major arsenic constituent. >50% of extractable arsenic; sig = significant arsenic constituent. >10–50% of extractable arsenic; minor = minor arsenic constituent. 1–10% of extractable arsenic constituent. <1% of extractable arsenic; nr = not recorded; nd = not detected.

It is only relatively recently that analytical systems have been able to determine arsenic compounds at such low levels, and these techniques have subsequently been applied to some interesting environmental samples, such as earthworms and ants (130).

Earthworms contain arsenite and arsenate as major or significant arsenicals (131). Some earthworm samples also contained appreciable quantities of arseno-sugar 1, and smaller quantities of the related arsenosugar 2 (see Fig. 2). Arsenobe-taine was also present at low levels in the earthworms. Ants contained mainly arsenate, with smaller amounts of arsenite and dimethylarsinate (132). Methyl-arsonate and arsenobetaine were also present in the ants as trace constituents.

Eggs from spoonbills contained dimethylarsinate as the major arsenical to-gether with significant amounts of arsenite (52). No other arsenical was detected in the spoonbill eggs. In contrast, sea gulls' eggs were reported to contain most of their arsenic as an unresolved mixture of trimethylarsine oxide and arsenobe-taine, with smaller quantities of dimethylarsinate, inorganic arsenic, and methyl-arsonate (113).

Interesting samples, albeit atypical, have been reported for sheep that graze on algae and hence have a very high intake of arsenosugars (78). The major compound in urine and blood serum was dimethylarsinate, an expected metabo-lite of arsenosugars. Control sheep also contained dimethylarsinate as the major urinary arsenical (but at 30-fold lower concentrations).

There have been many studies on arsenic species in human urine, and these data are covered in other parts of this book. The arsenic species present in urine depend largely on the diet, and some compounds contained in seafood (e.g., arse-nobetaine) are excreted essentially unchanged (12). Ingested inorganic arsenic is in part methylated and excreted as methylarsonate and dimethylarsinate in addi-tion to arsenite and arsenate (12). Interesting new urine metabolites are methylar-sonous acid and dimethylarsinous acid (133). These compounds had been re-ported earlier in lake water (68).

I. Arsenic Compounds in Soil and Sediments

The extraction efficiency of arsenic from soil and sediments is low (11,119). Because most analytical techniques for determining arsenic species are performed on water-based solutions of the analytes (extractable arsenic), the results obtained for soil and sediments represent only a small proportion of the total arsenic pres-ent. Of this extractable portion, the inorganic species arsenate and arsenite domi-nate (1), although methylarsonate and dimethylarsinate are also found as natural constituents in some soils (43). These four arsenicals are also commonly found in sediments or in the interstitial water (porewater) of the sediments (45,134,135), and a trimethylated arsenic species, possibly trimethylarsine oxide, has also been detected in some sediment porewater samples (135).

J. Arsenic Compounds in Air

The major sources of airborne arsenic are metal smelters, the burning of coal, and volcanoes (9). Particulate arsenic trioxide (As_2O_3) is the main form of arsenic released by these processes (9). Information on the various arsenic species in air is more limited than that for other environmental compartments. The reason for this is the greater analytical difficulty associated with the analysis of volatile arsenicals. Nevertheless, recent work has revealed the presence of arsine (AsH_3) and methylated arsines ($MeAsH_2$, Me_2AsH, and Me_3As) as trace constituents of air samples, particularly over sites where biological activity is high (14–16,28,29). More work is required in this important area. Further research should provide valuable information to enhance our understanding of the biogeochemical cycles for arsenic.

REFERENCES

1. WR Cullen, KJ Reimer. Arsenic speciation in the environment. Chem Rev 89:713–764, 1989.
2. JS Edmonds, KA Francesconi, RV Stick. Arsenic compounds from marine organisms. Nat Prod Rep 10:421–428, 1993.
3. DJH Phillips. The chemical forms of arsenic in aquatic organisms and their interrelationships. In: JO Nraigu, ed. Arsenic in the Environment. Part I: Cycling and Characterization. New York: Wiley, 1994, pp 263–288.
4. KA Francesconi, JS Edmonds. Arsenic and marine organisms. Adv Inorg Chem 44:147–189, 1997.
5. S Maeda. Biotransformation of arsenic in the freshwater environment. In: JO Nraigu, ed. Arsenic in the Environment. Part I: Cycling and Characterization. New York: Wiley, 1994, pp 155–187.
6. D Kuehnelt, W Goessler. Organoarsenic compounds in the terrestrial environment. In: PJ Craig, ed. Organometallic Compounds in the Environment. 2d ed. Chichester: Wiley, in press.
7. T Tanaka. Distribution of arsenic in the natural environment with an emphasis on rocks and soils. Appl Organomet Chem 2:283–295, 1988.
8. DK Bhumbla, RF Keefer. Arsenic mobilization and bioavailability in soils. In: JO Nraigu, ed. Arsenic in the Environment. Part I: Cycling and Characterization. New York: Wiley, 1994, pp 51–82.
9. A Léonard. Arsenic. In: E Merian, ed. Metals and Their Compounds in the Environment. Weinheim: VCH, 1991, pp 751–774.
10. E Smith, R Naidu, AM Alston. Arsenic in the soil environment: A review. Adv Agronom 64:149–195, 1998.
11. M Pantsar-Kallio, PKG Manninen. Speciation of mobile arsenic in soil samples as a function of pH. Sci Total Environ 204:193–200, 1997.
12. National Academy of Sciences. Chemistry and analysis of arsenic species in water

and biological materials. In: Arsenic in Drinking Water. Washington, DC: National Academy Press, 1999, pp 27–82.

13. JS Edmonds, KA Francesconi. Arsenic in seafoods: Human health aspects and regulations. Mar Pollut Bull 26:665–674, 1993.

14. J Feldmann, R Grümping, AV Hirner. Determination of volatile metal and metalloid compounds in gases from domestic waste deposits with GC-ICP-MS. Fresenius J Anal Chem 350:228–235, 1994.

15. I Koch, J Feldmann, L Wang, P Andrewes, KJ Reimer, WR Cullen. Arsenic in the Meager Creek hot springs environment, British Columbia, Canada. Sci Total Environ 236:101–107, 1999.

16. K Michalke, EB Wickenheiser, M Mehring, AV Hirner, R Hensel. Production of volatile derivatives of metal(loid)s by microflora involved in anaerobic digestion of sewage sludge. Appl Environ Microbiol 66:2791–2796, 2000.

17. JS Edmonds. Diastereoisomers of an "arsenomethionine"-based structure from *Sargassum lacerifolium*: The formation of the arsenic-carbon bond in arsenic-containing natural products. Bioorg Med Chem Lett 10:1105–1108, 2000.

18. J Gailer, KA Francesconi, JS Edmonds, KJ Irgolic. Metabolism of arsenic compounds by the blue mussel *Mytilus edulis* after accumulation from seawater spiked with arsenic compounds. Appl Organomet Chem 9:341–355, 1995.

19. JS Edmonds, Y Shibata, RIT Prince, KA Francesconi, M Morita. Arsenic compounds in tissues of the leatherback turtle *Dermochelys coriacea*. J Marine Biol Assoc UK 74:463–466, 1994.

20. Z Slejkovec, AR Byrne, T Stijve, W Goessler, KJ Irgolic. Arsenic compounds in higher fungi. Appl Organomet Chem 11:673–682, 1997.

21. KA Francesconi, JS Edmonds, RV Stick. Accumulation of arsenic in yellow-eye mullet (*Aldrichetta forsteri*) following oral administration of organoarsenic compounds and arsenate. Sci Total Environ 79:59–67, 1989.

22. F Challenger. Biological methylation. Chem Rev 36:315–361, 1945.

23. AR Byrne, Z Slejkovec, T Stijve, L Fay, W Goessler, J Gailer, KJ Irgolic. Arsenobetaine and other arsenic species in mushrooms. Appl Organomet Chem 9:305–313, 1995.

24. AR Byrne, M Tusek-Znidaric, BK Puri, KJ Irgolic. Studies of the uptake and binding of trace metals in fungi. Part II: Arsenic compounds in *Laccaria amethystina*. Appl Organomet Chem 5:25–32, 1991.

25. JS Edmonds, Y Shibata, KA Francesconi, RJ Rippington, M Morita. Arsenic transformations in short marine food chains studied by HPLC ICP MS. Appl Organomet Chem 11:281–287, 1997.

26. D Kuehnelt, J Lintschinger, W Goessler. Arsenic compounds in terrestrial organisms. IV: Green plants and lichens from an old arsenic smelter site in Austria. Appl Organomet Chem 14:411–420, 2000.

27. FB Whitfield, DJ Freeman, KJ Shaw. Trimethylarsine: An important off-flavour in some prawn species. Chem Ind 20:786–787, 1983.

28. J Feldmann, AV Hirner. Occurrence of volatile metal and metalloid species in landfill and sewage gases. Intern J Environ Anal Chem 60:339–359, 1995.

29. AV Hirner, J Feldmann, E Krupp, R Grümping, R Goguel, WR Cullen. Metal(loid)organic compounds in geothermal gases and waters. Org Geochem 29:1765–1778, 1998.

30. I Koch, L Wang, CA Ollson, WR Cullen, KJ Reimer. The predominance of inorganic arsenic species in plants from Yellowknife, Northwest Territories, Canada. Environ Sci Technol 34:22–26, 2000 and information from the supporting tables SI-1 and SI-2 available at http://pubs.acs.org.

31. JS Edmonds, KA Francesconi. Transformations of arsenic in the marine environment. Experientia 43:553–557, 1987.

32. A Geiszinger, W Goessler, SN Pedersen, KA Francesconi. Arsenic biotransformation by the brown macroalga Fucus serratus. Environ Toxicol Chem, in press.

33. KA Francesconi, JS Edmonds, RV Stick. Arsenic compounds from the kidney of the giant clam Tridacna maxima: Isolation and identification of an arsenic-containing nucleoside. J Chem Soc Perkin Trans I:1349–1357, 1992.

34. JS Edmonds, Y Shibata, KA Francesconi, J Yoshinaga, M Morita. Arsenic lipids in the digestive gland of the western rock lobster Panulirus cygnus: An investigation by HPLC ICP-MS. Sci Total Environ 122:321–335, 1992.

35. KA Francesconi, RV Stick, JS Edmonds. Glycerylphosphorylarsenocholine and phosphatidylarsenocholine in yellow-eye mullet (Aldrichetta forsteri) following oral administration of arsenocholine. Experientia 46:464–466, 1990.

36. EH Larsen. Speciation of dimethylarsinyl-riboside derivatives (arsenosugars) in marine reference materials by HPLC-ICP-MS. Fresenius J Anal Chem 352:582–588, 1995.

37. S Khokiattiwong, W Goessler, SN Pedersen, R Cox, KA Francesconi. Dimethylarsinoylacetate from microbial demethylation of arsenobetaine in seawater. Appl Organomet Chem 15:481–489, 2001.

38. JS Edmonds, Y Shibata, F Yang, M Morita. Isolation and synthesis of 1-deoxy-1-dimethylarsinoylribitol-5-sulfate, a natural constituent of Chondria crassicaulis and other red algae. Tertrahedron Lett 38:5819–5820, 1997.

39. M Morita, Y Shibata. Chemical form of arsenic in marine macroalgae. Appl Organomet Chem 4:181–190, 1990.

40. KA Francesconi, W Goessler, S Panutrakul, KJ Irgolic. A novel arsenic containing riboside (arsenosugar) in three species of gastropod. Sci Total Environ 221:139–148, 1998.

41. Y Shibata, M Morita. A novel, trimethylated arsenosugar isolated from the brown alga Sargassum thunbergii. Agric Biol Chem 52:1087–1089, 1988.

42. KA Francesconi, S Khokiattiwong, W Goessler, SN Pedersen, M Pavkov. A new arsenobetaine from marine organisms identified by liquid chromatography–mass spectrometry. Chem Comm:1083–1084, 2000.

43. RS Braman, CC Foreback. Methylated forms of arsenic in the environment. Science 182:1247–1249, 1973.

44. MO Andreae. Distribution and speciation of arsenic in natural waters and some marine algae. Deep-Sea Res 25:391–402, 1978.

45. MO Andreae. Arsenic speciation in seawater and interstitial waters: The influence of biological-chemical interactions on the chemistry of a trace element. Limnol Oceanogr 24:440–452, 1979.

46. SJ Santosa, S Wada, S Tanaka. Distribution and cycle of arsenic compounds in the ocean. Appl Organomet Chem 8:273–283, 1994.

47. E Gonzales Soto, E Alonso Rodriguez, D Prada Rodriguez, E Fernandez Fernandez.

Inorganic and organic arsenic speciation in seawaters by IEC-HG-AAS. Anal Lett 29:2701–2712, 1996.

48. AG Howard, SDW Comber. The discovery of hidden arsenic species in coastal waters. Appl Organomet Chem 3:509–514, 1989.

49. AMM de Bettencourt, MO Andreae. Refractory arsenic in estuarine waters. Appl Organomet Chem 5:111–116, 1991.

50. AG Howard, LE Hunt, C Salou. Evidence supporting the presence of dissolved dimethylarsinate in the marine environment. Appl Organomet Chem 13:39–46, 1999.

51. MO Andreae, TW Andreae. Dissolved arsenic species in the Schelde estuary and watershed, Belgium. Estuar Coast Shelf Sci 29:421–433, 1989.

52. JL Gomez-Ariza, D Sanchez-Rodas, I Giraldez, E Morales. A comparison between ICP-MS and AFS detection for arsenic speciation in environmental samples. Talanta 51:257–268, 2000.

53. T Kaise, M Ogura, T Nozaki, K Saitoh, T Sakurai, C Matsubara, C Watanabe, K Hanaoka. Biomethylation of arsenic in an arsenic-rich freshwater environment. Appl Organomet Chem 11:297–304, 1997.

54. XC Le, S Yalcin, M Ma. Speciation of submicrogram per liter levels of arsenic in water: On-site species separation integrated with sample collection. Environ Sci Technol 34:2342–2347, 2000.

55. T-H Lin, Y-L Huang, M-Y Wang. Arsenic species in drinking water, hair, fingernails, and urine of patients with blackfoot disease. J Toxicol Environ Health A 53: 85–93, 1998.

56. O Muñoz, D Vélez, R Montoro, A Arroyo, M Zamorano. Determination of inorganic arsenic [As(III) + As(V)] in water samples by microwave-assisted distillation and hydride generation atomic absorption spectrometry. J Anal At Spectrom 15: 711–714, 2000.

57. T Guerin, N Molenat, A Astruc, R Pinel. Arsenic speciation in some environmental samples: A comparative study of HG-GC-QFAAS and HPLC-ICP-MS methods. Appl Organomet Chem 14:401–410, 2000.

58. H Hasegawa, Y Sohrin, M Matsui, M Hojo, M Kawashima. Speciation of arsenic in natural waters by solvent extraction and hydride generation atomic absorption spectrometry. Anal Chem 66:3247–3252, 1994.

59. C-J Hwang, S-J Jiang. Determination of arsenic compounds in water samples by liquid chromatography–inductively coupled plasma mass spectrometry with an in situ nebulizer–hydride generator. Anal Chim Acta 289:205–213, 1994.

60. J Mattusch, R Wennrich. Determination of anionic, neutral, and cationic species of arsenic by ion chromatography with ICPMS detection in environmental samples. Anal Chem 70:3649–3655, 1998.

61. M Williams, F Fordyce, A Paijitprapaporn, P Charoenchaisri. Arsenic contamination in surface drainage and groundwater in part of the southeast Asian tin belt, Nakorn Soi Thammarat Province, southern Thailand. Environ Geol 27:16–33, 1996.

62. AG Howard, SC Apte. Seasonal control of arsenic speciation in an estuarine ecosystem. Appl Organomet Chem 3:499–507, 1989.

63. GE Millward, L Ebdon, AP Walton. Seasonality in estuarine sources of methylated arsenic. Appl Organomet Chem 7:499–511, 1993.

64. H Hasegawa. Seasonal changes in methylarsenic distribution in Tosa Bay and Uranouchi Inlet. Appl Organomet Chem 10:733–740, 1996.
65. H Hasegawa, M Matzui, S Okamura, M Hojo, N Iwasaki, Y Sohrin. Arsenic speciation including "hidden" arsenic in natural waters. Appl Organomet Chem 13:113–119, 1999.
66. JY Cabon, N Cabon. Speciation of major arsenic species in seawater by flow injection hydride generation atomic absorption spectrometry. Fresenius J Anal Chem 368:484–489, 2000.
67. DA Bright, M Dodd, KJ Reimer. Arsenic in subArctic lakes influenced by gold mine effluent: The occurrence of organoarsenicals and "hidden" arsenic. Sci Total Environ 180:165–182, 1996.
68. H Hasegawa. The behaviour of trivalent and pentavalent methylarsenicals in lake Biwa. Appl Organomet Chem 11:305–311, 1997.
69. Y Sohrin, M Matsui, M Kawashima, M Hojo, H Hasegawa. Arsenic biogeochemistry affected by eutrophication in lake Biwa, Japan. Environ Sci Technol 31:2712–2720, 1997.
70. VWM Lai, WR Cullen, CF Harrington, KJ Reimer. The characterization of arsenosugars in commercially available algal products including a nostoc species of terrestrial origin. Appl Organomet Chem 11:797–803, 1997.
71. JS Edmonds, KA Francesconi. Arseno-sugars from brown kelp (*Ecklonia radiata*) as intermediates in cycling of arsenic in a marine ecosystem. Nature 289:602–604, 1981.
72. JS Edmonds, KA Francesconi. Arsenic-containing ribofuranosides: Isolation from brown kelp *Ecklonia radiata* and NMR spectra. J Chem Soc Perkin Trans I:2375–2382, 1983.
73. VWM Lai, WR Cullen, CF Harrington, KJ Reimer. Seasonal changes in arsenic speciation in *Fucus* species. Appl Organomet Chem 12:243–251, 1998.
74. AD Madsen, W Goessler, SN Pedersen, KA Francesconi. Characterization of an algal extract by HPLC-ICPMS and LC electrospray MS for use in arsenosugar speciation studies. J Anal At Spectrom 15:657–662, 2000.
75. G Raber, KA Francesconi, KJ Irgolic, W Goessler. Determination of 'arsenosugars' in algae with anion-exchange chromatography and an inductively coupled plasma mass spectrometer as element-specific detector. Fresenius J Anal Chem 367:181–188, 2000.
76. SN Pedersen, KA Francesconi. Liquid chromatography electrospray mass spectrometry with variable fragmentor voltage gives simultaneous elemental and molecular detection of arsenic compounds. Rapid Commun Mass Spectrom 14:641–645, 2000.
77. JS Edmonds, M Morita, Y Shibata. Isolation and identification of arsenic-containing ribofuranosides and inorganic arsenic from Japanese edible seaweed *Hizikia fusiforme*. J Chem Soc Perkin Trans I:577–580, 1987.
78. J Feldmann, K John, P Pengprecha. Arsenic metabolism in seaweed-eating sheep from northern Scotland. Fresenius J Anal Chem 368:116–121, 2000.
79. Y Shibata, M Morita, JS Edmonds. Purification and identification of arsenic-containing ribofuranosides from the edible brown seaweed *Laminaria japonica* (Makonbu). Agric Biol Chem 51:391–398, 1987.
80. KA Francesconi, JS Edmonds, RV Stick, BW Skelton, AH White. Arsenic-

containing ribosides from the brown alga *Sargassum lacerifolium*: X-ray molecular structure of 2-amino-3-[5'-deoxy-5'-(dimethylarsinoyl)ribosyloxy]-propane-1-sulphonic acid. J Chem Soc Perkin Trans I:2707–2716, 1991.

81. K Jin, T Hayashi, Y Shibata, M Morita. Isolation and identification of arsenic-containing ribofuranosides from the edible brown seaweed *Sphaerotrichia divaricata* (Ishimozuku). Appl Organomet Chem 2:365–369, 1988.

82. M Morita, Y Shibata. Isolation and identification of arseno-lipid from a brown alga, *Undaria pinnatifida* (Wakame). Chemosphere 17:1147–1152, 1988.

83. Y Shibata, M Morita, K Fuwa. Selenium and arsenic in biology: Their chemical forms and biological functions. Adv Biophys 28:31–80, 1992.

84. S Branch, L Ebdon, P O'Neill. Determination of arsenic species in fish by directly coupled high-performance liquid chromatography–inductively coupled plasma mass spectrometry. J Anal At Spectrom 9:33–37, 1994.

85. EH Larsen, G Pritzl, SH Hansen. Arsenic speciation in seafood samples with emphasis on minor constituents: An investigation using high-performance liquid chromatography with detection by inductively coupled plasma mass spectrometry. J Anal At Spectrom 8:1075–1084, 1993.

86. JS Edmonds, KA Francesconi. Trimethylarsine oxide in estuary catfish (*Cnidoglanis macrocephalus*) and school whiting (*Sillago bassensis*) after oral administration of sodium arsenate and as a natural component of estuary catfish. Sci Total Environ 64:317–323, 1987.

87. J Albertí, R Rubio, G Rauret. Extraction method for arsenic speciation in marine organisms. Fresenius J Anal Chem 351:420–425, 1995.

88. W Maher, W Goessler, J Kirby, G Raber. Arsenic concentrations and speciation in the tissues and blood of sea mullet (*Mugil cephalus*) from Lake Macquarie, NSW, Australia. Marine Chem 68:169–182, 1999.

89. MB Amran, F Lagarde, MJF Leroy. Determination of arsenic species in marine organisms by HPLC-ICP-OES and HPLC-HG-QFAAS. Mikrochim Acta 127:195–202, 1997.

90. A El Moll, R Heimburger, F Lagarde, MJF Leroy, E Maier. Arsenic speciation in marine organisms: From the analytical methodology to the constitution of reference materials. Fresenius J Anal Chem 354:550–556, 1996.

91. EH Larsen, CR Quétel, R Munoz, A Fiala-Medioni, OFX Donard. Arsenic speciation in shrimp and mussel from the Mid-Atlantic hydrothermal vents. Marine Chem 57:341–346, 1997.

92. KA Francesconi, DA Hunter, B Bachmann, G Raber, W Goessler. Uptake and transformation of arsenosugars in the shrimp *Crangon crangon*. Appl Organomet Chem 13:669–679, 1999.

93. JS Edmonds, KA Francesconi, PC Healy, AH White. Isolation and crystal structure of an arsenic-containing sugar sulphate from the kidney of the giant clam *Tridacna maxima*: X-ray crystal structure of (2S)-3-[5-deoxy-5-(dimethylarsinoyl)-B-D-ribofuranosyloxy]-2-hydroxypropyl hydrogen sulphate. J Chem Soc Perkin Trans I:2989–2993, 1982.

94. K Shiomi, Y Kakehashi, H Yamanaka, T Kikuchi. Identification of arsenobetaine and a tetramethylarsonium salt in the clam *Meretrix lusoria*. Appl Organomet Chem 1:177–183, 1987.

95. WR Cullen, M Dodd. Arsenic speciation in clams of British Columbia. Appl Organomet Chem 3:79–88, 1989.
96. VWM Lai, WR Cullen, S Ray. Arsenic speciation in scallops. Marine Chem 66: 81–89, 1999.
97. JL Gomez-Ariza, D Sanchez-Rodas, I Giraldez, E Morales. Comparison of biota sample pretreatments for arsenic speciation with coupled HPLC-HG-ICP-MS. Analyst 125:401–407, 2000.
98. KA Francesconi, JS Edmonds, BG Hatcher. Examination of the arsenic constituents of the herbivorous marine gastropod *Tectus pyramis*: Isolation of tetramethylarsonium ion. Comp Biochem Physiol 90C:313–316, 1988.
99. W Goessler, W Maher, KJ Irgolic, D Kuehnelt, C Schlagenhaufen, T Kaise. Arsenic compounds in a marine food chain. Fresenius J Anal Chem 359:434–437, 1997.
100. W Goessler, A Rudorfer, EA Mackey, PR Becker, KJ Irgolic. Determination of arsenic compounds in marine mammals with high performance liquid chromatography and an inductively coupled plasma mass spectrometer as element specific detector. Appl Organomet Chem 12:491–501, 1998.
101. K Hanaoka, W Goessler, T Kaise, H Ohno, Y Nakatani, S Ueno, D Kuehnelt, C Schlagenhaufen, KJ Irgolic. Occurrence of a few organo-arsenicals in jellyfish. Appl Organomet Chem 13:95–99, 1999.
102. W Goessler, D Kuehnelt, C Schlagenhaufen, Z Slejkovec, KJ Irgolic. Arsenobetaine and other arsenic compounds in the National Research Council of Canada certified reference materials DORM 1 and DORM 2. J Anal At Spectrom 13:183–187, 1998.
103. Z Slejkovec, JT van Elteren, AR Byrne. Determination of arsenic compounds in reference materials by HPLC-(UV)-HG-AFS. Talanta 49:619–627, 1999.
104. JW McKiernan, JT Creed, CA Brockhoff, JA Caruso, RM Lorenzana. A comparison of automated and traditional methods for the extraction of arsenicals from fish. J Anal At Spectrom 14:607–613, 1999.
105. J Alberti, R Rubio, G Rauret. Arsenic speciation in marine biological materials by LC-UV-HG-ICP/OES. Fresenius J Anal Chem 351:415–419, 1995.
106. F Lagarde, MB Amran, MJF Leroy, C Demesmay, M Ollé, A Lamotte, H Muntau, P Michel, P Thomas, S Caroli, E Larsen, P Bonner, G Rauret, M Foulkes, A Howard, B Griepink, EA Maier. Certification of total arsenic, dimethylarsinic acid and arsenobetaine contents in a tuna fish powder (BCR-CRM 627). Fresenius J Anal Chem 363:18–22, 1999.
107. EH Larsen, GA Pedersen, JW McLaren. Characterization of national food agency shrimp and plaice reference materials for trace elements and arsenic species by atomic and mass spectrometric techniques. J Anal At Spectrom 12:963–968, 1997.
108. MA Suner, V Devesa, I Rivas, D Velez, R Montoro. Speciation of cationic arsenic species in seafood by coupling liquid chromatography with hydride generation atomic fluorescence detection. J Anal At Spectrom 15:1501–1507, 2000.
109. AA Benson, RE Summons. Arsenic accumulation in Great Barrier Reef invertebrates. Science 211:482–483, 1981.
110. K Hanaoka, W Goessler, K Yoshida, Y Fujitaka, T Kaise, KJ Irgolic. Arsenocholine- and dimethylated arsenic-containing lipids in starspotted shark *Mustelus manazo*. Appl Organomet Chem 13:765–770, 1999.

111. K Shiomi, Y Sugiyama, K Shimakura, Y Nagashima. Arsenobetaine as the major arsenic compound in the muscle of two species of freshwater fish. Appl Organomet Chem 9:105–109, 1995.

112. RA Schoof, LJ Yost, J Eickhoff, EA Crecelius, DW Cragin, DM Meacher, DB Menzel. A market basket survey of inorganic arsenic in food. Food Chem Toxicol 37:839–846, 1999.

113. Z Slejkovec, AR Byrne, B Smodis, M Rossbach. Preliminary studies on arsenic species in some environmental samples. Fresenius J Anal Chem 354:592–595, 1996.

114. D Kuehnelt, W Goessler, KJ Irgolic. Arsenic compounds in terrestrial organisms. II: Arsenocholine in the mushroom *Amanita muscaria*. Appl Organomet Chem 11: 459–470, 1997.

115. I Koch, L Wang, KJ Reimer, WR Cullen. Arsenic species in terrestrial fungi and lichens from Yellowknife, NWT, Canada. Appl Organomet Chem 14:245–252, 2000.

116. D Kuehnelt, W Goessler, KJ Irgolic. Arsenic compounds in terrestrial organisms. I: *Collybia maculata*, *Collybia butyracea* and *Amanita muscaria* from arsenic smelter sites in Austria. Appl Organomet Chem 11:289–296, 1997.

117. EH Larsen, M Hansen, W Goessler. Speciation and health risk considerations of arsenic in the edible mushroom *Laccaria amethystina* collected from contaminated and uncontaminated locations. Appl Organomet Chem 12:285–291, 1998.

118. LQ Ma, KM Komar, C Tu, W Zhang, Y Cai, ED Kennelley. A fern that hyperaccumulates arsenic. Nature 409:579, 2001.

119. KA Francesconi, P Visoottiviseth, W Sridokchan, W Goessler. Arsenic species in an arsenic hyperaccumulating fern, *Pityrogramma calomelanos*: A potential phytoremediator of arsenic-contaminated soils. Sci Total Environ, in press.

120. J Mattusch, R Wennrich, A-C Schmidt, W Reisser. Determination of arsenic species in water, soils and plants. Fresenius J Anal Chem 366:200–203, 2000.

121. A Geiszinger. Spurenelemente in Regenwuermern (Lumbricidae, Oligochaeta) und ihre oekologische Bedeutung. PhD dissertation, Karl-Franzens University, Graz, Austria, 1998.

122. A-C Schmidt, W Reisser, J Mattusch, P Popp, R Wennrich. Evaluation of extraction procedures for the ion-chromatographic determination of arsenic species in plant materials. J Chromatogr A 889:83–91, 2000.

123. RA Schoof, LJ Yost, E Crecelius, K Irgolic, W Goessler, HR Guo, H Greene. Dietary arsenic intake in Taiwanese districts with elevated arsenic in drinking water. Hum Ecol Risk Assess 4:117–135, 1998.

124. JA Caruso, DT Heitkemper, C B'Hymer. An evaluation of extraction techniques for arsenic species from freeze-dried apple samples. Analyst 126:136–140, 2001.

125. H Helgesen, EH Larsen. Bioavailability and speciation of arsenic in carrots grown in contaminated soil. Analyst 123:791–796, 1998.

126. EK Porter, PJ Petersen. Arsenic accumulation by plants on mine waste. Sci Total Environ 4:365–371, 1975.

127. T De Koe. *Agrostis castellana* and *Agrostis delicatula* on heavy metal and arsenic enriched sites in NE Portugal. Sci Total Environ 145:103–109, 1994.

128. J Bech, C Poschenrieder, M Llugany, J Barceló, P Tume, FJ Tobias, JL Barran-

zuela, ER Vásquez. Arsenic and heavy metal contamination of soil and vegetation around a copper mine in Northern Peru. Sci Total Environ 203:83–91, 1997.

129. RE Grissom, CO Abernathy, AS Susten, JM Donohue. Estimating total arsenic exposure in the United States. In: WR Chappell, CO Abernathy, RL Calderon, eds. Arsenic Exposure and Health Effects. Amsterdam: Elsevier, 1999, pp 51–60.

130. KJ Irgolic, W Goessler, D Kuehnelt. Arsenic compounds in terrestrial biota. In: WR Chappell, CO Abernathy, RL Calderon, eds. Arsenic Exposure and Health Effects. Amsterdam: Elsevier, 1999, pp 61–68.

131. A Geiszinger, W Goessler, D Kuehnelt, KA Francesconi, W Kosmus. Determination of arsenic compounds in earthworms. Environ Sci Technol 32:2238–2243, 1998.

132. D Kuehnelt, W Goessler, C Schlagenhaufen, KJ Irgolic. Arsenic compounds in terrestrial organisms. III: Arsenic compounds in *Formica* sp. from an old arsenic smelter site. Appl Organomet Chem 11:859–867, 1997.

133. XC Le, XF Lu, MS Ma, WR Cullen, HV Aposhian, BS Zeng. Speciation of key arsenic metabolic intermediates in human urine. Anal Chem 72:5172–5177, 2000.

134. L Ebdon, AP Walton, GE Millward, M Whitfield. Methylated arsenic species in estuarine porewaters. Appl Organomet Chem 1:427–433, 1987.

135. KJ Reimer, JAJ Thompson. Arsenic speciation in marine interstitial water. The occurrence of organoarsenicals. Biogeochemistry 6:211–237, 1988.

4
Arsenic Speciation in the Environment and Humans

X. Chris Le
University of Alberta, Edmonton, Alberta, Canada

I. INTRODUCTION

Humans are exposed to naturally occurring and anthropogenic sources of arsenic compounds in the environment. Health problems associated with exposure to arsenic (As) continue to command world attention. A wide variety of adverse health effects, including skin and several internal cancers and cardiovascular and neurological effects, have been attributed to chronic exposure to high levels of arsenic, primarily from drinking water (1). The U.S. Environmental Protection Agency (EPA) has proposed a revision of its maximum contaminant level (MCL) for arsenic in drinking water. After much scientific and public debate, the EPA is considering a reduction of its MCL from the current 50 µg/L to 10 µg/L (2). The World Health Organization (WHO) limit is also 10 µg/L (3). The Canadian guideline is 25 µg/L (4), which is under review by Health Canada.

Environmental fate and behavior, bioavailability, and toxicity of arsenic vary dramatically with the chemical forms (species) in which arsenic exists. While inorganic arsenite [As(III)] and arsenate [As(V)] are highly toxic, monomethylarsonic acid [MMA(V)] and dimethylarsinic acid [DMA(V)] are less toxic, and predominant arsenic species present in most crustacean types of seafood are essentially nontoxic (1,5–8). Thus, assessments of environmental impact and human health risk strictly based on measurements of total element concentration are not reliable. It is important to identify and quantify individual chemical species of the element (i.e., chemical speciation).

Arsenic compounds are abundant in environmental and biological systems (5–10). Arsenic is present in certain seafood at concentrations as high as several hundred micrograms per gram (µg/g). Different types of arsenic compounds in aquatic and terrestrial environments and the transformation of these arsenic compounds in the environment have been summarized previously in Chapter 3. The present chapter focuses on arsenic compounds in water and food, which are the main sources of human exposure to arsenic. This chapter also emphasizes arsenic speciation and biotransformation in humans because of the relevance to human health.

II. HUMAN EXPOSURE TO ARSENIC

Common routes of exposure to arsenic include ingestion and inhalation of arsenic compounds (1). Dermal absorption of arsenic compounds is generally considered a less significant route of human exposure to arsenic. Inhalation of airborne arsenic is commonly associated with the smelting and mining activities (e.g., copper smelting and gold mining) (11–15). High airborne exposures of arsenic (principally arsenic oxide) have been associated with increased risk of lung cancer in workers at copper smelting operations (16–20). Major inhalation exposure may also result from coal fly ash due to burning of high-arsenic content coals (21).

Exposure to arsenic by the general population occurs mainly through ingestion of arsenic present in drinking water and food. In several regions of Bangladesh, India, and China, the high natural arsenic content in the drinking water has caused endemic, chronic arsenic poisoning (22–25). Human epidemiological studies conducted in Taiwanese, Chilean, and Argentinean populations have demonstrated a direct association between elevated arsenic exposure via drinking water (several hundred µg/L) and the prevalence of skin, bladder, and lung cancers (26–32). However, estimates of cancer risk resulting from the exposure to low levels (e.g., at the current MCL) of arsenic are the subject of considerable debate. Assessment of cancer risk, based on extrapolations from epidemiological studies involving highly exposed populations, carries large uncertainties (1). Furthermore, the mechanism(s) of action responsible for arsenic carcinogenicity are not understood (1).

Arsenic poisoning from food ingestion is rare, but has been documented. In some regions of Guizhou Province in southwestern China, certain food items (such as corn and chili pepper) are dried over open fires fueled with high-arsenic coal. Chili peppers dried over the coal fires absorb 500 µg/g arsenic on average. Consumption of arsenic-contaminated foods causes chronic arsenic poisoning (33).

III. ARSENIC SPECIATION IN WATER

Arsenic concentration in seawater is typically 1–3 µg/L (5–7,34,35), although in surface seawater, arsenic species may be subject to some seasonal changes due to biological uptake, particularly in highly productive coastal regions. Inorganic As(V) is the major arsenic species; As(III), MMA(V), and DMA(V) are present at lower levels, usually under 10% of the total arsenic concentration (34–36).

In freshwater systems, the arsenic concentration varies considerably with the geological composition of the drainage area and the extent of anthropogenic input. A concentration range of 0.1–80 µg/L has been reported as typical (5,34,35). Inorganic As(III) and As(V) are the major arsenic species. Minor amounts of DMA(V), MMA(V), and methylated As(III) species can also be present in natural waters (5,36–39). The proportions of arsenic species vary with the extent of anthropogenic input and biological activities. For example, As(V) was the only detectable arsenic species in the Itchen estuary in southern England (40) for most of the year. However, 30% of the total arsenic was in the form of methylated species when the temperature was above 12°C and when productivity was at a maximum. In the river, DMA(V) was found only in May and June. In saline areas, the total arsenic concentration was 0.7–1.0 µg/L, and 0.2 µg/L at a freshwater site.

Studies conducted on Lake Biwa in Japan (38,39) revealed the presence of methylarsenic (III) species, possibly $(CH_3)As(OH)_2$ and $(CH_3)_2AsOH$, at low concentrations. Speciation and concentration of arsenic varied with the season, particularly in the euphotic southern basin of the lake where DMA(V) could be the dominant species. Similar compounds had been seen in sediment pore water from Yellowknife, Canada (41).

Inorganic As(III) and As(V) are the major arsenic species in groundwater. The devastating arsenic endemic episodes in India, Bangladesh, and China are attributed to the high levels of inorganic arsenic in well water (22–25). Inorganic arsenic concentrations in well waters from these regions are as high as several thousand µg/L. Earlier work on speciation of arsenic in groundwater mostly detected As(V) as the major water-soluble species. However, some of the procedures used previously for sample handling and analysis might have resulted in the oxidation of As(III) to As(V) (43). With improvements in methods of sampling, preservation, and analysis, there is increasing evidence (42) that As(III) might be more prevalent than previously determined. Methods of on-site separation of As(III) and As(V) species immediately after water sample collection using solid-phase cartridges are particularly useful. A recently developed method using disposable cartridges has been demonstrated for speciation of particulate and soluble arsenic (43). A measured volume of water sample is passed through a 0.45-µm membrane filter and a silica-based strong anion exchange cartridge connected in serial. The filter captures particulate arsenic, while the anion exchange cartridge retains As(V). Arsenite is not retained and is detected in the effluent. The anion-exchange cartridge is subsequently eluted with 1 M hydrochloric acid (HCl) and the eluent is analyzed for As(V) concentration.

The amount of particulate fraction of arsenic can be substantial (43,44). In well water samples, the particulate arsenic was found to be as high as 70% of the total arsenic. The particulate can be readily removed by filtration, thereby reducing arsenic levels in drinking water (44).

Hydride generation with various atomic spectrometry detectors is often used for the determination of low levels of arsenic in water (see Chap. 2). It is generally believed that the arsenic species detected by hydride generation are the oxy species $(CH_3)_x AsO(OH)_{3-x}$ (x = 0–3). However, arsenic/sulfur compounds might be expected to dominate in the reducing environments (5). The presence of the thio analogue [i.e., oxythioarsenate, $H_3As(V)O_3S$] in water from an arsenic-rich, reducing environment has been demonstrated (45).

In addition to the well-characterized arsenic species discussed above, there are also reports of unidentified arsenic species in water (37,41,46). Careful analysis of a reference material for trace metals revealed that 22% of the total arsenic was not identified (46). The water samples used to prepare the reference material were from the Ottawa River, Canada. The certified value for total arsenic in this reference water is 0.55 ± 0.08 µg/L. In coastal waters of southern England, 25% of total arsenic was not detected by hydride generation procedures (37). These hidden arsenic species were unidentified, although they appeared to contain dimethylarsenic moieties. Targus estuary water in Portugal contained hidden arsenic up to 25% of the total (47).

IV. ARSENIC SPECIATION IN FOOD

There is little information on the nature of arsenic species in the human diet, apart from seafood, although limited market basket surveys and duplicate diet surveys have provided useful information on total arsenic concentration (48–50). Comprehensive analyses of food collected in Canadian cities in the years 1985–1988 showed that the food groups containing the highest mean arsenic concentrations were fish (1662 ng/g), meat and poultry (24.3 ng/g), bakery goods and cereals (24.5 ng/g), and fats and oils (19.0 ng/g) (48). The average daily dietary intake of total arsenic by Canadians was estimated to be 38 µg and varied from 15 µg for children 1–4 yr old to 59 µg for males 20–39 yr old (49). Estimates of daily dietary intake of total arsenic from other countries include: 62 µg/day from the United States (50), 89 µg/day from the United Kingdom (51), 55 µg/day from New Zealand (52), and 160–280 µg/day from Japan (53).

Most dietary arsenic originates from fish, shellfish, and seaweed products. On the basis of the U.S. Food and Drug Administration (FDA) Total Diet Study for Market Baskets collected from 1990 through 1991, Adams et al. (54) estimated that food contributed 93% of the total daily intake of arsenic, with seafood accounting for 90%. The major arsenic species found in the fish and shellfish that are usually eaten is arsenobetaine (AB), a ubiquitous major arsenic species in crustaceans. Minor amounts of arsenocholine (AC), trimethylarsine oxide $[(CH_3)_3AsO]$, tetramethylarsonium ion $[(CH_3)_4As^+]$, DMA(V), MMA(V), As(V), and unidentified arsenicals have also been found in marine animals (5,9,10,55). Both arsenobetaine and arsenosugars are present in bivalves at comparable concentrations (55,56). The arsenic speciation in the marine environment has been comprehensively summarized (5,9,10,55). Arsenobetaine is not metabolized by humans. It is believed to have low or negligible toxicity (8). However, little work has been done on the toxicity of arsenosugars. Arsenosugars are metabolized by humans, mainly to DMA(V) (57,58), which is toxicologically relevant.

The presence of organoarsenicals in marine organisms is commonly assumed to be due to the accumulation of compounds that have been biotransformed from As(V) at low trophic levels. Arsenobetaine is believed to be the end product that is accumulated in higher trophic levels through the food chain, although its actual metabolic origin is not clear. Arsenosugars present in marine algae and mollusks are suspected to be involved in the production of arsenobetaine at some stage in the food chain. It is unclear, however, how arsenosugars are transformed into arsenobetaine within the higher trophic levels (see Chap. 3).

Although a high concentration of arsenate (38% of the total arsenic) is present in certain brown algae, generally only a small portion (usually under 10%) of the total arsenic present in the marine algae is in inorganic form. The rest is present as arsenosugars (9,10). Arsenosugars are present in seaweeds, oysters, mussels, and clams (10,55,56), which are common for human consumption. Recent studies also identified arsenosugars in commercial food products of freshwater algae (59), freshwater shellfish and terrestrial plants (60), and earthworms (61), confirming that arsenosugars are not limited to the marine environment and may be present in other food items common for human consumption.

The total arsenic concentrations are lower in freshwater fish than in those from the marine environment (62). Much less is known about the nature of arsenic species in freshwater fish. Recent speciation studies show that arsenobetaine is also present in freshwater fish, and arsenosugars in freshwater mussels (63). Approximately 50% of the soluble arsenic in salmon (0.31 μg/g) is arsenobetaine and the rest is unknown (62). Cultured rainbow trout and smelt also contain arsenobetaine at a concentration of approximately 1 μg As/g (63).

The only other food class for which significant speciation information is available is mushrooms (64–67). Most mushroom species grown on noncontaminated soil contain less than 0.1 μg As/g (65). However, arsenic-accumulating mushrooms contain a range of arsenicals depending on the fungus, as shown in Table 1. Most of these high-arsenic mushrooms grew on highly contaminated soil, such as from old mining and smelter sites (66,67). Many of these mushrooms, although not all, contain arsenobetaine. An aresenosugar, tetramethylarsonium $[(CH_3)_4As^+]$, and several unidentified arsenic species are also present in some mushrooms as trace levels.

Table 1 Arsenic Species in Arsenic-Accumulating Mushrooms

Mushroom species	Total extracted arsenic (μg/g dry weight)	Arsenic species (concentration) (μg/g dry weight)	Ref.
Sarcosphaera coronaria	2000	MMA(V) ($\sim$2000)	64
Entoloma lividum	40	As(V) (37); As(III) (3)	64
Agaricus sp.	8	AB ($\sim$8)	64
Laccaria amethystina	40	DMA(V) ($\sim$40)	66
Collybia maculata	43	AB (43)	66
Collybia butracea	11	AB (8.8), DMA(V) (1.9), unidentified (0.2)	66
Amanita muscaria	20	AB (15.1), AC (2.6), $(CH_3)_4As^+$ (0.8), unidentified (0.4)	66
Paxillus sp.	30	DMA(V) (16), As(III) (1.1), arsenosugar (0.8), $(CH_3)_4As^+$ (0.1), unidentified (11)	67
Psathyrella sp.	7	As(V) (4.4), As(III) (1.4), DMA(V) (0.6), AB (0.3), $(CH_3)_4As^+$ (0.2), MMA(V) (0.1)	67
Leccinum sp.	7	DMA(V) (6.5), As(III) (0.2)	67
Coprinus sp.	68	AB (60), As(V) (4), As(III) (2), DMA(V) (1.7), AC (0.3), unidentified (0.4)	67
Lycoperdon sp.	71	AB (30), MMA(V) (6), As(III) (3.3), As(V) (21), AC (0.6), $(CH_3)_4As^+$ (0.5)	67

Only a few studies have dealt with the speciation of arsenic in rice and yam (68), vegetables and fruits (68–71), and poultry (72). In meat and poultry, some of the high arsenic concentrations may be due to the use of arylarsonic acids as growth promoters in chicken and swine, although these species have not been specifically sought in these foods. It is also possible that the arsenic results from fish meal present in the animal feed.

V. ARSENIC SPECIATION IN HUMANS

Ingested arsenic compounds can be readily absorbed through the gastrointestinal tract into the bloodstream (57,58,73–80). The rate of absorption is dependent on the solubility and probably the chemical species of arsenic. Most of the arsenic compounds are metabolized in the body. Both parent arsenic compounds and their metabolites are further excreted into urine (11–15,57,58,76–80,92–96). The extent of metabolism and excretion depends on the chemical species of arsenic ingested.

A. Arsenic Speciation in Urine

Urinary excretion is the primary pathway for the elimination of arsenic compounds from the body (11–15,57,58,76–80,92–96). Examples of the reported concentrations of arsenic in human urine from the general population are (mean ± standard deviation, μg/L) 9 ± 7 from a U.S. population (81), 17 ± 11 and 11 ± 6 from European studies (82,83), 21 ± 7 from Taiwan (84), and 121 ± 101 from Japan (85). People exposed to higher levels of arsenic from drinking water and food have correspondingly higher levels of urinary arsenic, e.g., 56 ± 13 from blackfoot disease patients (84), 274 ± 98 from a highly exposed Argentina population (86), and 450–700 from a highly exposed Mexican group (87). Arsenic levels in urine have a short half-life and reflect recent exposure (57,77,79). Therefore, speciation of arsenic in urine is commonly used as a measure of recent exposure to arsenic.

Most of the inorganic arsenic is metabolized in the body to methylated arsenic species. Methylation of arsenic involves a stepwise process of two-electron reduction of the pentavalent arsenic species [e.g., As(V), MMA(V), and DMA(V)] to the trivalent arsenic species [e.g., As(III), MMA(III), and DMA(III)], followed by oxidative addition of a methyl group to the trivalent arsenic (Scheme 1) (5,88). Glutathione, cysteine, and dithiothreitol can act as reducing agents, and S-adenosylmethionine (SAM) is the methyl donor. Dimethylarsinous acid (V) [DMA(V)] is the usual end product detected in humans. Trimethylarsine oxide (V) [TMAO(V)] and trimethylarsine (III) [TMA(III)] are the end products produced by some microorganisms; TMAO(V) is also produced by rats and mice.

The stable metabolites, MMA(V) and DMA(V), have usually been determined in human urine (76–87). Figure 1A shows typical profiles from the speciation analyses of urine samples from three individuals. The major arsenic species in all urine samples is DMA(V); MMA(V), As(III), and As(V) are present at lower levels. Accurate measurement of inorganic As(III) and As(V) separately is more difficult because the concentrations of these arsenic species in urine sam-

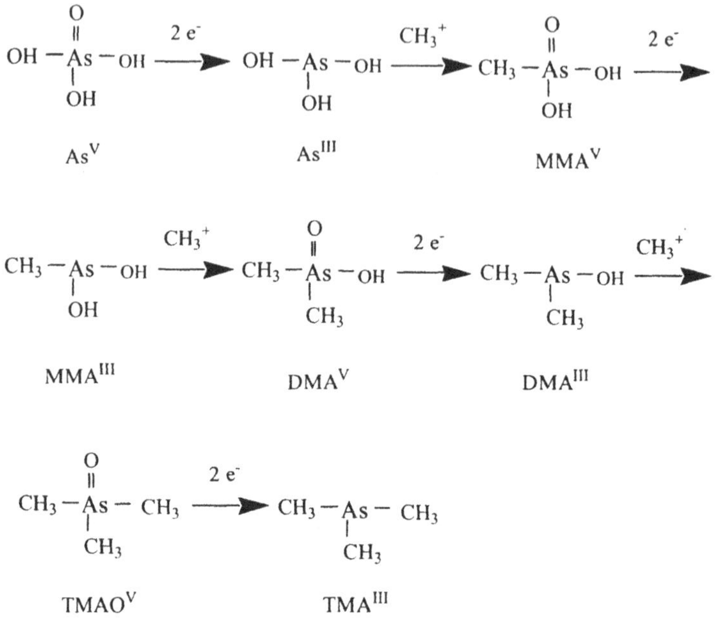

$$OH-\overset{\overset{O}{\|}}{\underset{\underset{OH}{|}}{As}}-OH \xrightarrow{2\ e^-} OH-\overset{}{\underset{\underset{OH}{|}}{As}}-OH \xrightarrow{CH_3^+} CH_3-\overset{\overset{O}{\|}}{\underset{\underset{OH}{|}}{As}}-OH \xrightarrow{2\ e^-}$$

$$As^V \qquad\qquad As^{III} \qquad\qquad MMA^V$$

$$CH_3-\overset{}{\underset{\underset{OH}{|}}{As}}-OH \xrightarrow{CH_3^+} CH_3-\overset{\overset{O}{\|}}{\underset{\underset{CH_3}{|}}{As}}-OH \xrightarrow{2\ e^-} CH_3-\overset{}{\underset{\underset{CH_3}{|}}{As}}-OH \xrightarrow{CH_3^+}$$

$$MMA^{III} \qquad\qquad DMA^V \qquad\qquad DMA^{III}$$

$$CH_3-\overset{\overset{O}{\|}}{\underset{\underset{CH_3}{|}}{As}}-CH_3 \xrightarrow{2\ e^-} CH_3-\overset{}{\underset{\underset{CH_3}{|}}{As}}-CH_3$$

$$TMAO^V \qquad\qquad TMA^{III}$$

Scheme 1 Pathway for biomethylation of arsenic.

ples are variable during sample storage (89). The proportion of the inorganic arsenic species and their methylation metabolites in urine is typically 55–80% DMA(V), 10–20% MMA(V), and 10–30% inorganic arsenic (76–80,90) in individuals who do not eat much food of marine origin, such as fish, shellfish, and algae.

The relative concentrations of the arsenic methylation metabolites in human urine have been used as a surrogate to compare methylation capacity between individuals and between populations (90). For example, a much lower portion (2.2%) of urinary MMA(V) was found in native Andean women (86), compared with 10–20% urinary MMA(V) in other populations. In another study of northern Argentina population, children were found to have a significantly higher percentage of inorganic arsenic (50%) in their urine samples than the adult women (32%) (91).

Ingestion of arsenosugar-containing food can dramatically change the proportion of DMA(V) in the urine because DMA(V) is one of the major human metabolites from arsenosugars. Figure 1B shows relative proportion of As(III), As(V), MMA(V), and DMA(V) in urine samples obtained from three subjects

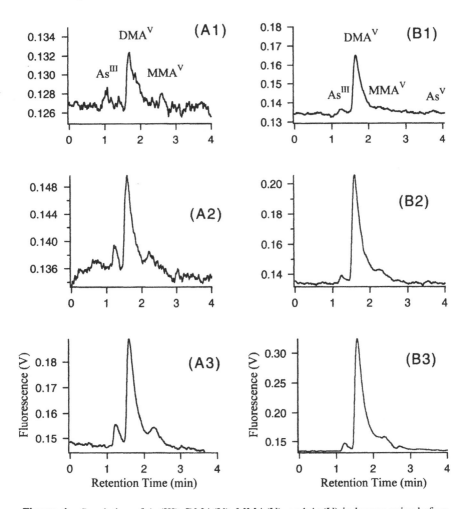

Figure 1 Speciation of As(III), DMA(V), MMA(V), and As(V) in human urine before (A) and after (B) the ingestion of 250 g of mussels. Typical arsenic speciation profiles are shown in urine samples (A1, A2, and A3) collected from three individuals (1, 2, and 3) before the ingestion of mussels. The concentrations of DMA(V) are markedly increased in the urine samples collected from the same three individuals after the ingestion of mussels (B1, B2, and B3). Arsenic compounds were separated on an ODS-3 column (15 cm × 4.6 mm, 3-μm particle size; Phenomenex, Torrance, CA), with a mobile phase (pH 5.8) containing 5 mM tetrabutylammonium hydroxide, 4 mM malonic acid, and 5% methanol. The flow rate of the mobile phase was 1.5 ml/min. The column temperature was 70°C. A hydride generation atomic fluorescence detector (Excalibur 10.003, P.S. Analytical, Kent, UK) was used for detection of arsenic. (Adapted from Ref. 92.)

approximately 15 hr after they ingested 250 g of mussels (92). The mussels contained arsenobetaine and two arsenosugars as the major arsenic species (Fig. 2a). While arsenobetaine is excreted unmodified, the arsenosugars are metabolized to DMA(V) and several not-yet-identified arsenic species that are excreted in the urine (Fig. 2b).

Until recently (93–96), little was known about the arsenic methylation intermediates, MMA(III) and DMA(III), in the human system, a result of the lack of techniques for the determination of these arsenic species. Recent developments of more sensitive and improved arsenic speciation techniques contribute to the discovery of these intermediary metabolites in human urine (93–96). Figure 3 shows typical chromatograms obtained from the analyses of arsenic compounds in deionized water and in urine samples. Coinjection of the urine sample with authentic MMA(III) standard (Fig. 3c) demonstrates the coelution of the suspected MMA(III) in the sample with that of the standard MMA(III), confirming the identity of MMA(III) in the urine sample. Similarly, coinjection of the urine sample with standard DMA(III) (Fig. 3d) and As(V) (Fig. 3e) confirms the presence of DMA(III) in the sample (96). Two other research groups have recently also found MMA(III) and DMA(III) in human urine samples (123,124).

Recent studies using cultured cells have shown that MMA(III) and DMA(III) are at least as toxic as the inorganic arsenic species (97–102). The methylated trivalent arsenic species are also proposed to be the proximate or ultimate genotoxic forms of arsenic (103). Thus, there has been much interest in the determination of these metabolites in humans. The observation of MMA(III) and DMA(III) species in human urine, together with these studies on arsenic toxic effects, indicates that methylation of arsenic may not be entirely a detoxification process for humans, as previously believed (104,105). Toxicological consequences of MMA(III) and DMA(III) in humans need to be further examined.

B. Arsenic Speciation in Blood

Table 2 summarizes literature data on arsenic species in blood. Limited speciation analyses revealed the presence of arsenobetaine (AB) and DMA(V) in human blood (106–108). Arsenobetaine and DMA(V) were detectable in serum of uremic (renal disease) patients, with mean values being 1 μg/L for DMA(V) and 3.5 μg/L for AB (107,108).

The majority of arsenic in blood is cleared with a half-time of about 1 hr. Concentrations of total arsenic in blood of people with no excess exposure to arsenic range from 0.3–2 μg/L (86,91,109,110). In populations exposed to higher concentrations of arsenic in drinking water and seafood, higher blood arsenic concentrations have been reported (Table 2). In people exposed to arsenic in drinking water (200 μg As/L) in northern Argentina, average blood arsenic con-

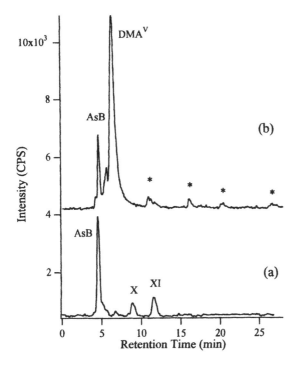

Figure 2 Arsenic speciation of a mussel extract (a) and a urine sample obtained from a 32-year-old male 42.5 hr after the ingestion of 250 g of mussels (b). Arsenic compounds were separated on an Inertsil ODS-2 column (4.6 × 250 mm; GL Sciences, Tokyo, Japan), with a mobile phase (pH 6.8) containing 10 mM tetraethylammonium hydroxide, 4.5 mM malonic acid and 0.1% methanol. The flow rate of the mobile phase was 0.8 ml/min. The column was maintained at 50°C. An inductively coupled plasma mass spectrometer (PlasmaQuad 2, VG Elemental) was used for detection. AsB, X, and XI stand for arsenobetaine and two arsenosugars, respectively. Peaks marked* indicate unidentified arsenic species from the metabolism of the arsenosugars. Structures of arsenosugars X and XI are shown below. Where R = OH for arsenosugar X, and R = $PO_4CH_2CH(OH)CH_2OH$ for arsenosugar XI. (Adapted from Ref. 92.)

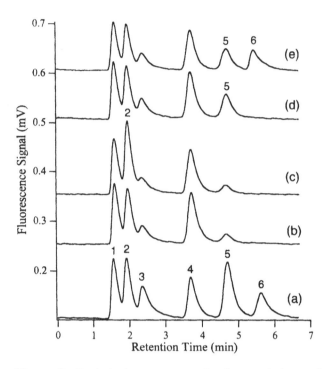

Figure 3 Typical chromatograms showing speciation analyses of As(III), As(V), MMA(V), DMA(V), MMA(III), and DMA(III) in deionized water (a), a urine sample (b), and the urine sample spiked with MMA(III) (c), DMA(III) (d), and As(V) (e). Separation was carried out on an ODS-3 column (15 cm × 4.6 mm, 3-μm particle size; Phenomenex) with a mobile phase (pH 5.95) containing 5 mM tetrabutylammonium hydroxide, 3 mM malonic acid, and 5% methanol. The flow rate of the mobile phase was 1.2 ml/min. The column was maintained at 50°C. A hydride generation atomic fluorescence detector was used for detection of arsenic. Peaks labeled 1–6 correspond to As(III), MMA(III), DMA(V), MMA(V), DMA(III), and As(V) respectively. The urine sample was collected from a person 4 hr after the administration of 300 mg sodium 2,3-dimercapto-1-propane sulfonate (DMPS). For clarity, chromatograms were manually shifted on vertical axis. (Adapted from Ref. 96.)

centration was 8 μg/L (86). People from the area in Taiwan with arsenic-rich water had blood arsenic of 22 μg/L (110). Blackfoot disease patients and their families had blood arsenic of 60 μg/L (110). The mean blood arsenic concentration in patients undergoing hemodialysis treatment was 8.5 μg/L compared with 10.6 μg/L in control patients (111).

Table 2 Arsenic Concentration and Species in Blood

Sources of sample	Sample type	Total arsenic conc. (µg/L) mean ± SD (range)	Arsenic species	Ref.
Belgium				107,108
Hemodialysis patients	Serum	6.5 ± 4.3	DMA(V) and AB	
Nonhemodialysis patient control	Serum	5.1 ± 5.6	DMA(V) and AB	
Healthy control	Serum	0.96 ± 1.5		
Argentina				86
Highly exposed Andean women	Blood	8 (2.7–18.3)		
Controls	Blood	1.5 (1.1–2.4)		
Belgium				82
Seafood-eating population	Blood	5.1 (0.5–32)		
Japan				106
A 29-year old male	Serum	4.6	AB	
	Plasma	3.3		
	Red blood cell	10.1		
Taiwan				110
Blackfoot disease families	Whole blood	60 ± 2		
	Plasma	38 ± 2		
	Red blood cell	93 ± 2		
Controls	Whole blood	22 ± 2		
	Plasma	15 ± 2		
	Red blood cell	33 ± 2		
California and Nevada, United States				112
Water arsenic 100–400 µg/L	Whole blood	3–4		

C. Arsenic Speciation in Hair and Nails

Arsenic is believed to accumulate in hair and nails because of the high content of keratin [and the corresponding high content of cysteine that might bind to As(III)]. In people with no known excess exposure to arsenic, the concentration of arsenic in hair is generally 0.02–0.2 µg/g (112–115). Reported normal values of arsenic in nails range from 0.02–0.5 µg/g (117). The concentrations of arsenic in hair and nails are elevated in cases of chronic poisoning (116,118). Hair and nail arsenic concentrations ranging from 3–10 µg/g have been reported in people exposed to high arsenic concentrations in drinking water in West Bengal, India (116).

Arsenic concentrations in hair and nails have been used as indicators of exposure to inorganic arsenic (112,113,121,122). However, arsenic in hair might be influenced by surface contamination via dust, water, soaps, and shampoos. Data on arsenic speciation in hair and nails is sparse. Only two studies reported the presence of dimethylated arsenic species in hair and nails (119,120).

VI. CONCLUDING REMARKS

Arsenic compounds are ubiquitous in the environment. Humans are exposed to arsenic from ingestion, inhalation, and less significantly, dermal absorption. Ingestion of arsenic from food and water is the main route of exposure to arsenic by the general population.

Inorganic As(III) and As(V) are the dominant arsenic species present in most natural waters. Methylated arsenic species may be present at lower concentrations that vary with biological and anthropogenic activities. Uncharacterized arsenic fractions have also been reported and research into their identification is needed.

Arsenic is abundant in seafood. Arsenobetaine is the major arsenic species present in fish and shellfish. Arsenosugars are the dominant arsenic species in seaweed. Both arsenobetaine and arsenosugars are present in mollusks (bivalves and gastropods) at comparable concentrations. Smaller amounts of arsenocholine, trimethylarsine oxide, tetramethylarsonium ion, DMA(V), MMA(V), inorganic arsenic, and unidentified arsenicals have also been found in seafood. Much less is known about the arsenic species in other food we eat.

Metabolism of arsenic in the body depends on the chemical species of arsenic absorbed. Inorganic As(III) and As(V) are metabolized in humans through a stepwise biomethylation pathway to form methylated arsenicals. Most of the arsenic compounds and their metabolites are readily excreted into the urine. Speciation of arsenic in human urine is the most suitable biomarker to assess recent exposure to arsenic.

ACKNOWLEDGMENTS

The author would like to thank Drs. Mingsheng Ma, Zhilong Gong, Serife Yalcin, and Ms. Xiufen Lu (University of Alberta), Dr. William R. Cullen and Ms. Vivian Lai (University of British Columbia), Dr. H. Vasken Aposhian (University of Arizona), Dr. Kenneth J. Reimer (Canadian Royal Military College), and Dr. Baoshan Zheng (Chinese Academy of Sciences) for their contributions. This work is supported by the Natural Sciences and Engineering Research Council of Canada.

REFERENCES

1. National Research Council. Arsenic in Drinking Water. Washington, DC: National Academy Press, 1999, pp 305.
2. United States Environmental Protection Agency. Fed Regist 66:6975–7066, 2001.
3. World Health Organization. Environmental Health Criteria 18: Arsenic. Geneva: WHO, 1981.
4. Health and Welfare Canada. Guidelines for Canadian Drinking Water Quality, Arsenic. Ottawa, Ontario: Ministry of Health, 1992.
5. WR Cullen, KJ Reimer. Arsenic speciation in the environment. Chem Rev 89:713–764, 1989.
6. JO Nriagu, ed. Arsenic in the Environment. Part II: Human Health and Ecosystem Effects. New York: Wiley, 1994.
7. S Tamaki, WT Frankenberger Jr. Environmental biochemistry of arsenic. Rev Environ Contam Toxicol 124:79–110, 1992.
8. T Kaise, S Fukui. The chemical form and acute toxicity of arsenic compounds in marine organisms. Appl Organomet Chem 6:155–160, 1992.
9. KA Francesconi, JS Edmonds. Arsenic in the sea. Oceanogr Marine Biol Annu Rev 31:111–151, 1993.
10. KA Francesconi, JS Edmonds. Arsenic and marine organisms. Adv Inorg Chem 44:147–189, 1997.
11. M Vahter, L Friberg, B Rahnster, A Nygren, P Nolinder. Airborne arsenic and urinary excretion of metabolites or inorganic arsenic among smelter workers. Int Arch Occup Environ Health 57:79–91, 1986.
12. TJ Smith, EA Crecelius, JC Reading. Airborne arsenic exposure and excretion of methylated arsenic compounds. Environ Health Perspect 19:89–93, 1977.
13. JA Offergelt, H Roels, JP Buchet, M Boeckx, R Lauwerys. Relation between airborne arsenic trioxide and urinary excretion of inorganic arsenic and its methylated metabolites. Br J Ind Med 49:387–393, 1992.
14. E Hakala, L Pyy. Assessment of exposure to inorganic arsenic by determining the arsenic species excreted in urine. Toxicol Lett 77:249–258, 1995.
15. JW Yager, JB Hicks, E Fabianova. Airborne arsenic and urinary excretion of arsenic metabolites during boiler cleaning operations in a Slovak coal-fired power plant. Environ Health Perspect 105:836–842, 1997.
16. SS Pinto, PE Enterline, V Henderson, MO Varner. Mortality experience in relation to a measured arsenic trioxide exposure. Environ Health Perspect 19:127–130, 1977.
17. K Welch, I Higgins, M Oh, C Burchfield. Arsenic exposure, smoking, and respiratory cancer in copper smelter workers. Arch Environ Health 37:325–335, 1982.
18. L Jarup, G Pershagen, S Wall. Cumulative arsenic exposure and lung cancer in smelter workers: A dose-response study. Am J Ind Med 15:31–41, 1989.
19. PE Enterline, VL Henderson, GM Marsh. Exposure to arsenic and respiratory cancer. A reanalysis. Am J Epidemiol 125:929–938, 1987.
20. PE Enterline, R Day, GM Marsh. Cancers related to exposure to arsenic at a copper smelter. Occup Environ Med 52:28–32, 1995.

21. JW Yager, HJ Clewell III, JB Hicks, E Fabianova. Airborne exposure to arsenic occurring in coal fly ash. In: WR Chappell, CO Abernathy, RL Calderon, eds. Arsenic Exposure and Health Effect. Amsterdam: Elsevier, 1999, pp 19–30.

22. A Chatterjee, D Dass, BK Mandal, TR Chowdhury, G Samanta, D Chakraborti. Arsenic in ground water in six districts of West Bengal, India: The biggest arsenic calamity in the world. Part 1. Arsenic species in drinking water and urine of the affected people. Analyst 120:643–650, 1995.

23. UK Chowdhury, BK Biswas, RK Dhar, G Samanta, BK Mandal, TR Chowdhury, D Chakraborti, S Kabir, S Roy. Groundwater arsenic contamination and suffering of people in Bangladesh. In: WR Chappell, CO Abernathy, RL Calderon, eds. Arsenic Exposure and Health Effects. Oxford: Elsevier, 1999, pp 165–182.

24. ZD Luo, YM Zhang, L Ma, GY Zhang, X He, R Wilson, DMA Byrd, JG Griffiths, S Lai, L He, K Grumski, SH Lamm. Chronic arsenicism and cancer in Inner Mongolia: Consequences of well-water arsenic levels greater than 50 μg/L. In: CO Abernathy, RL Calderon, WR Chappell eds. Arsenic Exposure and Health Effects. London: Chapman & Hall, 1997, pp 55–68.

25. C-J Chen, L-I Hsu, C-H Tseng, Y-M Hsueh, H-Y Chiou. Emerging epidemics of arseniasis in Asia. In: WR Chappell, CO Abernathy, RL Calderon, eds. Arsenic Exposure and Health Effects. Oxford: Elsevier, 1999, pp 113–121.

26. MM Wu, TL Kuo, YH Hwang, CJ Chen. Dose-response relation between arsenic concentration in well water and mortality from cancers and vascular diseases. Am J Epidemiol 130:1123–1132, 1989.

27. CJ Chen, CW Chen, MM Wu, TL Kuo. Cancer potential in liver, lung, bladder, and kidney due to ingested inorganic arsenic in drinking water. Br J Cancer 66: 888–892, 1992.

28. CJ Chen, CJ Wang. Ecological correlation between arsenic level in well water and age-adjusted mortality from malignant neoplasms. Cancer Res 50:5470–5474, 1990.

29. C Hopenhayn-Rich, ML Biggs, A Fuchs, R Bergoglio, EF Tello, H Nicolli, AH Smith. Bladder cancer mortality associated with arsenic in drinking water in Argentina. Epidemiology 7:117–124, 1996.

30. C Hopenhayn-Rich, ML Biggs, AH Smith. Lung and kidney cancer mortality associated with arsenic in drinking water in Cordoba, Argentina. Int J Epidemiol 27: 561–569, 1998.

31. AH Smith, M Goycolea, R Haque, ML Biggs. Marked increase in bladder and lung cancer mortality in a region of northern Chile due to arsenic in drinking water. Am J Epidemiol 147:660–669, 1998.

32. AH Smith, ML Biggs, L Moore, R Haque, C Steinmaus, J Chung, A Hernandez, P Lopipero. Cancer risks from arsenic in drinking water: Implications for drinking water standards. In: WR Chappell, CO Abernathy, RL Calderon, eds. Arsenic Exposure and Health Effects. Amsterdam: Elsevier, 1999, pp 191–199.

33. RB Finkelman, HE Belkin, B Zheng. Health Impacts of domestic coal use in China. Proc Natl Acad Sci USA 96:3427–3431, 1999.

34. MO Andreae. Distribution and speciation of arsenic in natural waters and some marine algae. Deep-Sea Res 25:391–402, 1978.

35. MO Andreae. Arsenic speciation in sea water and interstitial waters: The influence

of biological-chemical interactions on the chemistry of a trace element. Limnol Oceanogr 24:440–452, 1979.

36. RS Braman, CC Foreback. Methylated forms of arsenic in the environment. Science 182:1247–1249, 1973.
37. AG Howard, SDW Comber. The discovery of hidden arsenic species in coastal waters. Appl Organomet Chem 3:509–514, 1989.
38. H Hasegawa. The behavior of trivalent and pentavalent methyl arsenicals in Lake Biwa. Appl Organomet Chem 11:305–311, 1997.
39. Y Sohrin, M Matsui, M Kawashima, M Hojo, H Hasegawa. Arsenic biogeochemistry affected by eutrophication in Lake Biwa, Japan. Environ Sci Technol 31:2712–2720, 1997.
40. AG Howard, SC Apte. Seasonal control of arsenic speciation in an estuarine ecosystem. Appl Organomet Chem 3:499–507, 1989.
41. DA Bright, M Dodd, KJ Reimer. Arsenic in sub-arctic lakes influenced by gold mine effluent: The occurrence of organoarsenicals and "hidden" arsenic. Sci Total Environ 180:165–181, 1996.
42. NE Korte, Q Fernando. A review of arsenic(III) in groundwater. Crit Rev Environ Control 21:1–40, 1991.
43. XC Le, S Yalcin, M Ma. Speciation of submicrogram per liter levels of arsenic in water: On-site species separation integrated with sample collection, Environ Sci Technol 34:2342–2347, 2000.
44. H-W Chen, MM Frey, D Clifford, LS McNeill, M Edwards. Arsenic treatment considerations. J Am Water Works Assoc 91(3):74–85, 1999.
45. G Schwedt, M Reickhoff. Separation of thio- and oxothio-arsenates by capillary zone electrophoresis and ion chromatography. J Chromatogr A 736:341–350, 1996.
46. RE Sturgeon, MKW Siu, SN Willie, SS Berman. Quantification of arsenic species in a river water reference material for trace metals by graphite furnace atomic absorption spectrometric techniques. Analyst 114:1393–1396, 1989.
47. AH de Bettencourt, MH Florencio, MFN Duarte, MLR Gomes, LFC Vilas Boas. Refractory methylated arsenic compounds in esturine waters: Tracing back elusive species. Appl Organomet Chem 8:43–56, 1994.
48. RW Dabeka, AD McKenzie, GMA Lacroix. Dietary intakes of lead, cadmium, arsenic, and fluoride by Canadian adults: A 24-hour duplicate diet study. Food Addit Contam 4:89–101, 1987.
49. RW Dabeka, AD McKenzie, GM Lacroix, C Cleroux, S Bowe, RA Graham, HB Conacher, P Verdier. Survey of arsenic in total diet food composites and estimation of the dietary intake of arsenic by Canadian adults and children. J Assoc Off Anal Chem Internat 76:14–25, 1993.
50. MJ Gartrell, JC Craun, DS Podrebarac, EL Gunderson. Pesticides, selected elements, and other chemicals in adult total diet samples, October 1979–September 1980. J Assoc Off Anal Chem 68:862–875, 1985.
51. Food Additives and Contaminants Committee. Report on the Review of the Arsenic in Food Regulations. Ministry of Agriculture, Fisheries and Foods, FAC/REP/39. London: Her Majesty's Stationery Office, 1984.
52. GL Dick, JT Hughes, JW Mitchell, F Davidson. Survey of trace elements and pesti-

cide residues in the New Zealand diet. 1. Trace element content. NZ J Sci 21:57–69, 1978.

53. T Tsuda, T Inoue, M Kojima, S Aoki. Market basket and duplicate portion estimation of dietary intakes of cadmium, mercury, arsenic, copper, manganese, and zinc by Japanese adults. J Assoc Off Anal Chem Internat 78:1363–1368, 1995.

54. MA Adams, PM Bolger, EL Gunderson. Dietary intake and hazards of arsenic. In: WR Chappell, CO Abernathy, CR Cothern, eds. Arsenic: Exposure and Health. Northwood, UK: Science and Technology Letters, 1994, pp 41–49.

55. Y Shibata, M Morita, K Fuwa. Selenium and arsenic in biology: Their chemical forms and biological functions. Adv Biophys 28:31–80, 1992.

56. XC Le, WR Cullen, KJ Reimer. Speciation of arsenic compounds in some marine organisms. Environ Sci Technol 28:1598–1604, 1994.

57. XC Le, WR Cullen, KJ Reimer. Human urinary arsenic excretion after one-time ingestion of seaweed, crab, and shrimp. Clin Chem 40:617–624, 1994.

58. M Ma, XC Le. Effect of arsenosugar ingestion on urinary arsenic speciation. Clin Chem 44:539–550, 1998.

59. VW Lai, WR Cullen, CF Harrington, KJ Reimer. The characterization of arsenosugars in commercially available algal products including one of terrestrial origin, Nostoc sp. Appl Organomet Chem 11:797–803, 1997.

60. I Koch, L Wang, CA Ollson, WR Cullen, KJ Reimer. The predominance of inorganic arsenic species in plants from Yellowknife, Northwest Territories, Canada. Environ Sci Technol 34:22–26, 2000.

61. A Geiszinger, W Goessler, D Kuehnelt, K Francesconi, W Kosmus. Determination of arsenic compounds in earthworms. Environ Sci Technol. 32:2238–2243, 1998.

62. JF Lawrence, P Michalik, G Tam, HBS Conacher. Identification of arsenobetaine and arsenocholine in Canadian fish and shellfish by high-performance liquid chromatography with atomic absorption detection and confirmation by fast atom bombardment mass spectrometry. J Agric Food Chem 34:315–319, 1986.

63. K Shiomi, Y Sugiyama, K Shimakura, Y Nagashima. Arsenobetaine as the major arsenic compound in the muscle of two species of freshwater fish. Appl Organomet Chem 9:105–109, 1995.

64. AR Byrne, Z Slejkovec, T Stijve, L Fay, W Gössler, J Gailer, KJ Irgolic. Arsenobetaine and other arsenic species in mushrooms. Appl Organomet Chem 9:305–313, 1995.

65. J Vetter. Data on arsenic and cadmium contents of some common mushrooms. Toxicon 32:11–15, 1994.

66. DW Kuehnelt, W Goessler, KJ Irgolic. Arsenic compounds in terrestrial organisms. I. Collybia maculata, Collybia butyracea, and Amanita muscaria from arsenic smelter sites in Austria. Appl Organomet Chem 11:289–296, 1997.

67. I Koch, L Wang, KJ Reimer, WR Cullen. Arsenic species in terrestrial fungi and lichens from Yellowknife, NWT, Canada. Appl Organometal Chem 14:245–252, 2000.

68. RA Schoof, LJ Yost, E Crecelius, K Irgolic, W Goessler, HR Guo, H Greene. Dietary arsenic intake in Taiwanese districts with elevated arsenic in drinking water. Hum Ecol Risk Assess 4:117–135, 1998.

69. LJ Yost, RA Schoof, R Aucoin. Intake of inorganic arsenic in the North American diet. Hum Ecol Risk Assess 4:137–152, 1998.

70. H Helgesen, EH Larsen. Bioavailability and speciation of arsenic in carrots grown in contaminated soil. Analyst 123:791–796, 1998.

71. JA Caruso, DT Heitkemper, CB Hymer. An evaluation of extraction techniques for arsenic species from freeze-dried apple samples. Analyst 126:1–11, 2001.

72. JR Dean, L Ebdon, ME Foulkes, HM Crews, RD Massey. Determination of the growth promoter, 4-hydroxy-3-nitrophenylarsonic acid in chicken tissue by coupled high-performance liquid chromatography–inductively coupled plasma mass spectrometry. J Anal Atom Spectrom 9:615–618, 1994.

73. C Pomroy, SM Charbonneau, RS McCullough, GKH Tam. Human retention studies with [74]As. Toxicol Appl Pharmacol 53:550–556, 1980.

74. M Vahter, H Norin. Metabolism of [74]As-labeled trivalent and pentavalent inorganic arsenic in mice. Environ Res 21:446–457, 1980.

75. GB Freeman, RA Schoof, MV Ruby, AO Davis, JA Dill, SC Liao, CA Lapin, PD Bergstrom. Bioavailability of arsenic in soil and house dust impacted by smelter activities following oral administration in Cynomologus monkeys. Fundam Appl Toxicol 28:215–222, 1995.

76. JP Buchet, R Lauwerys, H Roels. Comparison of the urinary excretion of arsenic metabolites after a single dose of sodium arsenite, monomethylarsonate or dimethylarsinate in man. Int Arch Occup Environ Health 48:71–79, 1981.

77. JP Buchet, R Lauwerys. Inorganic arsenic metabolism in humans. In WR Chappell, CC Abernathy, CR Cothern, eds. Arsenic Exposure and Health. Norwood, UK: Science and Technology Letters, 1994, pp 181–189.

78. M Vahter. Metabolism of arsenic. In: BA Fowler, ed. Biological and Environmental Effects of Arsenic. Amsterdam: Elsevier, 1983, pp 171–198.

79. M Vahter. Species differences in the metabolism of arsenic compounds. Appl Organomet Chem 8:175–182, 1994.

80. M Vahter. Variation in human metabolism of arsenic. In: WR Chappel, CO Abernathy, RL Calderon, eds. Arsenic Exposure and Health Effects. Amsterdam: Elsevier, 1999, pp 267–279.

81. DA Kalman, J Hughes, G van Belle, T Burbacher, D Bolgiano, K Coble, NK Mottet, L Polissar. The effect of variable environmental arsenic contamination on urinary concentrations of arsenic species. Environ Health Perspect 89:145–151, 1990.

82. V Foà, A Colombi, M Maroni, M Burrati, G Calzaferri. The speciation of the chemical forms of arsenic in the biological monitoring of exposure to inorganic arsenic. Sci Total Environ 34:241–259, 1984.

83. JP Buchet, D Lison, M Ruggeri, V Föa, G Elia. Assessment of exposure to inorganic arsenic, a human carcinogen, due to the consumption of seafood. Arch Toxicol 70: 773–778, 1996.

84. TH Lin, YL Huang. Chemical speciation of arsenic in urine of patients with Blackfoot disease. Biol Trace Elem Res 48:251–261, 1995.

85. H Yamauchi, K Takahashi, M Mashiko, Y Yamamura. Biological monitoring of arsenic exposure of gallium arsenide– and inorganic arsenic–exposed workers by determination of inorganic arsenic and its metabolites in urine and hair. Am Ind Hyg Assoc J 50:606–612, 1989.

86. M Vahter, G Concha, B Nermell, R Nilsson, F Dulout, AT Natarajan. A unique metabolism of inorganic arsenic in native Andean women. Eur J Pharmacol Environ Toxicol 293:455–462, 1995.

87. LM Del Razo, GG Garcia-Vargas, H Vargas, A Albores, ME Gonsebatt, R Montero, P Ostrosky-Wegman, M Kelsh, ME Cebrian. Altered profile of urinary arsenic metabolites in adults with chronic arsenicism. A pilot study. Arch Toxicol 71:211–217, 1997.

88. WR Cullen, BC McBride, J Reglinski. The reduction of trimethylarsine oxide to trimethylarsine by thiols: A mechanistic model for the biological reduction of arsenicals. J Inorg Biochem 21:45–60, 179–194, 1984.

89. J Feldmann, VWM Lai, WR Cullen, M Ma, X Lu, XC Le. Sample preparation and storage can change arsenic speciation in human urine. Clin Chem 45:1988–1997, 1999.

90. C Hopenhayn-Rich, AH Smith, HM Goeden. Human studies do not support the methylation threshold hypothesis for the toxicity of inorganic arsenic. Environ Res 60:161–177, 1993.

91. G Concha, G Vogler, D Lezcano, B Nermell, M Vahter. Exposure to inorganic arsenic metabolites during early human development. Toxicol Sci 44:185–190, 1998.

92. XC Le, M Ma. Short-column liquid chromatography with hydride generation atomic fluorescence detection for the speciation of arsenic. Anal Chem 70:1926–1933, 1998.

93. HV Aposhian, B Zheng, MM Aposhian, XC Le, ME Cebrian, WR Cullen, RA Zakharyan, M Ma, RC Dart, Z Cheng, P Andrews, L Yip, GF O'Malley, RM Maiorino, W Van Voorhies, SM Healy, A Titcomb. DMPS-Arsenic challenge test: Modulation of arsenic species, including monomethylarsonous acid, excreted in human urine. Toxicol Appl Pharmacol 165:74–83, 2000.

94. VH Aposhian, ES Gurzau, XC Le, A Gurzau, SM Healy, X Lu, M Ma, L Yip, RA Zakharyan, RM Maiorino, RC Dart, MG Tirus, D Gonzalez-Ramirez, DL Morgan, D Avram, MM Aposhian. Occurrence of monomethylarsonous acid (MMA^{III}) in urine of humans exposed to inorganic arsenic. Chem Res Toxicol 13:693–697, 2000.

95. XC Le, M Ma, X Lu, WR Cullen, V Aposhian, B Zheng. Determination of monomethylarsonous acid, a key arsenic methylation intermediate, in human urine. Environ Health Perspect 108:1015–1018, 2000.

96. XC Le, X Lu, M Ma, WR Cullen, V Aposhian, B Zheng. Speciation of key arsenic metabolic intermediates in human urine. Anal Chem 72:5172–5177, 2000.

97. JS Petrick, F Ayala-Fierro, WR Cullen, DE Carter, HV Aposhian. Monomethylarsonous acid (MMA^{III}) is more toxic than arsenite in Chang human hepatocytes. Toxicol Appl Pharmacol 163:203–207, 2000.

98. JS Petrick, B Jagadish, EA Mash, HV Aposhian. Methylarsonous acid (MMA^{III}) and arsenite: LD$_{50}$ in hamsters and in vitro inhibition of pyruvate dehydrogenase. Chem Res Toxicol 14:651–656, 2001.

99. M Styblo, DJ Thomas. In vitro inhibition of glutathione reductase by arsenotriglutathione. Biochem Pharmacol 49:971–974, 1995.

100. M Styblo, SV Serves, WR Cullen, DJ Thomas. Comparative inhibition of yeast

glutathione reductase by arsenicals and arsenothiols. Chem Res Toxicol 10:27–33, 1997.

101. S Lin, WR Cullen, DJ Thomas. Methylarsenicals and arsinothiols are potent inhibitors of mouse liver thioredoxin reductase. Chem Res Toxicol 12:924–930, 1999.

102. S Lin, LM Del Razo, M Styblo, C Wang, WR Cullen, DJ Thomas. Arsenicals inhibit thioredoxin reductase in cultured rat hepatocytes. Chem Res Toxicol 14: 305–311, 2001.

103. MJ Mass, A Tennant, RC Roop, WR Cullen, M Styblo, DJ Thomas, AD Kligerman. Methylated trivalent arsenic species are genotoxic. Chem Res Toxicol 14:355–361, 2001.

104. RA Goyer. Toxic effects of metals. In: CD Klaassen, ed. Casarett and Doull's Toxicology: The Basic Science of Poisons. 5th ed. New York: McGraw-Hill, 1996, pp 696–698.

105. H Yamauchi, BA Fowler. Toxicity and metabolism of inorganic and methylated arsenicals. In: JO Nriagu, ed. Arsenic in the Environment. Part II: Human Health and Ecosystem Effects. New York: Wiley, 1994, pp 35–43.

106. Y Shibata, J Yoshinaga, M Morita. Detection of arsenobetaine in human blood. Appl Organomet Chem 8:249–251, 1994.

107. X Zhang, R Cornelis, J De Kimpe, L Mees, V Vanderbiesen, A De Cubber, R Vanholder. Accumulation of arsenic species in serum of patients with chronic renal disease. Clin Chem 42:1231–1237, 1996.

108. X Zhang, R Cornelis, L Mees, R Vanholder, N Lameire. Chemical speciation in serum of uraemic patients. Analyst 123:13–17, 1998.

109. EI Hamilton, E Sabbioni, MT van der Venne. Element reference values in tissues from inhabitants of the European community. IV. Review of elements in blood plasma and urine and a critical evaluation of reference values for the United Kingdom population. Sci Total Environ 158:165–190, 1994.

110. K Heydorn. Environmental variation of arsenic levels in human blood determined by neutron activation analysis. Clin Chim Acta 28:349–357, 1970.

111. DR Mayer, W Kosmus, H Pogglitsch, D Mayer, W Beyer. Essential trace elements in humans: Serum arsenic concentrations in hemodialysis patients in comparison to healthy controls. Biol Trace Elem Res 37:27–38, 1993.

112. JL Valentine, HK Kang, G Spivey. Arsenic levels in human blood, urine, and hair in response to exposure via drinking water. Environ Res 20:24–32, 1979.

113. A Olguin, P Jauge, M Cebrian, A Albores. Arsenic levels in blood, urine, hair, and nails from a chronically exposed human population. Proc West Pharmacol Soc 26: 175–177, 1983.

114. M Wolfsperger, G Hauser, W Gössler, C Schlagenhaufen. Heavy metals in human hair samples from Austria and Italy: Influence of sex and smoking habits. Sci Total Environ 156:235–242, 1994.

115. CE Rogers, AV Tomita, PR Trowbridge, JK Gone, J Chen, P Zeeb, HF Hemond, WG Thilly, I Olmez, JL Durant. Hair analysis does not support hypothesized arsenic and chromium exposure from drinking water in Woburn, Massachusetts. Environ Health Perspect 105:1090–1097, 1997.

116. D Das. A Chatterjee, BK Mandal, G Samanta, D Chakraborti, B Chanda. Arsenic in ground water in six districts of West Bengal, India: The biggest arsenic calamity

in the world. Part 2. Arsenic concentration in drinking water, hair, nails, urine, skin-scale and liver tissue (biopsy) of the affected people. Analyst 120:919–924, 1995.

117.   Y Takagi, S Matsuda, S Imai, Y Ohmori, T Masuda, JA Vison, MC Mehra, BK Puri, A Kaniewski. Survey of trace elements in human nails: An international comparison. Bull Environ Contam Toxicol 41:690–695, 1988.

118.   CA Pounds, EF Pearson, TD Turner. Arsenic in fingernails. J Forensic Sci Soc J 19:165–174, 1979.

119.   N Yamato. Concentration and chemical species of arsenic in human urine and hair. Bull Environ Contam Toxicol 40:633–640, 1988.

120.   H Yamauchi, Y Yamamura. Metabolism and excretion of orally administered dimethylarsinic acid in the hamster. Toxicol Appl Pharmacol 74:134–140, 1984.

121.   MR Karagas, TD Tosteson, J Blum, B Klaue, JE Weiss, V Stannard, V Spate, JS Morris. Measurement of low levels of arsenic exposure: A comparison of water and toenail concentrations. Am J Epidemiol 152:84–90, 2000.

122.   MR Karagas, TA Stukel, JS Morris, TD Tosteson, JE Weiss, SK Spencer, ER Greenberg. Skin cancer risk in relation to toenail arsenic concentrations in a US population-based case-control study. Am J Epidemiol 153:559–565, 2001.

123.   BK Mandal, Y Ogra, KT Suzuki. Identification of dimethylarsinous and monomethylarsonous acids in human urine of the arsenic-affected areas in West Bengal, India. Chem Res Toxicol 14:371–378, 2001.

124.   LM Del Razo, M Styblo, WR Cullen, DJ Thomas. Determination of trivalent methylated arsenicals in biological matrices. Toxicol Appl Pharmacol 2001, in press.

5
Bioavailability and Risk of Arsenic Exposure by the Soil Ingestion Pathway

Nicholas T. Basta
Oklahoma State University, Stillwater, Oklahoma

Robin R. Rodriguez
Stratum Engineering, Inc., Bridgeton, Missouri

Stan W. Casteel
University of Missouri, Columbia, Missouri

I. INTRODUCTION

Remedial investigations (RIs) are conducted on hazardous waste sites ("Superfund" sites) under the guidance of the Comprehensive Environmental Response, Compensation and Liability Act (CERCLA) to determine two overall important issues: first, to determine the nature and extent of contamination that exists and, second, to determine the extent to which some level of cleanup must be performed to be protective of human health and the environment. The typical RI includes the collection and chemical analyses of site media, including surface soils, subsurface soils, groundwater, surface water, sediment, and biota (plant and animal species). In some instances, air monitoring may be conducted to determine airborne concentrations of contaminants. For a comprehensive investigation, analytical testing for the full list of priority pollutants is typically performed. Priority pollutants include the major chemical classes of inorganics, volatile and semivolatile organics, pesticides, polychlorinated biphenyls (PCBs), and dioxins and furans.

An integral component of the RI is the development of the Human Health

Baseline Risk Assessment. The risk assessment is the foundation upon which site remediation goals are determined and is developed following two fundamental assessments. A toxicity assessment is performed to collect the most recent and pertinent toxicity data for carcinogenic and noncarcinogenic effects of chemical contaminants detected in site media. An exposure assessment is performed to quantify human intake of contaminated media. Subsequently, by measuring the concentrations of chemicals detected in the site media, the dose of chemical intake can then be quantified to complete the exposure assessment. Risk characterization is the final step, performed by coupling the results of the toxicity assessment with those of the exposure assessment to obtain an overall cumulative site risk.

Criticisms have been put forth regarding the conservative nature of current risk assessment methodology. Some have criticized the development of toxicity factors (such as the cancer slope factors) or the definition of acceptable risk (1,2), while others find fault with methodology used to assess exposure (3). The focus of this chapter is to evaluate the methodology of assessing risk from exposure to arsenic (As) in soils.

II. ASSESSING RISK OF INCIDENTAL INGESTION PATHWAY

A. Exposure to Contaminated Soil

Humans may come into contact with and may be exposed to chemicals in soil via a number of pathways, including inhalation, dermal absorption, and ingestion. In most cases, the significant pathway in regard to exposure to metal-contaminated soil is ingestion. Except for the rare condition of soil pica in children (cases where children intentionally ingest large quantities of soil), soil ingestion results from hand-to-mouth type activities whereby soil is accidentally, or incidentally, ingested.

Incidental soil ingestion by children is an important pathway in assessing public health risks associated with exposure to arsenic-contaminated soils. Incidental ingestion of soil represents the principal direct pathway for exposure to nondietary sources of As in contaminated areas. The importance of soil ingestion by children as a health issue has been reported by numerous researchers and fully illustrates the importance of this pathway in terms of subsequent chemical exposure (4–8).

B. Quantifying Arsenic Exposure

The first step of the exposure assessment is to quantify the amount of chemical received as a dose following exposure to contaminated site soils. Exposure must be quantified considering the magnitude, frequency, and duration of exposure for the receptors and pathways selected for quantitative evaluation. For incidental ingestion, the following formula is used to quantify average daily chemical intake (9):

$$CDI = \frac{(CS)(IR)(CF)(FI)(EF)(ED)}{(BW)(AT)} \qquad (1)$$

where

CDI = chemical daily intake (mg kg^{-1} day^{-1})
CS = chemical concentration in soil (mg kg^{-1})
IR = ingestion rate (mg soil day^{-1})
CF = conversion factor (10^{-6} kg mg^{-1})
FI = fraction ingestion from contaminated source (unitless)
EF = exposure frequency (days yr^{-1})
ED = exposure duration (yr)
BW = body weight (kg)
AT = averaging time (period over which exposure is averaged—days, typically 70 yr)

The CS variable, chemical concentration in soil, is a site-specific value measured by performing U.S. Environmental Protection Agency (USEPA) SW-846 Method 3050 (10) on site soil samples for metal analysis. This is a destructive method involving a hot acid digestion with nitric acid (HNO_3) and water, and results in a total metals analysis rather than a determination of a specific species or soil fraction of metal. The underlying assumption, in quantifying metal intake by the above formula, is that all of the As measured by the total metal analysis is quantified as the absorbed dose.

However, there is an inherent problem with the above assumption. For an adverse health effect to be realized, the chemical toxicant (in this case, the metal) must be dissolved for As absorption to occur. Forms of As found in soils and waste materials may not be soluble under conditions associated with human ingestion. Arsenic, for example, may exist in many different forms that have a wide range of solubilities. Arsenic typically exists in soils in the (III) and (V) oxidation states and may be present in the ($-III$) and (0) oxidation states in strongly reduced soils and sediments (11). Arsenic in the (III) form, arsenite, may be found as $As(OH)_3$, $As(OH)_4^-$, and $AsO_2(OH)^{2-}$. Arsenic in the (V) form, arsenate, is found as AsO_4^{3-} (the oxidized state) and is stable in aerobic soils (11). In addition to multiple possible oxidation states, As may also be found in a variety of matrices resulting from industrial processes, such as the mining, milling, and smelting of copper, lead, and zinc sulfide ores (12,13); raw and spent oil shale (14); and coal fly ash (15,16). The combination of various chemical species with different soil/solid matrices of As produces a wide range of As solubility in contaminated media. Most metal and metalloid sulfides, for example, are less soluble than their respective oxidized compounds; for As, the solubility of As_2S_3 in water is 0.005 g L^{-1}, while the solubility of As_2O_3 is 37 g L^{-1}. Differences in solubility will certainly have a significant impact on the dose absorbed from ingestion of contaminated soil.

Quantification of the absorbed dose involves As dissolution and biological uptake. It is not yet fully understood whether absorption of soluble As is an active or passive mechanism. To further complicate the issue, the rate-limiting step (dissolution kinetics vs. absorption mechanism) has not yet been determined.

An understanding of the potential for As in site soils to be absorbed by human receptors will have a significant impact on the resulting site baseline risk. The bioavailability, in terms of a fraction of the total As that may be absorbed, could be used to reduce the uncertainty and overly conservative nature of risk assessments in general and provide risk managers with a more reasonable, site-specific baseline risk estimate. In the following section, methods of obtaining site-specific bioavailability data are presented.

III. METHODS FOR DETERMINING BIOAVAILABILITY OF ARSENIC FROM SOIL INGESTION

Bioavailable As is the portion of an As dose from soil ingestion that enters the systemic circulation. Bioavailability of As in soil can be divided into two kinetic steps: dissolution of As in gastrointestinal fluids and absorption across the gastrointestinal epithelium into the bloodstream. Arsenic in contaminated soil may exist in many geochemical forms (e.g., oxides, sulfides, arsenates) and physical forms (e.g., flue dust, slag, tailings, waste ore). Therefore, combining the variability of geochemical forms of As in contaminated soil with dissolution chemistry and biological absorption processes in the gastrointestinal tract results in a complex system. Controlled dosing studies are required to accurately determine the bioavailability of As in this complex system.

Human subjects cannot be used in dosing studies for ethical reasons. Therefore, appropriate animal models are often used to determine bioavailability of As in contaminated soil. Recently reviewed in vivo models used to measure bioavailable As include juvenile swine, monkey, rabbit, and dog (17,18). In these in vivo dosing trials, soil As bioavailability is evaluated by measuring As in urine, blood, feces, and/or storage tissues (bone, skin, nails, hair). Most As is measured in urine following oral or intravenous doses of soluble As considered 100% bioavailable. Ingestion of soluble As by human volunteers resulted in 64–69% of dosed As received in urine. Similar As recoveries of highly soluble forms of dosed As have been reported in monkey (19), juvenile swine (20), and dog

(21). Measuring systemic As in urine is a convenient method of evaluating bioavailable As.

Regardless of the method (i.e., urine, blood, tissue) used to estimate systemic As or the animal model used, most studies show that As bioavailability in contaminated soil is much lower than the bioavailability of soluble inorganic As (i.e., sodium arsenate) used for assessing risk from As in drinking water (17,18). Bioavailability of As in soil contaminated by ore mining and smelting relative to sodium arsenate (i.e., relative bioavailability) ranged from 0–98% with a median value of 35.5% for 16 contaminated soils and media (17) and from 4.07–42.9% with a median value of 25.5% for 14 contaminated soils and media (22). Most contaminated soils had relative bioavailability of <50% showing clearly that As was less bioavailable in soil than when dissolved in water.

Juvenile swine and monkey are commonly used animal models to obtain site-specific bioavailability of soil As for use in risk assessment at Superfund sites. Use of an appropriate animal model for investigating the enteric bioavailability of As in humans, and especially children, necessitates selection based on similar age and anatomical and physiological characteristics. Both monkey and swine (23) are remarkably similar to humans with respect to their digestive tract, nutritional requirements, bone development, and mineral metabolism. Juvenile swine are commonly used because of several factors, including the economics of husbandry, ease of dose delivery, and the concern of animal rights' groups regarding animal model selection. Young swine are considered to be a good physiological model for gastrointestinal absorption in children (24,25). Feeding behavior and its connection to the presence of ingesta in the stomach have a significant impact on the enteric bioavailability of As. The presence of ingesta in the stomach clearly reduces the absorption of metals by pigs as it does in humans (26). Pigs, like humans, tend to ingest food intermittently allowing the stomach to evacuate periodically. This physiology is consistent with the way children most likely ingest As-contaminated materials—between meals when the gastric pH is lowest. Modeling the maximal exposure that might reasonably be expected by assessing bioavailability of As-laden soil or other material on an empty stomach is possible only in species with periodic feeding behavior (24). Use of the swine model for bioavailability determinations has been thoroughly validated during the testing of more than a dozen lead (Pb)-contaminated materials and one As-contaminated material for the USEPA Region VIII office. Standard operating procedures (SOPs) have been developed for all aspects of the project (17). Recently, SOPs have been reported (20) for the use of juvenile swine for determination of bioavailable As in soil.

Several disadvantages in conducting animal studies include expense, specialized facilities and personnel requirements, and time required to measure contaminant bioavailability. In order to overcome some of the difficulties and expenses associated with animal dosing trials used to assess bioavailability of contaminants in soil, research efforts have been directed toward development of in vitro chemical methods that simulate the gastrointestinal environment. The gastrointestinal digestive processes are quite complicated and difficult to simulate in vitro. Several studies in the area of human nutrition have reported in vitro methods to assess bioavailable iron (Fe) in foodstuffs (28–30). Many of these procedural steps are based upon the medical and biochemical scientific literature to gain an understanding of the digestive process, especially in terms of digestive solution volumes produced in response to food intake volume, pH conditions during digestive phases, and quantities of digestive juices and enzymes produced, such as pepsin, bile acids, pancreatin, etc. (31,32). Most in vitro methods are sequential extractions with two distinct extraction steps: a gastric phase extraction that simulates the acidic conditions of the stomach, and a subsequent intestinal phase extraction that simulates the biochemical environment of the small intestine. In vitro gastrointestinal simulation methods are usually conducted at 37°C. The amount of As and/or other contaminants dissolved in the extraction solution at the end of the gastric and intestinal extraction steps is used to estimate potential bioavailability (i.e., availability or bioaccessability) of the contaminant in soil. Many studies have shown Pb extracted by the gastric phase of the physiologically based extraction test (PBET) (33) is strongly correlated with bioavailable Pb measured from animal dosing trials (17). However, application of PBET to As-contaminated materials has been limited to only a small number of materials. The ability of an in vitro gastrointestinal (IVG) method to estimate bioavailable As in 15 contaminated soils and media was reported by Rodriguez et al. (34). Results from the IVG method were strongly correlated with bioavailable As measured by juvenile swine dosing studies.

A comprehensive review of in vivo and in vitro methods is beyond the scope of this chapter. An overview of the commonly used in vivo juvenile swine method used to measure bioavailable As and an overview of an in vitro method for estimating available As (34) in contaminated soil is presented in the following sections.

A. Juvenile Swine In Vivo Method

Accurate assessment of human health risks associated with oral exposure to metals requires knowledge of the fraction of the dose absorbed into the blood. This information is important for As-contaminated environmental media, such as soil and mine waste, because metal contaminants exist in a variety of soluble and insoluble forms and may be contained within particles of inert matrix, such as rock or slag. Physicochemical properties such as these influence the enteric absorption fraction (bioavailability) of ingested metals. Therefore, site-specific data on metal bioavailability in the environmental media of concern will increase the accuracy and decrease the uncertainty in human health risk estimates.

The enteric bioavailability of As can be described in absolute (ABA) or relative (RBA) terms. The ABA is also referred to as the oral absorption fraction (AF$_o$) and is equal to absorbed dose/ingested dose.

$$ABA = \frac{\text{Absorbed dose}}{\text{Ingested dose}} \tag{2}$$

Relative bioavailability (RBA) is the ratio of the ABA of As present in some test material (contaminated soil) compared with the ABA of As in some reference material, usually the chemical dissolved in water or some fully soluble form that completely dissolves in the digestive tract.

$$RBA = \frac{\text{ABA (test material)}}{\text{ABA (reference material)}} \tag{3}$$

For example, if 100 μg of As dissolved in drinking water were ingested and a total of 80 μg entered the body, the ABA would be 0.80 (80%). Likewise, if 100 μg of As contained in soil were ingested and 20 μg entered the body, the ABA for the soil As would be 0.20 (20%). If the As dissolved in water was used as the reference substance (e.g., sodium arsenate) for describing the relative amount of As absorbed from soil, the RBA would be 0.20/0.80 = 0.25 (25%).

The toxicokinetic fate of ingested As can be conceptualized in the following manner. In most animals and humans, absorbed As is excreted mainly in the urine. Consequently, the urinary excretion fraction (UEF), defined as the amount excreted in the urine divided by the amount dosed, is a reasonable approximation of the oral absorption fraction or ABA. However, this ratio will underestimate total absorption because some absorbed As is excreted back into the intestine via the biliary mechanism, and some absorbed As enters tissue compartments (e.g., liver, kidney, skin, and hair) from which it is cleared very slowly. Therefore, the UEF should not be equated with the absolute absorption fraction.

The conceptual model for As toxicokinetics requires further clarification. Salient features of the model include the following: (1) absorbed As is primarily excreted in the urine. Thus, the UEF, defined as the amount of As excreted in urine divided by the dosed amount, can be used to estimate the ABA. (2) Absolute bioavailability (ABA = AF$_o$) of As from a test material can be estimated from the ratio of UEF of As from test material compared with intravenously dosed As. (3) The RBAs of two orally dosed materials (e.g., a test soil and sodium arsenate) can be calculated from the ratio of their UEFs. This calculation is independent of the extent of tissue binding and biliary excretion:

$$RBA \text{ (test vs. ref)} = \frac{AF_o \text{ (test)}}{AF_o \text{ (ref)}} = \frac{(D)(AF_o, \text{test})(K_u)}{(D)(AF_o, \text{ref})(K_u)} = \frac{UEF \text{ (test)}}{UEF \text{ (ref)}} \tag{4}$$

Where D is the dose of As and K_u is the fraction of absorbed As excreted in the urine. Based on the conceptual model above, raw data can be reduced and analyzed. The amount of As excreted in urine by each animal over each collection period (48 hr) is calculated by multiplying the urine volume by the urine concentration:

$$\text{Excreted As } (\mu\text{gAs}/48 \text{ hr}) = (\text{urine } \mu\text{gAs/L})(\text{urine volume in L}/48 \text{ hr}) \quad (5)$$

For each test material, the amount of As excreted by each animal is plotted as a function of the amount administered (μg As/48 hr), and the best-fit straight line (calculated by linear regression) through the data (μg excreted vs. μg administered) is used as the best estimate of the UEF. The relative bioavailability of As in test material is calculated as:

$$\text{RBA} = \frac{\text{UEF (test)}}{\text{UEF (sodium arsenate)}} \quad (6)$$

where sodium arsenate is used as the soluble reference form of As.

As noted above, each RBA value is calculated as the ratio of two slopes (UEFs), each of which is estimated by linear regression through a set of data points. Because of the variability in the data, uncertainty exists in the estimated slope (UEF) for each material. This uncertainty in the slope is described by the standard error of the mean (SEM) for the slope parameter.

A repeated dose protocol is employed in juvenile swine (10–20 kg) for 15 days (20). The doses of As administered are below the toxic level for swine. Groups of four or five pigs are dosed with either sodium arsenate (reference material) or test material (e.g., soil, mine tailings, or slag). The general study design is presented in Table 1. Twenty-four-hour urine samples are collected on several days during each study. Following the 15-day dosing period, pigs are euthanized and tissue samples collected. Tissue and urine samples were analyzed

Table 1 General Experimental Design for Determination of Bioavailable Arsenic in Site Soil and Solid Media Using the Juvenile Swine Model

Group	n	Treatment	Arsenate/soil arsenic	Arsenic dose (μg kg^{-1}d^{-1})
1	5	Control	Weight adjusted	0
2	5	Na$_2$AsO$_4$7H$_2$O	Weight adjusted	25
3	5	Na$_2$AsO$_4$7H$_2$O	Weight adjusted	50
4	5	Na$_2$AsO$_4$7H$_2$O	Weight adjusted	100
5	5	Site$_1$ media	Mass and weight adjusted	60
6	5	Site$_1$ media	Mass and weight adjusted	120
7	5	Site$_1$ media	Mass and weight adjusted	240

for As by atomic absorption spectrophotometry or inductively coupled plasma atomic emission spectroscopy with hydride generation (AA-HG or ICP-HG). Arsenic, introduced into the spectrophotometer as arsine gas free from interelement and matrix interferences, can accurately be quantified at $\mu g \ L^{-1}$ concentrations in urine.

Arsenic excretion in urine is a linear function of dose for all treatments. Urinary excretion also is independent of time after study day 5, thereby allowing the UEF for each treatment group to be estimated by the slope of the best-fit linear regression line through dose-response data after the fifth day of dosing.

When exposure begins, the animals are about 5–6 wk old and weigh about 9–10 kg. Animals are weighed every 3 days during the course of each study. On average, animals gain about 0.4–0.5 kg day^{-1}, and the rate of weight gain is comparable in all groups. Doses and feed are adjusted every 3 days. Each day, every animal is given an amount of standard swine chow (University Feed Mill S 2 starter ration without added antibiotics) equal to 5% of the mean body weight of all animals on study. Feed is administered in two equal portions (2.5% of the mean body weight) at 11:00 AM and 5:00 PM daily. Drinking water is provided ad libitum via self-activated watering nozzles within each cage. Water and feed samples are analyzed during each study. Based on these data, estimated intake of As in unexposed animals is <0.1 $\mu g \ kg^{-1} \ day^{-1}$ via water and ~10 $\mu g \ kg^{-1} day^{-1}$ via the diet. This routine is followed sequentially for each pig on a daily basis for 15 consecutive days for all groups.

Dose material is placed in the center of a small portion (about 5 g) of moistened feed which has doughlike consistency, and this is administered to the animals by hand. The dose level administered is based on the As content of the test material, with target doses of 25–100 $\mu g \ kg^{-1} \ day^{-1}$ for Na_2AsO_4 and approximately double this range for each test material.

Although there is variability in the data, most dose-response curves are approximately linear, with the slope of the best-fit straight line being equal to the best estimate of the UEF. This finding is consistent with results from both animals and humans, which suggest that there is no threshold for As absorption or excretion up to doses of at least 5000 $\mu g \ day^{-1}$.

The RBA results for different test materials demonstrate that As in most soils and mine wastes is not as well absorbed as soluble As. Because As in test materials is less extensively absorbed than soluble As, use of default toxicity factors for assessing human health risk will lead to an overestimate of hazard. Therefore, application of site-specific RBA estimates is expected to increase the accuracy and decrease the uncertainty in human health risk assessments for As.

B. In Vitro Gastrointestinal Method

Gastrointestinal absorption of metal contaminants is a dynamic process involving dissolution and absorption (Fig. 1). In vitro gastrointestinal methods based solely

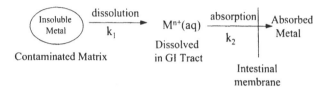

Figure 1 Diagram of metal contaminant dissolution and absorption processes that affect the metal contaminant bioavailability of soil in the gastrointestinal environment. k_1 and k_2 are rate constants for respective kinetic processes.

on measuring contaminant (metal) solubility can only be accurate estimators of contaminant bioavailability if dissolution of the contaminant matrix is the rate-limiting step in this kinetic process. The rate of dissolution of many Pb minerals is controlled by surface-controlled and not transport-controlled kinetic processes (35). The ability of the PBET method to provide accurate estimates of Pb bioavailability may be evidence that mineral dissolution is the rate-limiting step of Pb bioavailability in ingested soil (17,36).

Rodriguez et al. (34) evaluated the ability of the two in vitro gastrointestinal methods to estimate bioavailable As in contaminated soil and media. One method incorporated an adsorbent in a permeable membrane to act as a sink for dissolved As to mimic gastrointestinal absorption (IVG-AB). The other method (IVG) did not use an adsorbent and simply relied on dissolution of As from soil. In vitro results were compared with in vivo relative bioavailable As determined from dosing trials using immature swine. Fifteen contaminated soils collected from mining/smelter sites ranging from 401 to 17,500 mg As kg^{-1} were evaluated.

A schematic diagram depicting the in vitro reactor design is illustrated in Figure 2. Glass jars (1 L) were used as reactor vessels because of their wide mouth and heavy glass composition. All in vitro procedures were conducted in a water bath at body temperature (37°C), anaerobic conditions were maintained by constantly diffusing argon gas through the solution, and the pH of the in vitro solutions was monitored constantly and adjusted as necessary throughout the procedure. Constant mixing was maintained throughout the procedures (to simulate gastric mixing) by use of individual paddle stirrers at a speed of ~100 rpm. In the IVG method, As was sequentially extracted from contaminated soil with simulated gastric and intestinal solutions (Fig. 2). Simulated gastrointestinal solution pH and composition was adopted from Crews et al. and Magagelada et al. (28,31). The gastric solution was 0.15 M sodium chloride (NaCl) and 1% porcine pepsin (Sigma Chemical Co., St. Louis, MO, Cat. No. 146518). Soil (4 g) was added to 600 ml of gastric solution. An equivalent amount of the dosing vehicle (200 g of feed) was added to the gastric solution to mimic the in vivo dosing of 100 mg of soil in 5 g of doughlike feed. Gastric solution pH was

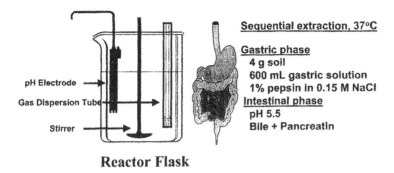

Reactor Flask

Figure 2 Gastrointestinal in vitro reactor and simulated gastrointestinal extraction solutions.

adjusted to 1.8 following the addition of soil. The gastric solution was modified to the intestinal solution by adjusting the pH to 5.5 with a saturated solution of sodium bicarbonate ($NaHCO_3$) followed by addition of porcine bile extract (2.10 g; Sigma Chemical Co., St. Louis, MO, Cat. No. B8631) and porcine pancreatin (0.21 g; Cat. No. P1500). Gastric or intestinal solution samples were collected using a syringe at the end of each phase (1 hr). Samples were centrifuged and the supernatant was filtered through a 0.45-μm filter, acidified to pH 2.0 with concentrated hydrochloric acid (HCl), and analyzed for As. The in vitro procedure with absorption simulation (IVG-AB) is the same as the IVG method de-

Table 2 Comparison of Methods Used to Measure Bioavailable Arsenic in Contaminated Soils and Solid Media[a]

| | In vitro arsenic method | | | | | | |
| | IVG | | | PBET | | | |
Samples	Gastric	Intestinal	IVG-AB intestinal	Gastric	Intestine	In vivo	Critical value[b]
All media ($n = 14$)	19.6ab	17.3ab	18.3ab	16.7b	9.78c	22.9a	6.4
Calcine ($n = 4$)	2.68b	2.55b	2.90b	0.97b	1.24b	11.2a	4.2
All media except calcine ($n = 10$)	26.4a	23.2a	24.9a	22.9a	13.2b	28.0a	6.4

[a] Values reported are mean percent relative bioavailable arsenic for that group. Mean separation statistics were generated using Duncan's multiple range test (37). Multiple comparison of mean values are made between bioavailable As method (horizontally). Mean values with the same letter designation indicate no difference between groups at $P < 0.05$.
[b] Quantitative difference between means necessary for methods significantly different at $P < 0.05$.

scribed above with the exception of adding freshly prepared amorphous Fe hydroxide gel during the intestinal phase as an adsorbent. The total amount of As measured by the IVG-AB method is the summed mass of As in the intestinal phase solution and the As dissolved from the Fe hydroxide gel.

The IVG in vitro methods (gastric, intestinal, and IVG-AB intestinal) were equivalent with the in vivo method ($P < 0.05$) across all media (Table 2). Evaluating the contaminated media separately, the noncalcinated waste materials (i.e., slags and soils) tested by the IVG methods were statistically equivalent to the in vivo method. However, the in vitro methods underestimated bioavailable As in calcine materials. However, when only the noncalcinated materials were evaluated (all media except the calcines), close agreement between the IVG and PBET gastric methods and in vivo bioavailable As was obtained. The IVG methods and

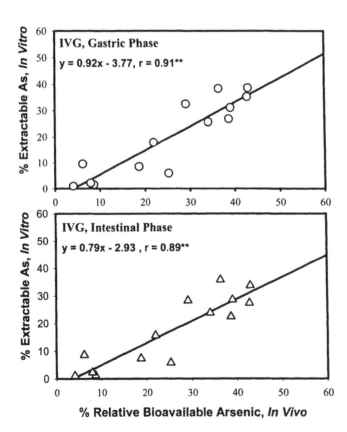

Figure 3   Linear regression of in vitro gastrointestinal (IVG) gastric or intestinal phases vs. relative bioavailable As, in vivo. $**P < 0.01$.

the PBET gastric phase were accurate estimators of bioavailable As for non-calcinated slags and soils. Few statistical differences between the IVG gastric phase, IVG intestinal phase, and IVG-AB intestinal phase were found for most groups of material. In other words, extending the in vitro method beyond the gastric phase did not improve the ability of the method to measure bioavailable As. Results suggest dissolution of As solid phases in studied contaminated media by simulated gastrointestinal solutions appears to be the rate-limiting step rather than the subsequent Fe gel adsorption step in controlling dissolved As in the IVG method. Therefore, adding a step to the IVG method that mimics gastrointestinal absorption did not increase the accuracy of the IVG method to assess the bioavailability of As in contaminated media.

Results of the IVG gastric phase were linearly correlated ($r = 0.92$) with in vivo bioavailable As ($P < 0.01$) (Fig. 3). The IVG intestinal phase was also linearly correlated with in vivo As with an r value of 0.91 ($P < 0.01$). The PBET

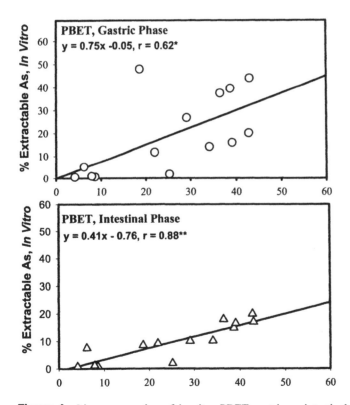

Figure 4 Linear regression of in vitro PBET gastric or intestinal phases vs. relative bioavailable As, in vivo. *$P < 0.05$; **$P < 0.01$.

gastric phase results were linearly correlated with in vivo As ($P < 0.05$), and the PBET intestinal phase was strongly correlated with an r of 0.88 ($P < 0.01$) (Fig. 4).

The IVG and PBET in vitro gastrointestinal methods were correlated with bioavailable As in contaminated media. Recent studies have shown the IVG method of Rodriguez et al. (34) can be used to estimate Pb bioavailability in contaminated soil and media (38).

IV. RISK CHARACTERIZATION OF ARSENIC IN SOILS

Risk to human receptors from exposure to contaminated soils is evaluated for carcinogenic effects by multiplying the quantified chemical daily intake by the cancer slope factor. The slope factor (SF) converts estimated daily intakes averaged over a lifetime of exposure directly to incremental risk of an individual developing cancer (9). Slope factors can be found in toxicological databases, such as the on-line USEPA Integrated Risk Information System (IRIS). For Superfund sites, the dose-response relationship is generally assumed to be linear in the low-dose portion of the multistage model dose-response curve, and risks presented will be <0.01 (9). Using these assumptions, carcinogenic risk is calculated by the following equation (9):

$$\text{Risk} = \text{CDI} \times \text{SF} \tag{7}$$

where

Risk = a unitless probability (e.g., 1×10^{-5} or 1E-05, meaning 1 in 10,000)
CDI = chronic daily intake averaged over 70 yr (mg kg^{-1}day^{-1})
  SF = cancer slope factor (mg kg^{-1}day^{-1})$^{-1}$

Noncancer effects are not expressed in terms of a probabilistic approach. Instead, the potential for noncarcinogenic effects is evaluated by comparing an exposure level over a specified period (typically, a lifetime) with a reference dose (RfD) derived for a similar exposure period (9). Reference doses may also be found in toxicological databases, such as IRIS. Noncancer risk is presented as a ratio of exposure to toxicity and is called a hazard quotient (HQ) (9). Using these assumptions, noncarcinogenic risk is calculated by the following equation (9):

$$\text{Hazard quotient} = \frac{\text{CDI}}{\text{RfD}} \tag{8}$$

where

CDI = chronic daily intake averaged over 70 yr (mg kg^{-1}day^{-1})
RfD = reference dose (mg kg^{-1}day^{-1})

Toxicity data (the SF and RfD) for As have been derived from toxicological studies performed using soluble forms of As. Therefore, site-specific bioavailability data obtained by in vitro or in vivo methods require conversion to relative bioavailability (RBA), as presented in the previous section, prior to its use in risk assessment. Bioavailability data can be used to provide more accurate exposure assessments that will result in more reasonable and site-specific risk estimates. Using RBA values, adjustments to toxicity values can be made as follows:

$$RfD_{adjusted} = \frac{RfD_{IRIS}}{RBA} \tag{9}$$

$$SF_{adjusted} = SF_{IRIS} \times RBA \tag{10}$$

Alternatively, making adjustments to the dose rather than the toxicity values is just as acceptable and mathematically correct as using the following equation:

$$CDI_{adjusted} = CDI \times RBA \tag{11}$$

V. ARSENIC BIOAVAILABILITY AND CLEANUP LEVELS

Excretion of absorbed As is mainly via the urine. Short-term exposure of humans to As can be evaluated by measurements of urinary As as a biomarker and to provide an indication of not only exposure but also, more importantly, absorption. The biological half-life of ingested inorganic As in humans is about 10 hr, and 50–80% of As absorbed is excreted in about 3 days (39).

Several studies have been conducted to evaluate the relationship between soil As and urinary As for individuals living near As-contaminated sites. However, significant variability exists in terms of correlation between soil and urinary As. For example, elevated urinary As levels have been demonstrated in two studies conducted near copper smelters. The urinary As means of children, ages 1–5, living in three towns near a copper smelter (''exposed'') were compared with those of children living in three towns without smelters (''unexposed'') (40). Soil As levels were not reported except to note concentrations up to 100 mg kg^{-1}. The urinary As geometric mean of children exposed was 18.7 µg L^{-1}, while the geometric mean of children unexposed was 5.8 µg L^{-1}. Eighteen of the 19 t values evaluated were >1 when the means were compared. Another study (41) evaluated soil As and urinary As for three communities near a copper smelter and one community without a smelter. Soil As means for the three communities near the smelter ranged from 136–398 mg kg^{-1}, while the soil As mean for the community without a smelter was 44 mg kg^{-1}. Urinary As means for the commu-

Table 3 Range of Remediation Levels for Soil Arsenic at USEPA Superfund Sites

Ref.	State	Soil remediation level (mg kg^{-1})
USEPA, 1992 Vol 1 (49)	AL	10
	CA	0.005 (subsurface soil), 2.0 (surface soil)
	MI	0.0004, 0.05,
	NE	11 (background)
	NC	2.7
	OH	0.56
	PA	0.21, 450 (unsaturated soil)
	SC	18
	TN	0.3–1.0
USEPA, 1993 (50)	CO	160
	FL	16
	ID	<100
	LA	0.016
	MA	12
	MI	1.7
	NM	0.37
	NY	19
	OK	25 (surface soil)
	SD	100
	TX	30 (off-site soil)
	UT	35
USEPA, 1992 Vol 2 (49)	CA	15 (background), 50
	CO	5–10 (background)
	FL	3.5, 5.0
	ME	60
	MA	250
	NC	48 (background), 94
	TX	300 (on-site soil)
	UT	70
	WA	200
USEPA, 1990 (51)	NJ	20
USEPA, 1996 (52)	MT	250 (residential)
USEPA, 1994 (53)	MT	500 (industrial), 1000 (recreational)
ODEQ, 1994 (54)	OK	600 (industrial)
USEPA, 1998 (55)	UT	960 (industrial)
USEPA, 1993 (56)	WA	230 (residential)

nities near the smelter ranged from 58–264 µg L^{-1} and was only 4 µg L^{-1} for the control community. Statistical significance was not reported.

Other studies reported in the literature show no correlation between soil As and urinary As for receptors living near smelters and mining sites. For example, neither soil nor indoor dust As levels were correlated with total urinary As levels (42–45) for receptors living near mining or smelter sites. In two other studies (46,47), urinary As levels for individuals living near smelter sites were not statistically different from individuals living in control areas without a smelter. Moreover, in a study reported by Gebel et al. (48), urinary As levels for receptors living in a control area with no exposure to elevated soil As were actually higher than receptors living near a smelter site in Germany. The authors concluded that dietary sources of As in seafood had a more significant impact on urinary As levels than soil As sources.

Accurately quantifying exposure is critical in developing site-specific risk information that in turn provides relevant and reasonable information to develop site cleanup goals. Table 3 demonstrates the high variability in remediation goals that have been developed for numerous Superfund sites in the United States as described in the Records of Decision. The high variability reflects both variation in site characteristics, acceptable levels of risk, and differences in risk assessment methodology.

To further understand the importance of collecting site-specific bioavailability data and their impact on cleanup levels, consider the following case study. Site-specific data collected at the Anaconda Superfund Site, Montana, was evaluated to determine the effect specific data had on cleanup levels as compared with using USEPA defaults (55). Site-specific soil As bioavailability was determined using cynomolgus monkeys and was found to be 25% for indoor dust and 18.3% for soil (19). Walker and Griffin (55) estimate that using the USEPA default of 100% bioavailability, rather than site-specific bioavailability, resulted in a 200% overestimation of As exposure. Further, they found that risk from exposure to As-contaminated soils was reduced by an order of magnitude from 1.7E-04 to 4E-05 (central tendency analysis) using site-specific bioavailability instead of the standard USEPA default.

VI. FUTURE DIRECTIONS AND NEEDS

Many data gaps and uncertainties remain when considering risk from As-contaminated soils: from disagreements with established As toxicity values to the methodology of developing soil As cleanup levels that are adequate and yet economic to be protective of human health and the environment.

There is a wide diversity of opinion regarding the methodology used to establish As toxicity values. Most disagreements arise from human epidemiological data used to establish the carcinogenic dose-response curve. Limited epidemiological data reported for very low doses show uncertainty as to whether (1) the dose-response relationship is linear thereby allowing for extrapolation from high dose data; (2) a threshold (dose level below which there is no effect) exists; or (3) As is relatively nontoxic at low levels (57). Also, only water-soluble As toxicological data has been used to establish As toxicity values (e.g., SF and RfD). As it is understood that few species of As in soils are soluble to any great extent, As health risk assumptions related to soluble forms of As in water cannot be directly applied to As in soils (18).

As seen in studies to evaluate the correlation of soil As with human As biomarkers (e.g., urinary As), As from other sources has a significant impact. There appears to be an inconsistency in federal regulation of As contributions from other sources. For example, food As is unregulated, although food is the major source of As for many Americans (57). The U.S. Food and Drug Administration regards As as a natural constituent in food and thus acceptable by default. However, the USEPA is legally required to set allowable standards for As in drinking water and in water and soil at Superfund sites (57). We have also seen that for sites where soil As and urinary As correlations have been investigated, and food As is not assumed to have a significant impact, most of these studies do not demonstrate a correlation, suggesting two possible conclusions. One possible conclusion may be that the study itself was flawed or not enough data were collected with respect to the receptors and their potential for exposure to the As-contaminated soils. An alternate conclusion may be that exposure to low doses of As in soil does not result in significant absorption, therefore indicating low bioavailability and, subsequently, low risk.

Methodologies presently used to evaluate risk to contaminated soils at Superfund sites often result in findings of unacceptable levels of risk from As found at naturally occurring background levels. For example, the soil screening level (SSL) approach derives soil chemical concentrations below which there is no concern for human health (58). Exceeding the SSL does not trigger the automatic need for cleanup; rather it presents a need to retain the chemical for further evaluation. For carcinogenic chemicals, the USEPA sets the SSL at the soil concentration associated with a 10^{-6} risk. For As, the SSL is 0.4 mg kg^{-1} (59,60), which is well below the geometric mean As concentration of 5 mg kg^{-1} for soils in many regions of the United States (61). By this approach, most soils in the United States would appear to have some level of unacceptable risk associated with them.

To properly understand environmental As health risks, the relationship of soil As to ingested As, absorbed As, and urinary As should be further evaluated. Research and development of in vivo and in vitro methodologies is needed to provide a comprehensive understanding of measuring As bioavailability. Several animal models have been used to measure bioavailable As, but few studies have compared results obtained with animal models with human subjects. The number of studies using animal models to evaluate bioavailable As are few, and different animals are often used. The in vivo As bioavailability database for soil materials is very limited and further dosing trials are needed. Perhaps an interspecies (monkey, swine) dosing study would provide information on the possibility of combining or pooling in vivo data. For example, data could be pooled into one in vivo database for soil material if similar As bioavailabilities were measured by different animal models.

The ability of in vitro gastrointestinal methods to estimate bioavailable As is promising. However, few studies have been conducted on a very limited number of contaminated soil and solid media. More studies that compare in vitro results with in vivo bioavailable As are needed over a wide range of sample matrices (soil, slag, etc). In short, the limitations (sample matrix, form or contaminant, competitive effects of other ions) of the in vitro methods are unknown. In vitro gastrointestinal methods may not provide highly accurate estimates of As for all contaminated media. The human digestive system is too complex and dynamic to simulate in the laboratory. A more reasonable approach may be by the use of in vitro methods as screening tools to provide rapid estimates of As bioavailability at a contaminated site. An estimate of variability of As bioavailability in many samples collected from heterogeneous contaminated sites can be obtained rapidly and inexpensively by in vitro methods. Bioavailable contaminant "hot spots" can readily be identified and aid in the design and cost effectiveness of remedial strategies at contaminated sites.

Use of site-specific As bioavailability data in risk assessment may dramatically lower cleanup costs. The demand for site-specific soil bioavailability data will likely increase because in vivo and in vitro bioavailability methods are inexpensive compared with site cleanup based on overly conservative soil bioavailability As levels. Site-specific As availability information will lower the degree of uncertainty in risk assessment and provide scientifically derived data to aid in the selection of appropriate remedies that are cost effective and protective of human health and the environment.

REFERENCES

1. JH Lehr. Toxicological risk assessment distortions: Part 1. Groundwater 26:2–8, 1990.
2. CC Travis, HA Hatterner-Fray. Determining an acceptable level of risk. Environ Sci Technol 22:873–876, 1987.
3. DE Burmaster, RH Harris. The magnitude of compounding conservatisms in Superfund risk assessments. Risk Anal 13:131–134, 1993.
4. S Binder, D Sokal, D Maughan. Estimating the amount of soil ingested by young children through tracer elements. Arch Environ Health 41:341–345, 1986.
5. E Calabrese, R Barnes, EJ Stanek. How much soil do young children ingest: An epidemiologic study. Reg Toxicol Pharmacol 10:123–137, 1989.

6. P Clausing, B Brunekreff, JH van Wijnen. A method for estimating soil ingestion in children. Int Arch Occup Environ Med 59:73–82, 1987.

7. S Davis, P Waller, R Buschom, J Bailou, P White. Quantitative estimates of soil ingestion in normal children between the ages of 2 and 7 years: Population-based estimates using aluminum, silicon, and titanium as soil tracer elements. Arch Environ Health 45:112–122, 1990.

8. JH van Wijnen, P Clausing, B Brunekreef. Estimated soil ingestion by children. Environ Res 51:147–162, 1990.

9. United States Environmental Protection Agency (USEPA). Risk Assessment Guidance for Superfund (RAGS). Vol. I. Human Health Evaluation Manual (Part A). Washington, DC: Office of Emergency and Remedial Response, EPA/540/1-89/002, 1989.

10. United States Environmental Protection Agency (USEPA). Test Methods for Evaluating Solid Wastes. 3rd ed. SW-846. Washington DC, 1986.

11. MB McBride. Environmental Chemistry of Soils. New York: Oxford University Press, 1994.

12. L Lindau. Emissions of arsenic in Sweden and their reduction. Environ Health Perspect 19:25–30, 1977.

13. KW Nelson. Industrial contributions of arsenic to the environment. Environ Health Perspect 19:31–34, 1977.

14. AD Shendrikar, GB Faudel. Distribution of trace metals during oil shale retorting. Environ Sci Technol 12:332–334, 1978.

15. LD Hansen, D Silberman, GL Fisher, DJ Eatough. Chemical speciation of elements in stack-collected, respirable-size, coal fly ash. Environ Sci Tech 18:181–186, 1984.

16. A Wadge, M Hutton. The leachability and chemical speciation of selected trace elements in fly ash from coal combustion and refuse incineration. Environ Pollut 48:85–99, 1987.

17. MV Ruby, R Schoof, W Brattin, M Goldade, G Post, M Harnois, DE Mosby, SW Casteel, W Berti, M Carpenter, D Edwards, D Cragin, W Chappell. Advances in evaluating oral bioavailability of inorganics in soil for use in human health risk assessment. Environ Sci Technol 33:3697–3705, 1999.

18. PA Valberg, BD Beck, TS Bowers, JL Keating, PD Bergstrom, PD Boardman. Issues in setting health-based cleanup levels for arsenic in soil. Reg Toxicol Pharmacol 26:219–229, 1997.

19. GB Freeman, RA Schoof, MV Ruby, AD Davis, JA Dill, SC Liao, CA Lapin, PD Borgstrom. Bioavailability of arsenic in soil and house dust impacted by smelter activities following oral administration in Cynomolgus monkeys. Fundam Appl Toxicol 28:215–222, 1995.

20. SW Casteel, T Evans, ME Dunsmore, CP Weis, B Lavelle, WJ Brattin, TL Hammon. Relative bioavailability of arsenic in VB170 site soils. Report prepared for USEPA Region VIII, February 2000.

21. K Groen, HAMG Vaessen, JJG Kleist, JLM deBoer, T van Ooik, A Timmerman, RF Vlug. Bioavailability of inorganic arsenic from bog ore-containing soil in the dog. Environ Health Perspect 102:182–184, 1994.

22. NT Basta, RR Rodriguez, SW Casteel. Development of chemical methods to assess

bioavailability of arsenic in contaminated media. Final Report prepared for USEPA Office of Research and Development, Washington, DC, February 2001.

23. JW Dodds. The pig model for biomedical research. Fed Proc 41:247–256, 1982.

24. CP Weis, JM LaVelle. Characteristics to consider when choosing an animal model for the study of lead bioavailability. Chem Spec Bioavail 3:113–119, 1991.

25. SW Casteel, RP Cowart, GM Henningsen, E Hoffman, WJ Brattin, MF Starost, JT Payne, SL Stockham, SV Becker, JR Turk. A swine model for determining the bioavailability of lead from contaminated media. In: ME Tumbleson, LD Schook, eds. Advances in Swine in Biomedical Research. Vol. 2. Proceedings of International Symposium on Swine in Biomedical Research. New York: Plenum Press, 1996, pp 637–646.

26. MB Rabinowitz, DC Bellinger. Soil-lead blood-lead relationship among Boston children. Bull Environ Contam Toxicol 41:791–797, 1988.

27. SW Casteel. Project Manual for Systemic Availability of Lead to Young Swine from Subchronic Administration of Lead-Contaminated Soil. Columbia, MO: Veterinary Medical Diagnostic Laboratory, University of Missouri—Columbia, 1995.

28. HM Crews, JA Burrell, DJ McWeeney. Preliminary enzymolysis studies on trace element extractability from food. J Sci Food Agric 34:997–1004, 1983.

29. ER Miller, DE Ullrey. The pig as a model for human nutrition. Ann Rev Nutr 7: 361–382, 1987.

30. R Schwartz, Z Belko, EM Wien. An in vitro system for measuring intrinsic dietary mineral exchangeability alternative to intrinsic isotopic labeling. J Nutr 112:497–504, 1982.

31. JR Magagelada, GG Lonstreth, WHJ Summerskill, VLW Go. Measurement of gastric functions during digestion of ordinary solid meals in man. Gastroenterology 70: 203–210, 1976.

32. JM Orten, OW Neuhaus. Human biochemistry. St. Louis: Mosby, 1975.

33. MV Ruby, A Davis, R Schoof, S Eberle, CM Sellstone. Estimation of lead and arsenic bioavailability using a physiologically based extraction test. Environ Sci Technol 30:422–430, 1996.

34. RR Rodriguez, NT Basta, SW Casteel, and LW Pace. An in vitro gastrointestinal method to assess bioavailable arsenic in contaminated soils and solid media. Environ Sci Technol 33:642–649, 1999.

35. DL Sparks. Environmental Soil Chemistry. New York: Academic, 1995.

36. MV Ruby, A Davis, JH Kempton, JW Drexler, PD Bergstrom. Lead bioavailability: Dissolution kinetics under simulated gastric conditions. Environ Sci Technol 26: 1242–1248, 1992.

37. SAS. SAS/STAT User's Guide, Release 6.03 ed. Cary, NC: SAS Institute Inc., 1988.

38. JL Schroder, NT Basta, DC Ward, RR Rodriguez, SW Casteel. Estimating bioavailable As, Pb, and Cd in contaminated soil by an in vitro gastrointestinal method. Proc Agron Soc Am Soil Sci Soc of Am, Minneapolis, MN, 2000, p. 235.

39. RA Goyer. Toxic effects of metals. 4th ed. In: MO Amdur, J Doull, CD Klaasser, eds. Casarett and Doull's Toxicology: The Basic Science of Poisons. New York: McGraw-Hill, 1991.

40. EL Baker, CG Hayes, PJ Landrigan, JL Handke, RT Leger, MJ Housworth, JM

Harrington. A nationwide survey of heavy metal absorption in children living near primary copper, lead, and zinc smelters. Am J Epidemiol, 106:261–273, 1977.

41. S Binder, D Forney, W Kaye, D Paschal. Arsenic exposure in children living near a former copper smelter. Bull Environ Contam Toxicol 39:114–121, 1987.

42. Butte-Silver Bow Department of Health and University of Cincinnati Department of Environmental Health. The Butte-Silver Bow County Environmental Health Lead Study. Final Report. February 1992.

43. Colorado Department of Health, Division of Disease Control and Environmental Epidemiology, University of Colorado at Denver, Center for Environmental Sciences, and Agency for Toxic Substances and Disease Registry (ATSDR). Leadville Metals Exposure Study. Denver: State of Colorado, April 1990.

44. DJ Hewitt, GC Millner, AC Nye, M Webb, RG Huss. Evaluation of residential exposure to arsenic in soil near a Superfund site. Human Ecol Risk Assess 1:323–335, 1995.

45. YH Hwang, RL Bornschein, J Grote, W Menrath, S Roda. Environmental arsenic exposure of children around a former copper smelter. Environ Res 72:72–81, 1997.

46. L Polissar, K Lowry-Coble, DA Kalman, JP Highes, G van Belle, DS Covert, TM Burbacher, D Bolgiano, NK Mottet. Pathways of human exposure to arsenic in a community surrounding a copper smelter. Environ Res 53:29–47, 1990.

47. F Diaz-Barriga, MA Santos, JJ Mejia, L Batres, L Yanez, L Carrizales, E Vera, LM Del Razo, ME Cebrian. Arsenic and cadmium exposure in children living near a smelter complex in San Luis Potosi, Mexico. Environ Res 62:242–250, 1993.

48. TW Gebel, RH Suchenwirth, C Bolten, HH Dunkelberg. Human biomonitoring of arsenic and antimony in case of an elevated geogenic exposure. Environ Health Perspect 106:33–39, 1998.

49. United States Environmental Protection Agency (USEPA). ROD Annual Report FY 1991. Volumes 1, 9355.6-05-1 and Volume 2, 9355.6-05-2. Washington, DC, 1992.

50. United States Environmental Protection Agency (USEPA). Superfund Record of Decision: Commencement Bay—Nearshore/Tideflats, WA. EPA/ROD/R10-93/062. Region 10, Seattle, WA, 1993.

51. United States Environmental Protection Agency (USEPA). Superfund Record of Decision: Meyers Property, NJ. EPA/ROD/R02-90/115, 1990.

52. United States Environmental Protection Agency (USEPA). Record of Decision: Community Soils Operable Unit, Anaconda Smelter NPL Site, Anaconda, MT. 7760-037-DD-DNJY. Helena, MT: Region VIII and Montana Department of Environmental Quality, 1996.

53. United States Environmental Protection Agency (USEPA). Record of Decision: Old Works/East Anaconda Development Area Operable Unit, Anaconda Smelter NPL Site, Anaconda, MT. EPA/ROD/R08-94/083. Region VIII, Helena, MT, 1994.

54. Oklahoma Department of Environmental Quality. Record of Decision for Operable Unit One of the National Zinc Site. Oklahoma City, 1994.

55. S Walker, S Griffin. Site-specific data confirm arsenic exposure predicted by the U.S. Environmental Protection Agency. Environ Health Perspect 106:133–139, 1998.

56. United States Environmental Protection Agency (USEPA). ROD Annual Report FY 1992, 9355.6-6-06. Washington, DC, 1993.

57. WR Chappel, BD Beck, KG Brown, R Chaney, C Richard Cothern, KJ Irgolic, DW

North, I Thornton, TA Tsongas. Inorganic arsenic: A need and an opportunity to improve risk assessment. Environ Health Perspect 105:1060–1067, 1997.

58. United States Environmental Protection Agency (USEPA). Soil Screening Guidance: Technical Background Document. EPA/540/R95/128. Washington, DC: Office of Solid Waste and Emergency Response, 1996.

59. United States Environmental Protection Agency (USEPA), Region 3. Risk Based Concentration (RBC) Tables. http://www.epa.gov/reg3hwmd/risk/rbc.pdf., 2000.

60. United States Environmental Protection Agency (USEPA), Region 9. Preliminary Remediation Goals (PRGs). http://www.epa.gov/Region9/waste/sfund/prg/index.htm., 2000.

61. JJ Connor, HT Shacklette. Background geochemistry of some rocks, soils, plants, and vegetables in the conterminous United States. U.S. Geological Survey (USGS) Professional Paper 574-F. Washington, DC, 1975.

6

Aspects of Arsenic Chemistry in Relation to Occurrence, Health, and Treatment

Laurie S. McNeill
Utah State University, Logan, Utah

Hsiao-wen Chen
University of Colorado, Boulder, Colorado

Marc Edwards
Virginia Polytechnic Institute and State University, Blacksburg, Virginia

I. INTRODUCTION

Spurred by increasing concern about the adverse health effects from exposure to low levels of arsenic in drinking water (1), the U.S. Environmental Protection Agency (USEPA) has proposed lowering the Maximum Contaminant Level (MCL) for arsenic from 50 μg/L down to 5 μg/L (2). Total compliance costs for a regulation of 5 μg/L have been estimated at \$1.47 billion per year (3).

This chapter details several experimental and field results regarding arsenic occurrence and speciation that may have important implications for health effects and for arsenic removal during water treatment.

II. ARSENIC OCCURRENCE IN THE UNITED STATES COMPARED WITH TAIWAN: IMPLICATIONS FOR HEALTH

Many of the studies on the human health effects from arsenic exposure have been conducted in Taiwan. These results are often extrapolated to estimate exposure effects on U.S. populations (4,5), although there is some debate regarding the validity of this approach in terms of genetics and assumed dose-response impacts (6). This study highlights some other differences in arsenic exposure and occurrence that may be relevant to this debate.

A. Arsenic Concentrations and Co-Occurring Constituents

Water samples were taken from three deep wells in the blackfoot disease endemic area in Taiwan. An aliquot of each sample was filtered and speciated in the field to separate particulate arsenic, arsenate [As(V)] and arsenite [As(III)] (7). Arsenic and co-occurring elements were measured with inductively coupled plasma mass spectrometry (ICP-MS) or ICP emission spectroscopy (ICP-ES). These Taiwanese well samples were compared with results from the National Arsenic Occurrence Survey (NAOS), which obtained raw water samples from 428 water supplies in all 50 states of the United States (8,9).

The average total arsenic concentration in the Taiwanese well samples was nearly 690 μg/L, which is significantly higher than the USEPA MCL of 50 μg/L. These levels were nearly six times higher than the highest U.S. surface water concentration and more than 10 times higher than the highest U.S. groundwater concentration measured in the NAOS (Fig. 1). The Taiwanese samples also con-

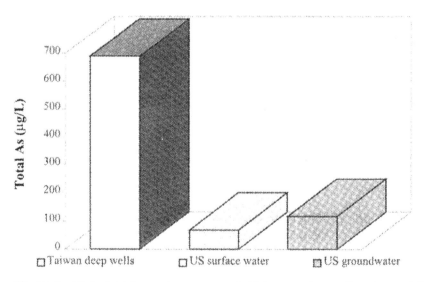

Figure 1 Highest total arsenic occurrence in Taiwanese deep well waters and U.S. drinking water sources.

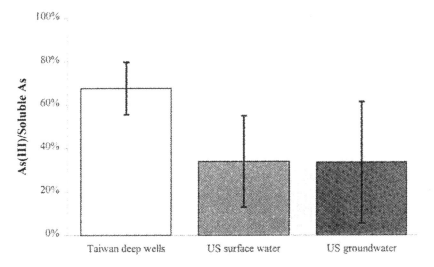

Figure 2 Average percentage As(III) occurrence in Taiwanese deep well waters and U.S. drinking water sources. The error bars indicate the standard deviation.

tained a much greater proportion of As(III) (Fig. 2). As(III) is considered the more toxic of the arsenic species, and is generally more difficult to remove during treatment.

The Taiwanese well samples also contained higher average concentrations of phosphorus and boron compared with the U.S. water samples (Fig. 3). Phosphorus has similar chemistry to arsenic and can interfere with arsenic detection and removal. The higher levels of boron may be of interest to health researchers, because boron strongly increases retention of phosphate species in the human body (10), and might be expected to impact metabolism and retention of chemically similar arsenate species as well.

B. Effect of Boiling on Arsenic Speciation

The Taiwanese almost invariably boil their drinking water. In addition to the change in temperature, during boiling the dissolved oxygen concentration goes to nearly zero, so it is possible that any As(V) present may be reduced to As(III), the more toxic species [Eq. (1)].

$$2AsO_4^{3-} \xrightarrow[\substack{\Delta \\ O_{2(aq)} \approx 0}]{\leftarrow} 2AsO_3^{3-} + O^2 \tag{1}$$

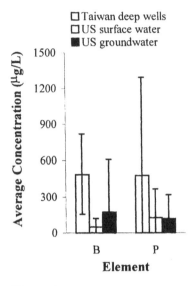

Figure 3 Elements co-occurring with arsenic in Taiwanese deep well waters and U.S. drinking water sources. The error bars represent the standard deviation.

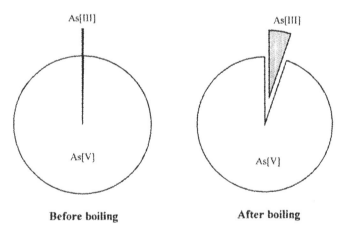

Figure 4 Arsenic speciation in Taiwanese deep well water before and after boiling.

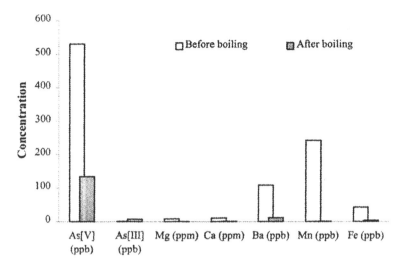

Figure 5 Concentrations of arsenic and metals in Taiwanese deep well water before and after boiling.

One of the Taiwanese deep well samples was boiled to investigate whether adverse changes in arsenic speciation could be attributed to this practice. The water was boiled for 30 min and cooled at room temperature for 1 hr. Subsequently, an aliquot was filtered and speciated.

After boiling, the concentration of As(V) dramatically decreased while the concentration of As(III) increased (Figs. 4 and 5). This trend was confirmed for a range of boiled waters, including samples from the United States, and the percentage conversion of As(V) to As(III) was as high as 83%. For the Taiwanese sample, the concentrations of soluble barium (Ba), manganese (Mn), and iron (Fe) also decreased significantly (Fig. 5), along with the overall arsenic concentration, presumably due to precipitation of solids onto the sides of the container. While the change from As(V) to As(III) species is surely an adverse consequence of boiling given current understanding of health impacts, the possible ingestion of arsenic precipitates produced during boiling has uncertain health implications, and is deserving of future research.

III. IMPLICATIONS FOR WATER TREATMENT

This study uncovered several aspects of arsenic occurrence and chemistry that may be important to drinking water utilities trying to meet the newly lowered arsenic MCL.

A. Co-Occurrence of Silica

The National Arsenic Occurrence Survey found that the average silica concentration in U.S. groundwaters was more than twice as high as in surface waters (25 vs. 15 mg/L). The majority of water utilities affected by the new arsenic MCL will be groundwater systems (11) and many of these systems will install point-of-use or point-of-entry treatment systems. Silica has been shown to interfere with arsenic sorption onto iron oxides and hydroxides (12,13), which is a common arsenic removal mechanism for both conventional and point-of-use/point-of-entry treatment systems. Thus, higher silica levels could prove problematic for utilities wishing to improve arsenic removal. Indeed, it is possible that higher concentrations of silica in the water, coupled with higher pH, could cause mobilization of arsenic from sediments and soil (14,15).

B. Variations in Arsenic Concentration

Several studies have reported daily and seasonal variations in raw water arsenic concentration. For example, one study reported diurnal variations in arsenic of 39–59 and 57–83 µg/L in a stream (16). Seasonal concentration variations have also been reported (17–20), including 11–21 (21) and 16–63 µg/L (22) in surface waters and 10–220 µg/L (23) in a groundwater. Variations in arsenic concentration and speciation will become increasingly important to water utilities trying to meet a lowered MCL.

1. Diurnal Variation

A field sampling was conducted at the Madison River in Montana to test for diurnal variations in arsenic concentration. The Madison River contains nearly 300 µg/L arsenic at its source (24), mostly due to geothermal contributions from Yellowstone National Park. In fact, a grab sample from the runoff of the Old Faithful geyser, which eventually flows into this river, contained 1.96 mg/L total arsenic. Grab samples were taken each hour from a single location in the Madison River and analyzed for arsenic and pH for one diurnal cycle. The arsenic concentration varied from 45–52 µg/L over 24 hr and correlated with changes in pH (Fig. 6). This is consistent with previous studies that showed diurnal arsenic variations were due to release and uptake of arsenic by river sediments as the pH and temperature changed (22,25–27).

2. Seasonal Variation

One possible reason for seasonal arsenic concentration variations is uptake and release of arsenic by aquatic river plants. These plants may absorb arsenic in the spring and summer (28,29), causing a very slight decrease in arsenic concentra-

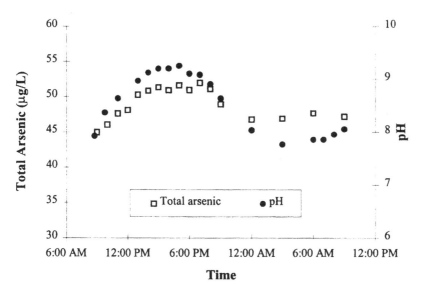

Figure 6 Diurnal variation of the total arsenic concentration and pH in the Madison River, Montana.

tions in the water. In fact, one study found that periphyton absorbed arsenic as they grew and contained as much as 2572 µg of arsenic per gram of plant tissue (28). In the fall, it might be expected that the plants could release a significant amount of arsenic, and it is also possible that this release could occur in a relatively short time span.

To examine this possibility, native macrophytes taken from the Madison River were placed in a container filled with river water. The container was then sealed and covered to exclude all light. Over a period of 1 mo, the concentration of As(III) in the water increased (Fig. 7), indicating that the plants were releasing arsenic as they decayed and could contribute to seasonal variations in influent arsenic concentration experienced by downstream water treatment facilities. Moreover, the released arsenic was in the As(III) form, which is not only more toxic to humans, but generally more difficult to remove during treatment.

One additional curious result was observed during this field sampling. During the sampling, the river water was filtered through 0.025-µm Anodisc® filters. Previous laboratory QA/QC showed that these filters retained at least 16% of As(V) that was passed through them, presumably due to sorption on the aluminum oxide filter material. The presence of other obvious anions, such as carbonate, sulfate, or NOM, did not prevent arsenic sorption by the filter in the laboratory. However, when used in the field, these filters did not remove any arsenic

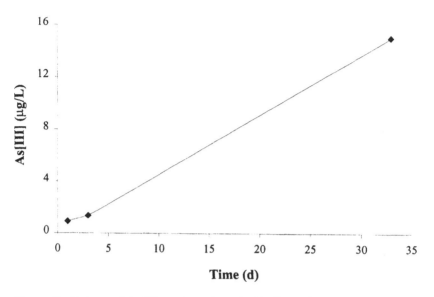

Figure 7   Release of As(III) as plants from the Madison River decayed.

from the river water, even though the arsenic was identified as As(V) by the field speciation method. Although this result has not been explained, it appears that either the As(V) in the river water was associated with some other species that prevented it from being sorbed by the filter, or some species in the water caused erroneous speciation between As(III) and As(V) during the field speciation method. Although previous research also showed that arsenic can be methylated by biological activity (30,31), it is not likely that the arsenic in the Madison River was an organic form, because organic arsenic forms would be erroneously categorized as As(III) by the ion-exchange speciation method (32). In this study, to within the limits of detection, all of the arsenic in this river was classified as As(V) by the ion-exchange speciation method, although it did not stick as strongly to the oxide filter media. This issue is deserving of additional study.

C.   Arsenic Removal at Conventional Coagulation Utilities

A simplified linear isotherm [Eq. (2)] can be used to describe arsenate removal by sorption to iron or aluminum hydroxides at coagulation water treatment plants (33):

$$\text{As sorbed } (\%) = 100\% \, \frac{K^*[Fe + Al] \text{ mM}}{1 + K^*[Fe + Al] \text{ mM}} \tag{2}$$

where

As sorbed (%) = arsenate removed from solution by metal hydroxide solids
[Fe + Al] = the amount of iron and aluminum hydroxide formed (in mM)
K = characteristic sorption constant

A general value of $K = 78$ mM^{-1} was derived that provided excellent fit to the data from 25 sampling events from full-scale utilities (33). In addition, several utilities used either jar test or full-scale data to determine a K value specific to their individual treatment conditions (34).

An interesting example of the application of the isotherm model comes from a case study of two coagulation utilities that utilize the same river as their water source, with the plant intakes only a few miles apart. One utility practices alum coagulation while the other uses ferric chloride. These utilities participated in a year-long project where once a week they reported raw water quality (pH and temperature) and treatment conditions (coagulant dose and pH) and collected influent and treated water samples for arsenic analysis.

The influent water quality was very similar for both of these utilities, although there was a great deal of seasonal variation through the year, including a variation in raw water arsenic concentrations of 4.6–14.4 µg/L (Fig. 8). The period from mid-May to early June represents a time when both utilities experienced severe flooding due to high runoff, and treatment effectiveness was "lost." The high arsenic concentrations in this river are again due to geothermal contributions from Yellowstone National Park.

Although they were treating similar water, the arsenic removal performance varied widely between the two plants (Fig. 8). In general, the utility using iron coagulant achieved better removal than the utility using alum. Because only total influent and effluent arsenic concentrations were measured at these facilities, these data were not ideal to predict soluble arsenic as a function of coagulant dose [Eq. (2)] although the model can still be applied if it is assumed that all soluble Al or Fe added to water formed an insoluble precipitate. Moreover, the model is only for As(V) sorption, so the fit of the data could be affected by the presence of any As(III) in the raw water. In spite of these limitations, the long-term data from each utility are usually predicted reasonably by the simplified model (Fig. 9), with a K value of 28 mM^{-1} for the alum plant and 364 mM^{-1} for the iron plant. The open points represent data from the high runoff/flooding period, when influent arsenic was diluted and the influent turbidities were quite high, which were not included in the model fit. The data for the alum utility would undoubtedly show a better fit to the model if the actual amount of aluminum hydroxide formed was quantified, a factor shown to be critical at alum utilities (33).

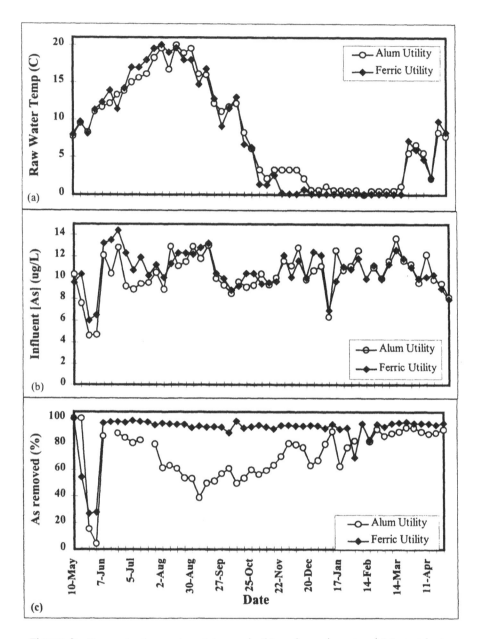

Figure 8 Raw water temperature (a), arsenic (b), and arsenic removal (c) over 1 yr.

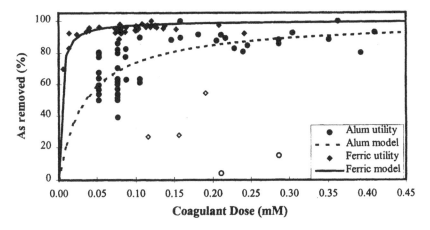

Figure 9 Modeling arsenic removal data using simplified isotherm.

IV. SUMMARY

Arsenic occurrence in the United States vs. Taiwan: implications for health effects:

- Waters sampled from the blackfoot disease endemic area in Taiwan had much higher average arsenic concentrations than the U.S. water samples. Taiwan waters also contained a higher percentage of As(III).
- The Taiwanese water samples had higher average concentrations of boron and phosphorus.
- Boiling water increased the As(III) concentration and decreased concentration of As(V), barium, manganese, and iron.

Implications for water treatment:

- U.S. groundwaters contained higher levels of silica compared with U.S. surface waters, which may interfere with As removal during water treatment.
- Significant diurnal and seasonal variations in arsenic concentration were observed in two U.S. surface water supplies.
- Native river plants are known to uptake arsenic when they grow, and they were shown to release it as As(III) when they decayed.
- A simplified sorption isotherm may be useful to utilities for predicting arsenic removal throughout the year as water quality changes, although it is severely limited during some time periods when treatment is especially difficult.

REFERENCES

1. National Research Council. Arsenic in Drinking Water. Washington, DC: National Academy Press, 1999.
2. United States Environmental Protection Agency Proposed Arsenic Rule. Fed Regist 65:38887–38983, 2000.
3. MM Frey, J Chwirka, R Narasimhan, S Kommineni, Z Chowdhury. Cost Implications of a Lower Arsenic MCL. Denver, CO: American Water Works Association Research Foundation, 2000.
4. AH Smith, C Hopenhayn-Rich, MN Bates, HM Goeden, I Hertz-Picciotto, HM Duggan, R Wood, MJ Kosnett, MT Smith. Cancer risks from arsenic in drinking water. Environ Health Perspect 97:259–267, 1992.
5. KH Morales, L Ryan, T-L Kuo, M-M Wu, C-J Chen. Risk of internal cancers from arsenic in drinking water. Environ Health Perspect 108:655–661, 2000.
6. P Mushak, AF Crocetti. Risk and revisionism in arsenic cancer risk. Environ Health Perspect 103:684–689, 1995.
7. M Edwards, S Patel, LS McNeill, H-W Chen, MM Frey, AD Eaton, RC Antweiler, H Taylor. Considerations in As analysis and speciation. J Am Water Works Assoc 90:103–113, 1998.
8. MM Frey, M Edwards, G Amy, D Owen, Z Chowdhury. National Compliance Assessment and Costs for the Regulation of Arsenic in Drinking Water. Denver, CO: American Water Works Association, 1996.
9. MM Frey, M Edwards. Surveying arsenic occurrence in US drinking water. J Am Water Works Assoc 89:105–117, 1997.
10. SL Meacham, IJ Taper, SL Volpe. Effect of boron supplementation on blood and urinary calcium, magnesium, and phosphorus, and urinary boron in athletic and sedentary women. Am J Clin Nutr 61:341–345, 1995.
11. United States Environmental Protection Agency Proposed Arsenic in Drinking Water Rule Regulatory Impact Analysis. Washington DC: USEPA, EPA 815-R-00-013, 2000.
12. CC Davis, H-W Chen, M Edwards. Re-examining the role of silica sorption in iron hydroxide surface chemistry. Environ Sci Technol, submitted for publication.
13. CC Davis, WR Knocke, M Edwards. Implications of aqueous silica sorption to iron hydroxide: Mobilization of iron colloids and interference with sorption of arsenate and humic substances. Environ Sci Technol, in press, 2001.
14. JE Davis, M Edwards. Role of sulfide and silica in mobilization of sorbed arsenate from iron and aluminum hydroxides. Proceedings of American Geophysical Union Annual Meeting, San Francisco, CA, 2000.
15. C Davis, H-W Chen, JE Davis, M Edwards. The impact of silica and other competing ions on arsenic removal during water treatment. Proceedings of 31st International Geological Congress, Pre-Congress Workshop on As in Groundwater or Sedimentary Aquifers, Rio de Janeiro, Brazil, 2000.
16. CC Fuller, JA Davis, GW Zellwegger, KE Goddard. Coupled chemical, biological and physical processes in Whitewood Creek, South Dakota: Evaluation of the controls of dissolved arsenic. Crit Rev Environ Control 21:235–246, 1991.

17. M Cebrian, A Albores, M Aguilar, E Blakely. Chronic Arsenic Poisoning in the North of Mexico. Hum Toxicol 2:121–133, 1983.

18. J Nadakavukaren, R Ingermann, G Jeddeloh, S Falkowski. Seasonal variation of arsenic concentration in well water in Lane County, Oregon. Bull Environ Contam Toxicol 33:264–269, 1984.

19. GE Mallard, SE Ragone. U.S. Geological Survey Toxic Substances Hydrology Program, Proceedings of the Technical Meeting. USGS, Water Resources Investigations Report 88-4220, 1988.

20. F Frost, D Frank, K Pierson, L Woodruff, B Raasina, R Davis, J Davies. A seasonal study of arsenic in groundwater, Snohomish County, Washington, USA. Environ Geochem Health 15:209–213, 1993.

21. LA Baker, T Quereshi, L Farnsworth. Sources and fate of arsenic in the Verde and Salt Rivers, Arizona. Proceedings of WEFTEC'94, Chicago, IL, 1994, pp 239–248.

22. SJ McLaren, ND Kim. Evidence for a seasonal fluctuation of arsenic in New Zealand's longest river and the effect of treatment on concentrations in drinking water. Environ Pollut 90:67–73, 1995.

23. JW Gibbs, LP Scanlan. Arsenic removal in the 1990s: Full scale experience from Park City, Utah. Proceedings of AWWA WQTC, San Francisco, CA, 1994.

24. C Flaherty. Montana's Water: The Good, the Bad and the Beautiful. Montana State University Communications Services. 6/5/1996, 1996.

25. CC Fuller, JA Davis. Influence of coupling of sorption and photosynthetic processes on trace element cycles in natural waters. Nature 340:52–54, 1989.

26. WM Mok, MW Chien. Distribution and mobilization of arsenic and antimony species in the Coeur D'Alene River, Idaho. Environ Sci Technol 24:102–108, 1990.

27. WM Mok, CM Wai. Mobilization of arsenic in contaminated river waters. In: JO Nriagu, ed. Arsenic in the Environment. Part I: Cycling and Characterization. New York: J Wiley, 1994, pp 99–117.

28. JS Kuwabara, CCY Chang, SP Pasilis. Periphyton Effects on Arsenic Transport in Whitewood Creek, South Dakota. Reston, VA:U.S. Geological Survey, Water Resources Investigations Report 88-4220, 1988.

29. Y Shibata, M Sekiguchi, A Otsuki, M Morita. Arsenic compounds in zoo- and phytoplankton of marine origin. Appl Organomet Chem 10:713–719, 1996.

30. SJ Santosa, S Wada, S Tanaka. Distribution and cycle of arsenic compounds in the ocean. Appl Organomet Chem 8:273–283, 1994.

31. GE Millward, HJ Kitts, SDW Comber, L Ebdon, AG Howard. Methylated arsenic in the southern North Sea. Estuar Coast Shelf Sci 43:1–18, 1996.

32. GP Miller, DI Norman, PL Frisch. A comment on arsenic species separation using ion exchange. Water Res 34:1397–1400, 2000.

33. LS McNeill, M Edwards. Predicting arsenate removal during metal hydroxide precipitation. J Am Water Works Assoc 89:75–86, 1997.

34. LS McNeill, M Edwards. Arsenic removal via softening and conventional treatment. Proceedings of AWWA Inorganic Contaminants Workshop, San Antonio, TX, 1998.

7

Biogeochemical Controls on Arsenic Occurrence and Mobility in Water Supplies

Janet G. Hering and Penelope E. Kneebone*
California Institute of Technology, Pasadena, California

I. INTRODUCTION

The occurrence of arsenic (As) in potable water supplies is increasingly a cause of concern worldwide. Significant health effects (including cancer of the skin and internal organs) have been linked to chronic exposure to arsenic in drinking water (1). A large-scale shift in water resource allocation from surface water to groundwater in West Bengal, India, and Bangladesh and the inadvertent exposure of local populations to groundwater containing arsenic at concentrations of several hundred µg/L has resulted in an extreme example of environmental health effects (2,3). Although arsenic concentrations in potable water in the United States are low compared with those regions (e.g., Taiwan, West Bengal, Bangladesh, Chile, etc.) where health effects have been studied, the U.S. Environmental Protection Agency (USEPA) is, in response to these health concerns, re-evaluating the current enforceable standard for arsenic in drinking water of 50 µg/L (0.67 µM) (4). A standard of 10 µg/L (0.13 µM) was promulgated in January 2001 (5); the effective date of the rule has been delayed (pending reconsideration of a standard between 3 and 20 µg/L) but the original compliance date in 2006 remains in force (4).

* *Current affiliation*: ENVIRON International Corporation, Emeryville, California

Since arsenic is characterized as a "minor" constituent of surface water and groundwater (i.e., an element having a (naturally occurring) median concentration between 1 μg/L and 1 mg/L) (6), a standard in the range of 3 to 20 μg/L will pose a substantial challenge to water suppliers. The USEPA estimates that the source waters for 5.4% of groundwater-based community water systems (CWSs) and 0.8% of surface-water–based CWSs would require treatment to comply with an arsenic standard of 10 μg/L (7). Since, in most cases, the least expensive avenue for compliance would involve modification of existing treatment systems to accomplish arsenic removal (8), a standard at this level would be particularly burdensome for groundwater-based CWSs without existing treatment, that is, for 50% of very small systems (serving populations of 25–100) and approximately 25% of small systems (serving populations of 101–10,000) (7). Although the occurrence of arsenic in the source water will determine whether treatment would be required to meet a new standard, other water quality parameters will also influence the feasibility of various treatment technologies. Both the oxidation state of arsenic and the concentrations of co-occurring source water constituents will affect the efficiency of arsenic removal (see Chaps. 6 and 9).

From the perspective of water supply, it is the occurrence and mobility of naturally occurring arsenic in groundwater and (to a lesser extent) fresh, surface water that are chiefly of concern. Although substantial contamination of surface water and groundwater has resulted from mining and smelting, manufacture of arsenical pesticides, and other human activities, such contamination rarely affects potable water supplies in the United States (9). This is not, however, necessarily the case in developing countries; in Thailand, for example, elevated concentrations of arsenic in drinking water are linked to mining activities (10).

The concentration of naturally occurring arsenic in groundwater and surface water can vary quite widely with the most elevated concentrations on the order of tens of mg/L (11). This variability can be attributed both to the variations in the arsenic content of the soils, rocks, and aquifer minerals that serve as the proximate source of arsenic in groundwater and surface water and to the variable extent of processes that release or sequester arsenic from or to solid phases. Numerous reviews of arsenic geochemistry are available (9,11–16) and this chapter seeks to complement them by focusing on the following questions:

- How are surface water and groundwater different or similar with regard to arsenic occurrence, speciation, and cycling?
- How does the proximate source of arsenic influence dissolved arsenic concentrations?
- Are arsenic occurrence, speciation, and mobility dominated by equilibrium or kinetic controls?

II. COMPARISON OF GROUNDWATER AND SURFACE WATER SUPPLIES

The concentration of arsenic in precipitation is quite low, averaging 0.019 µg/L for precipitation originating from unpolluted ocean air masses and 0.46 µg/L for that from terrestrial air masses (17). As rain or snowmelt infiltrates through soils, its composition is altered by the chemical weathering of soils and aquifer minerals. The differences in the chemical composition of surface water and groundwater reflect the progress of chemical weathering under varying conditions and for varying durations that are characteristic of these types of natural waters. For a given potable water source (i.e., a surface water intake or groundwater well), the quality of the water supply derives both from the characteristics of the inputs to the river, lake, reservoir, or aquifer and from the evolution of chemical composition during the residence time of water in that system.

A. Discharge to Surface Waters and Groundwater Recharge

Inputs to surface waters include baseflow (i.e., groundwater inputs), surface flow, and interflow occurring during (or soon after) precipitation (or snowmelt) events, and direct precipitation (18). The chemical composition of the interflow component reflects relatively short-term interactions with the soil matrix; surface flows have even less contact with soils and thus correspondingly less alteration of their chemical composition. Inputs to groundwaters include recharge from surface waters and percolation of water through the unsaturated (or vadose) zone of soils in recharge areas. Significant changes in the composition of soil moisture (such as several-fold increases in the total dissolved solids (TDS) concentration relative to that of average precipitation) occur within meters of the surface due both to evapotranspirative concentration and chemical weathering (6). The accumulation of trace constituents is likely to parallel, to some extent, that of the more well-studied major ions in soil waters.

For both surface waters and groundwaters, input of or mixing with geothermal waters can have dramatic impacts on both major ion and trace element composition because of the enrichment of geothermal waters in many chemical constituents. In particular, geothermal waters are often enriched in arsenic (11) and contribute significant amounts of arsenic to surface waters, for example, in the eastern Sierra Nevada in California (19,20).

B. Evolution of Water Quality

The chemical compositions (including arsenic concentrations and speciation) of surface waters and groundwaters evolve under varying conditions during residence times that range from means of 2 wk for rivers, 10 yr for lakes, and 1700 yr for groundwater. Actual residence times for a given system can, however, differ substantially from the mean; groundwater ages, in particular, can vary from less than decades for shallow aquifers to tens of thousands of years (6). Because of their longer residence times and isolation from the earth's surface, deep groundwaters are subject to only minor seasonal effects, which can be quite pronounced for surface waters and even shallow aquifers. The extent of mixing and heterogeneity can also be quite different in surface water and groundwater systems. Surface waters are subject to large-scale mixing (e.g., by wind) though stratification may develop seasonally and heterogeneities arise at sediment–water interfaces (21). Although mechanical mixing during fluid advection and molecular diffusion does result in some hydrodynamic dispersion of solute concentrations in groundwaters (22), signals of local heterogeneities in the porous medium can be preserved to a significant extent in the contacting groundwater.

Surface waters are exposed to sunlight and the atmosphere, which support the growth of photosynthetic organisms. ''Nutrient-like'' behavior of arsenic (i.e., surface depletion in its total concentration) has been observed in surface seawater (23) and in some lakes (24), though, in other cases, surface depletions were not observed (25–28). The occurrence of arsenic in the (thermodynamically unstable) +III oxidation state and of monomethylated and dimethylated arsenic species in oxic surface waters has been attributed to the activity of phytoplankton (25–27,29).

Exchange with the atmosphere can be limited, thus allowing depletion of dissolved oxygen, even in some surface waters (e.g., hypolimnetic water) as well as in groundwater. Changes in redox potential then occur as oxygen and other terminal electron acceptors (such as nitrate, manganese oxides, etc.) are consumed by both microbial and abiotic reactions (see Sec. IV.A). Microbial processes in groundwater systems can have a significant impact on the distribution of arsenic between solid and aqueous phases (see Sec. IV and Chaps. 8 and 11). The uptake of arsenic by the biota is, however, less likely to affect dissolved arsenic concentrations in groundwater than in surface waters where uptake by biota provides a mechanism for transport of arsenic from the water column to the sediments.

C. Comparison of Arsenic Occurrence in Surface Water and Groundwater

Several surveys have been performed in the United States to determine the occurrence of arsenic in potable water supplies and the likelihood of exceedances of drinking water standards set at various levels (7,9,11,30–32). Of these, Frey and Edwards (31) provided data specifically for source waters (rather than finished waters, whose compositions reflect effects of water treatment) and obtained samples representative of various regions of the United States. As shown in Figure 1, distinct regional patterns are observed in arsenic occurrence with the lowest concentrations found in the Mid-Atlantic and Southeast regions. In all regions except North Central, arsenic concentrations above 5 μg/L occur more frequently in groundwater than in surface waters. Only one region (North Central) has any occurrence of arsenic above 20 μg/L in surface waters. The data set presented by Frey and Edwards (31), however, represents only about 500 of the 54,432 CWSs in the United States. Furthermore, since potable water supplies at the time of the survey were subject to the arsenic standard of 50 μg/L and substituting a lower arsenic source water would have been an attractive alternative to treatment, this survey of CWSs may have excluded source waters with arsenic in excess of 50 μg/L. As an extreme example, evaporative concentration has been shown to result in elevated concentrations of arsenic (up to mg/L levels) in groundwater in the arid southwestern United States (33) and in terminal alkaline lakes (e.g., 17 mg/L arsenic in Mono Lake) (25). In these cases, however, the high levels of TDS preclude these waters from use as potable water supplies.

D. Surface Water and Groundwater as a Coupled System

As mentioned above, groundwater can be recharged from and also discharge into surface waters and thus surface water and groundwater cannot be considered independently. In Montana, arsenic concentrations in groundwater of 16 to 176 μg/L have been attributed in part to direct aquifer recharge by Madison River water with concentrations of geothermally derived arsenic as high as 100 μg/L (as well as to leaching of arsenic from aquifer sediments) (34). Groundwater inputs can contribute substantially to streamflow; in a U.S. survey, groundwater contributions ranged from 14–90%, with a median value of 55% (35). Elevated concentrations of chemical constituents in groundwater can influence the composition of surface waters though these impacts may be modified by processes occurring in the hyporheic zone (36).

Artificial recharge of aquifers with surface water, which offers a mechanism for water storage in the arid Southwest, also raises issues of water quality. For example, arsenic occurs in groundwater in the Kern Fan element of the Kern water bank (a proposed groundwater recharge project) at concentrations up to 211 μg/L (37). The extent to which the composition of recharged surface water will evolve to resemble the ambient groundwater during storage is not known.

III. PROXIMATE SOURCES OF ARSENIC

In both surface waters and groundwaters, arsenic in the aqueous (mobile) phase is derived from the arsenic contained in soils, rocks, and aquifer minerals. Arsenic

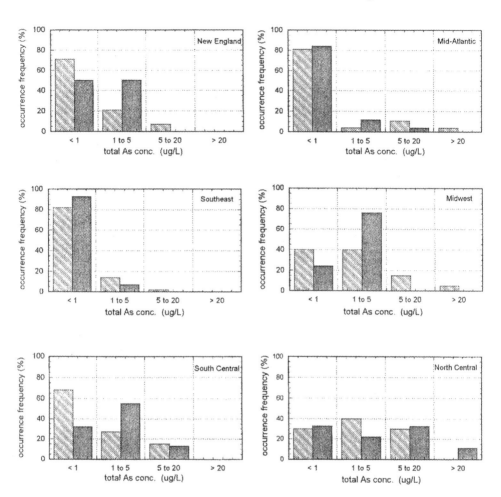

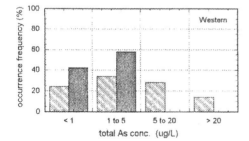

is a widely distributed constituent of geological materials, with an average crustal abundance of 1.8 mg/kg. Relatively low concentrations of arsenic (<0.1 to not more than 20 mg/kg) occur in igneous rocks, schists, carbonates, and sandstones. Elevated concentrations of arsenic (ranging up to hundreds of mg/kg) occur in basalts, slate, marine shale, and phosphorites (11). The arsenic content of soils ranges from <0.1 to hundreds of mg/kg; elevated concentrations are found in soils derived from shales and in soils in mineralized areas or areas of volcanic or geothermal activity (11,38,39).

Elevated arsenic concentrations in soils or aquifer minerals are not, however, required to support dissolved arsenic concentrations in the range of a few to hundreds of µg/L. For a solid with an arsenic content (M) of 1.8 mg/kg (i.e., equal to the crustal abundance), the fraction (f) of arsenic that would need to be released to support a given dissolved arsenic concentration in the contacting porewater ($[As]_{diss}$ in µg/L) can be calculated from the equation:

$$f = \frac{[As]_{diss}(\phi)}{(M)(D)(1 - \phi)(1000)} \tag{1}$$

where D is the specific gravity of the solid (in g/cm^3) and ϕ is porosity (11). For typical values of specific gravity (D = 2.6 g/cm^3 for plagioclase) and porosity (ϕ = 0.3) and M = 1.8 mg/kg, it requires the solubilization of only 0.09% of the arsenic contained in the solid to support a dissolved arsenic concentration of 10 µg/L.

Despite this, a substantial proportion of U.S. groundwaters [42% in the Frey and Edwards (31) study] have arsenic concentrations below 0.5 µg/L. In addition, the variability of arsenic concentrations in groundwaters suggests that arsenic is differentially mobilized from various source materials under different environmental conditions. Thus the speciation of arsenic in the solid phase and its susceptibility to leaching by contacting porewater as well as mechanisms for the sequestration of arsenic from the aqueous to solid phases must be considered.

Arsenic in soils, sediments, and rocks subject to weathering is present largely in association either with sulfides or iron oxyhydroxides. In mineralized areas, arsenic is associated with sulfide minerals (i.e., as arsenopyrite and as a trace element in pyrite and other minerals) and with secondary arsenate and arse-

---

Figure 1 Arsenic occurrence in U.S. source waters in groundwater (diagonal hashed bars) and surface water (shaded bars). Regions: New England (ME, NH, VT, MA, RI, CT, NY, NJ), Mid-Atlantic (PA, MD, DE, VA, DC, WV, KY, NC, SC), Southeast (TN, GA, FL, AL, MS), Midwest (OH, IN, MI, IL, WI, MN, IA), South Central (MO, AR, LA, TX, OK, NM, KS, NE, CO), North Central (MT, ND, SD, WY), Western (AZ, CA, NV, ID, OR, WA, AK, HA). (Data from Ref. 31.)

nate-sulfate minerals (e.g., bukovskyite, kankite, pitticite, and scorodite) formed by weathering processes (40). Authigenic precipitation of arsenic in sulfide solids can occur as a result of bacterial sulfate and/or arsenate reduction (41,42). In contrast, elevated arsenic contents in alluvial sediments has been attributed to its deposition in association with iron oxides (43) and arsenic and iron contents are also correlated in phosphate rock (44). The enrichment of (easily leachable) arsenic in pegmatites has been related to elevated concentrations of arsenic in groundwater from drilled bedrock wells in New Hampshire (45). Clearly, these different associations of arsenic in solid phases will confer differing susceptibility to mobilization.

IV. EQUILIBRIUM VERSUS KINETIC CONTROLS OF ARSENIC REMOBILIZATION AND SEQUESTRATION

Dissolved arsenic concentrations can be limited either by the solubility of minerals containing arsenic as a constituent element (or in solid solution) or by sorption of arsenic onto various mineral phases. For both the precipitation–dissolution of arsenic-containing minerals and sorption–desorption of arsenic onto solid phases, equilibrium calculations can indicate the level of control over dissolved arsenic concentrations that can be exerted by these processes. However, neither of these types of reactions is necessarily at equilibrium in natural waters. The kinetics of these reactions can be very sensitive to a variety of environmental parameters and to the level of microbial activity. In particular, a pronounced effect of the prevailing redox conditions is expected because potentially important sorbents (e.g., Fe(III) oxyhydroxides) are unstable under reducing conditions and because of the differing solubilities of As(V) and As(III) solids.

A. Redox Controls

In aquatic environments with restricted contact with the atmosphere, sub-oxic and anoxic conditions are generated by the biological oxidation of organic carbon and concomitant reduction of oxygen, nitrate, manganese and iron oxides, and sulfate. Although the concentrations of redox-active species formally define the redox potential (pe or E_H) of the system, it is often the case that redox potentials calculated from different redox couples do not agree with each other or with the measured redox potentials (23,46) unless extraordinary care is exercised (47). These discrepancies indicate that concentrations of redox-active species reflect the kinetics of operative redox reactions in the system (which establish steady-state concentrations) rather than redox equilibrium. Modeling of the distribution of porewater and sediment constituents in coastal marine sediments suggests that the rates of microbial and abiotic redox reactions can be comparable (48).

Arsenic occurs in aquatic environments in multiple oxidation states, most commonly +III and +V (12,13,16). Although As(V) is thermodynamically favored in oxic waters and As(III) under reducing conditions, coexistence of As(III) and As(V) has been observed in both oxic surface waters and anoxic groundwaters and hypolimnetic waters (24,25,49). Such coexistence has been attributed to biologically mediated interconversions of As(III) and As(V) and to the inertness of arsenic species toward reaction with chemical oxidants, particularly oxygen (50,51), and reductants, such as sulfide at circumneutral pH values (52). Microorganisms can directly mediate reduction of As(V) to As(III) (41,42,53,54) and also oxidation of As(III) to As(V) (55–58). However, abiotic oxidation of As(III) by manganese oxides can also be rapid (59–64). Oxidation of As(III) in lake sediments has been attributed to reactions with manganese oxides (65) and the heterogeneous oxidation of As(III) by soils has been shown to be related to the level of oxalate-extractable manganese (66). Abiotic reduction of As(V) by sulfide may be important at low pH values where sulfide is present as H_2S (52).

Although coexistence of As(III) and As(V) in solution (inconsistent with the calculated or measured redox potential) is frequently observed, speciation of arsenic in the solid phase is closely coupled to the redox status of the sediments (67–71). Sequential extraction of arsenic-contaminated sediments has shown that the solid-phase speciation of arsenic shifts from the iron and manganese oxyhydroxide fraction in the oxidized region to the sulfide fraction in the reduced zone (69). In soils amended with arsenate and subjected to flooding (under N_2) and subsequent aeration, arsenopyrite was identified in the flooded soils but not after aeration. During aeration, arsenic appeared to be sorbed and/or coprecipitated with iron oxyhydroxides (72). Release of arsenic from sediments has been found to be related to reductive dissolution of Fe(III) oxides and oxidative dissolution of sulfide minerals as well as to the redox cycling of arsenic (67–69,71,73–77). Correlation of sediment profiles also suggests some control of arsenic mobility by manganese, possibly due to redox reactions of manganese oxides with As(III) (74,78). Redox conditions (whether subject to equilibrium or kinetic controls) will thus profoundly affect the distribution of arsenic between the aqueous and solid phases.

B. Dissolution–Precipitation

Limits on the dissolved concentrations of arsenic due to precipitation of authigenic minerals (with arsenic as a constituent ion) are unlikely in oxic systems. Most arsenate minerals are too soluble to precipitate under environmental conditions. Scorodite, $FeAsO_4$(s), is observed as a weathering product of arsenopyrite and is likely to have formed under conditions of elevated concentrations of dissolved As(V) and Fe(III) and at low pH (40). At circumneutral pH, scorodite is unstable with respect to transformation to Fe(III) oxyhydroxide phases, which may then serve as sorbents for As(V) solubilized by scorodite dissolution (see Sec. IV.C). Calcium arsenates have been identified at a contaminated site where acid mine drainage water with very elevated As(V) concentrations (tens to hundreds of mg/L) interact with a limestone substratum (79). Although Wagemann (80) calculated that barium arsenate, $Ba_3(AsO_4)_2(s)$, was oversaturated in natural waters, redetermination of the solubility product for this solid by Essington (81) demonstrated that these calculations were based on a grossly incorrect constant. Based on a reinterpretation of previously reported solubility data, Essington (81) also reported a solubility product for $BaHAsO_4(s)$, $K_{sp} = 10^{-26.64}$. Figure 2 shows that the concentration of total, dissolved As(V) in equilibrium with both Ba-$HAsO_4(s)$ and barite, $BaSO_4(s)$, at fixed pH varies inversely with the sulfate concentration. Thus water with, for example, a sulfate concentration of 50 mg/L would be oversaturated with respect to $BaHAsO_4(s)$ (at pH 8.3) if the total, dissolved As(V) concentration exceeded about 30 µg/L. However, the kinetics of solid precipitation at a low degree of oversaturation can be quite slow and barium arsenates have not been identified in natural environments.

Under reducing conditions, As(III) may precipitate with sulfide as orpiment, $As_2S_3(s)$, realgar, $As_4S_4(s)$, or arsenopyrite, FeAsS(s). However, even if such solids form, the dissolved As(III) concentrations in equilibrium with them are not necessarily low. A study of orpiment solubility demonstrated that, in

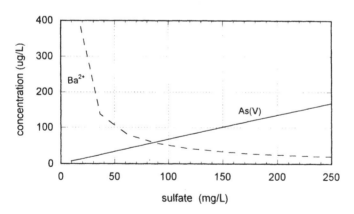

Figure 2   Dissolved concentrations of barium and As(V) (in µg/L) in equilibrium with the solids $BaHAsO_4(s)$ and $BaSO_4(s)$ calculated as a function of the dissolved sulfate concentration (mg/L) using MINEQL$^+$ (133). Calculations were performed at pH 8.3 with ionic strength fixed at 0.01 M and major ion composition in mg/L (mM in parentheses): SiO_2 9.3 (0.16), Ca 72 (1.8), Mg 29 (1.2), Na 93 (4.1), K 4.5 (0.12), HCO_3 154 (2.5), Cl 87 (2.4), F 0.29 (0.015), B 0.14 (0.013). The solubility constants for $BaHAsO_4(s)$ and $Ba_3(AsO_4)_2(s)$ were taken from Essington(81).

the absence of excess sulfide, the concentrations of dissolved As(III) (present as H_3AsO_3) in equilibrium with orpiment were about 750 µg/L over a pH range of about 2–6 and were predicted to increase with pH above pH 6. In the presence of excess sulfide, solubility was lower at pH <3 but substantially higher at circumneutral pH (in the tens of mg/L); this increased solubility was attributed to the formation of the dissolved complex $H_2As_3S_6^-$ (82). The presence of this species in solutions saturated with orpiment was supported by analysis of solubility data by Helz et al. (83) who also provided spectroscopic evidence for the monomeric dissolved species $H_2AsS_2^-$ and $HAsS_2^{2-}$ in undersaturated alkaline solutions. Formation of carbonato complexes of As(III) has also been proposed (84). Dissolved As(III) concentrations in equilibrium with mixed metal sulfides can be lower than those controlled by orpiment solubility as demonstrated for the assemblage CuS(covellite)-$Cu_{1.8}S$(digenite)-Cu_3AsS_4 (85). Even in this case, however, dissolved As(III) concentrations were between 12 and 160 µg/L for pH values of 8–8.5 and varying concentrations of dissolved sulfide.

Such sulfide phases are unstable in the presence of dissolved oxygen. In eastern Wisconsin, release of arsenic from a sulfide-bearing secondary cement horizon has been attributed to introduction of oxygen in water supply wells. The intersection of static water levels in residential wells with this horizon was found to correlate strongly with concentrations of arsenic in groundwater, which were as high as 12 mg/L (86). An x-ray photoelectron spectroscopic study of arsenopyrite oxidation showed that arsenic diffuses from the interior of the mineral to the surface during oxidation, suggesting that arsenic would be preferentially leached from the solid (87).

Under either reducing or oxidizing conditions, the solubilization of arsenic from sulfide phases can be subject to kinetic limitations. Mass transfer constraints, particularly in porous media, can result in localized saturation conditions near the surface of the solid. For oxidative dissolution, depletion of dissolved oxygen may limit dissolution kinetics. Microorganisms may also play a role in catalyzing such oxidative dissolution as has been demonstrated for pyrite oxidation (88) and thus dissolution rates may reflect the level of microbial activity (which may be subject, for example, to nutrient limitation). Thus, although equilibrium calculations indicate solubility constraints on dissolved arsenic concentrations, actual concentrations may be lower than the predicted equilibrium values due to slow dissolution kinetics or greater due to slow precipitation kinetics.

C. Sorption–Desorption

Unlike authigenic mineral precipitation, sorption is clearly an important mechanism controlling dissolved As(V) concentrations. Arsenic adsorption onto oxide minerals (specifically iron oxides and hydroxides) has been invoked in interpreting the occurrence and mobility of arsenic in lake sediments (68,69,73,74,89,90), observed diel cycles of dissolved arsenic concentrations in streamwater (91), scavenging of arsenic from seawater at hydrothermal vents (92,93), and the association of iron and arsenic in alluvial sediments (43,94–96). Arsenate sorbs strongly on many mineral phases, particularly Fe(III) and Al(III) oxyhydroxides and clay minerals (97–107). Retention of arsenate by soils has been correlated with the amounts of oxalate-extractable aluminum and (108) and with citrate-dithionite extractable iron (104). Sorption is also the basis for many water treatment technologies for arsenic removal, such as coagulation and packed-bed media filtration (see Chaps. 6 and 9).

Sorption of As(III) to Fe(III) oxyhydroxides has been widely observed (101,105–107,109). At high pH values, As(III) can be sorbed to a greater extent than As(V). Sorption of As(III) on clays and amorphous aluminum hydroxide has been reported (102). However, removal of As(III) by coagulation with alum is generally poor (110,111). Adsorption experiments conducted with hydrous aluminum oxide under conditions comparable to water treatment with coagulants (pH 6, 120 μM total Al) showed 99% removal of As(V) but only 3% removal of As(III) at initial arsenic concentrations of 10 and 100 μg/L (112).

The extent of arsenic sorption in natural waters will be influenced by many factors, relating to both the sorbent and the water composition. As(V) and As(III) have different affinities for various sorbent phases that may be present in sediment, soils, and aquifers. Thus the redox speciation of arsenic and the characteristics of available sorbents will strongly affect the extent of arsenic sorption as will the pH and concentrations of co-occurring inorganic and organic solutes in the aqueous phase. Since sorption is a surface phenomenon and is limited by the availability of surface sites on the sorbing phase(s), the extent of competition between arsenic and other sorbates will depend not only on the affinity of each sorbate for the surface but also on their concentrations relative to each other and to the surface site concentration. Elevated concentrations of phosphate have been used to desorb arsenic from clays (51) and from soils contaminated with arsenical pesticides (113).

Precipitation reactions cannot decrease dissolved arsenic concentrations below that in equilibrium with the solid. In contrast, the dissolved arsenic concentrations controlled by equilibrium sorption will decrease with increasing sorbent concentration. This effect is illustrated in Figure 3, which shows the calculated distribution of As(V) between sediment and porewater where this distribution is controlled by As(V) adsorption onto hydrous ferric oxide (HFO). Increasing the concentration of iron (present as HFO) in the sediment from 1 to 1.5 mg/g significantly decreases the predicted concentration of As(V) in the porewater. This modeling follows the approach used by Welch and Lico (33) except that we have used a published constant for the sorption of silica on ferrihydrite (114) rather than the estimated constants used by Welch and Lico (33). The choice of the

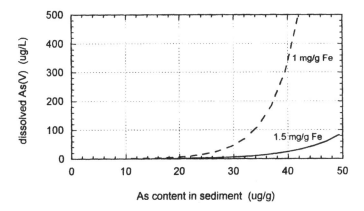

Figure 3 Dissolved As(V) concentrations in porewater (μ/L) as a function of the arsenic content of the sediment (μg/g) where partitioning between the solid and solution is controlled by sorption onto hydrous ferric oxide (HFO) with 1 or 1.5 mg/g Fe(III) (as HFO) present in the sediments. Modeling was performed using MINEQL$^+$ (133) which incorporates constants for sorption onto HFO from Dzombak and Morel (134). Constants for sorption of silica from Hansen et al. (114) and of fluoride estimated by Dzombak and Morel (134) were included in the modeling. Calculations were performed at pH 8 with ionic strength fixed at 0.01. Dissolved concentrations of major ions were fixed at the values reported for Figure 2 (i.e., sorption of these ions onto HFO was not allowed to deplete their concentration in the porewater). Contents of As and Fe in the sediments were calculated based on Eq. (1) using D = 2.6 g /cm^3 and ϕ = 0.3.

constant for silica sorption does significantly affect the results of the calculations (for the conditions used here). Consistent with our choice of constants, previous studies of As(V) removal by sorption/coprecipitation with ferric chloride in electrolyte solutions amended with silica (115) or in whole water samples (110) indicate that silica has only a modest effect on the sorption of As(V). However, Welch and Lico (33) did find that, for sediment samples obtained from the Carson Desert in Nevada, the predicted concentration of arsenic adsorbed to HFO agreed with the concentrations obtained from sediment extraction within a factor of 10. Miller (116) has also successfully described the observed partitioning of arsenic in sediment water systems with an adsorption model.

The equilibrium distribution of arsenic between the surface and solution can be destabilized by changes in environmental conditions. Changes in pH or in the concentration of competing ions could shift a pre-existing sorption equilibrium to release arsenic into the contacting fluid. Consistent with competitive desorption, correlations between dissolved arsenic and other (competing) anions have been observed in oxic groundwater (96). Changes in environmental conditions that result in dissolution of the sorbing phase could also increase the dissolved arsenic concentration. Enhanced release of arsenic associated with dissolution of iron oxides has been demonstrated in selective leaching experiments with the strong organic complexing agent EDTA (ethylenediaminetetraacetic acid) (117).

Redox conditions and pH have been shown to be crucial factors influencing the release of arsenic from contaminated sediments (75,78). Changes in redox conditions, specifically a decreasing redox potential, can affect arsenic sorption both by altering the redox speciation of arsenic and by favoring reductive dissolution of some sorbents, particularly Fe(III) oxyhydroxides. These effects are illustrated in studies of microbial reduction. When a microorganism capable of reducing both Fe(III) and As(V) was incubated with a coprecipitate of As(V) and Al(III) oxyhydroxide, all the As(III) produced was released into solution. With a coprecipitate of As(V) and Fe(III) oxyhydroxide, most of the As(III) produced remained associated with the solid, though some Fe(II) was also released onto solution (54). When an Fe(III)-reducing bacterium was incubated with scorodite or As-contaminated sediments, Fe(II) and As(V) were released into solution and no As(III) was detected (118). The importance of the redistribution of arsenic between sorbent and solution as the sorbing phase is progressively dissolved is illustrated by a calculation of dissolved arsenic concentrations in groundwater by Welch et al. (9).

A role of microbial processes in release of arsenic into groundwater concomitant with the reductive dissolution of Fe(III) oxyhydroxides has been suggested based on the observed correlation between dissolved arsenic and bicarbonate concentrations (94,95). Increased bicarbonate concentrations are attributed to the oxidation of organic matter with Fe(III) oxyhydroxides as the terminal electron acceptor. Like oxidative dissolution, reductive dissolution may be kinetically limited. Rates of microbial reduction may be limited by the supply (and nature) of organic carbon.

D. Sequestration and Release of Arsenic in the Los Angeles Aqueduct System

The importance of the availability of Fe(III) oxyhydroxide as a sorbent and of the diagenetic alteration of this sorbent under changing environmental conditions on the mobility of arsenic is illustrated by the Los Angeles Aqueduct (LAA) system.

Elevated arsenic concentrations in the LAA water supply derive from geothermal inputs of arsenic in the Owens Valley. Eccles (19) estimated that about 60% of the arsenic input to Crowley Lake, the first reservoir in the LAA system resulted from inputs of geothermal water with arsenic concentrations of approximately 1 mg/L in Hot Creek Gorge. Although rapid oxidation of As(III) to As(V) was observed downstream of the geothermal inputs, total arsenic concentrations remained constant (119). This observation is consistent with an earlier study that showed that comparable discharges of geothermal water in Hot Creek Gorge could be calculated based on arsenic, chloride, and boron (120). In contrast, losses of arsenic up to 50% were reported for geothermally influenced streams in Lassen Volcanic National Park; the difference between the conservative behavior of arsenic in Hot Creek Gorge and its nonconservative behavior at Lassen was explained by the different inputs of iron in these systems (121). The paucity of iron and the abundance of phosphate are also likely factors affecting the conservative transport of arsenic through Crowley Lake (27).

The conservative transport of arsenic through the LAA system is deliberately perturbed as part of the interim arsenic management plan implemented in 1996 by the Los Angeles Department of Water and Power (LADWP) in order to provide LADWP customers with drinking water that contains less than 10 μg/L arsenic (as compared with the historical annual average of 20 μg/L) (122). Under the interim arsenic management plan, the Cottonwood Treatment Plant (originally built in 1973 and operated intermittently to reduce turbidity) has been operated nearly continuously since March 1996 for arsenic removal. For this purpose, ferric chloride (at an average dose of 5.7 mg $FeCl_3$/L) and cationic polymer (at an average dose of 1.7 mg/L) are introduced into the LAA through diffusers at the Cottonwood Treatment Plant. Flocculated solids resulting from this treatment are deposited in North Haiwee Reservoir, 27 km downstream of the treatment plant.

A field study conducted in December 1999 (123,124) confirmed the efficacy of this treatment process. The LAA discharges into North Haiwee Reservoir in a channel about 25 m wide, bordered on the eastern side by a small peninsula. Samples of overlying water, sediment, and sediment porewaters were collected in this channel and in a control site on the eastern side of the peninsula. In the channel, an orange-brown floc was visible floating in the water and was observed to settle near the banks, where flow velocity was reduced. In the water sample collected closest to the LAA discharge, the comparison between arsenic and iron concentrations in unfiltered samples (17 μg/L As and 1.9 mg/L Fe) and samples filtered through 0.45-μm membrane filters (3.3 μg/L As and 0.13 mg/L Fe) indicated that both As and Fe were largely associated with the particulate phase. In contrast, at the control site, the concentrations of arsenic in unfiltered and filtered samples were quite similar; unfiltered samples contained 5.7 and filtered samples 4.6 μg/L As.

The sediments collected in the channel were enriched in both arsenic and iron relative to the sediments collected at the control site. A 20-cm core was collected at the control site and a 40-cm core in the channel; sediments were digested using a modification of the U.S. Environmental Protection Agency

(EPA) Method 6010. Based on analysis of these digests, the average concentrations in the sediments (on a dry weight basis) were 240 and 8 µg/g for arsenic and 4.4 and 0.2% for iron in cores from the channel and control sites, respectively.

Sediment porewaters were collected to a depth of 28 cm at a vertical resolution of 0.5 cm with a gel probe device modified from Krom et al. (125). The patterns in the concentrations of manganese, iron, and arsenic in sediment porewater indicated that the sediments became more reducing with depth and were consistent with a shift in terminal electron acceptors from O_2 to Mn oxides to Fe oxides. Dissolved manganese concentrations increased below 5 cm. Dissolved iron and arsenic concentrations increased below 10 cm and were closely correlated. The maximum dissolved arsenic concentration observed in the porewater was 1.3 mg/L.

These data can be used to examine the delivery of iron and arsenic to the sediments and their diagenetic behavior within the sediments. Based on 4-yr average values of the arsenic concentrations above Cottonwood (24.8 µg/L) and below Haiwee (8.3 µg/L) and the ferric chloride dose (5.7 mg/L), the molar ratio of iron to arsenic in the floc is 160. The same value is observed in the particulate fraction in the water samples collected below the LAA discharge. Figure 4 shows the molar ratios of iron to arsenic in the sediments from the channel core and in

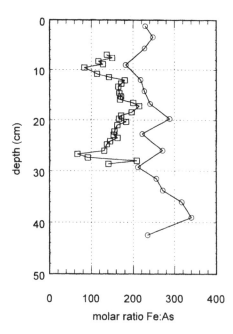

Figure 4 Molar ratios of arsenic and iron in porewaters (□) and sediments (○) as a function of depth.

the porewaters (excluding the near-surface samples that are very low in both arsenic and iron). The average values of these molar ratios are 250 for the sediment and 150 for the porewaters; neither the sediment nor porewaters exhibit any clear trend in the Fe/As molar ratio with depth. The correspondence of these values for the initial floc, particulate material in the overlying water, sediments, and porewaters suggests that the sediment composition at this site is dominated by the effects of the coagulant input at Cottonwood and that iron and arsenic are released congruently from the sediments during diagenesis.

The fractions of iron and arsenic in the sediments that would have to be released to support the observed concentrations of dissolved iron and arsenic in the porewaters can be calculated using Eq. (1) with appropriate values of specific gravity, $D = 2.5$ g/cm^3 for coagulant sludge (126), and porosity ($\phi = 0.9$). Figure 5 illustrates that the calculated fractional solubilization is quite low, less than 2.5% for all samples, and is similar for iron and arsenic for most samples. The

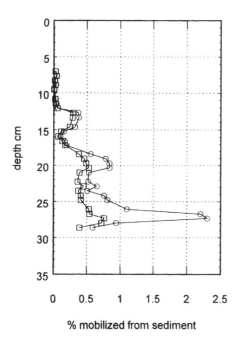

% mobilized from sediment

Figure 5   Fractions of arsenic (○) and iron (□) that would need to have been mobilized from the sediment to support the observed concentrations of dissolved arsenic and iron in the porewaters. Note that the porewater data had finer vertical resolution than the sediment samples. Arsenic and iron contents in the sediment sampled at a given depth interval were applied for all porewater data within that interval. Calculations were based on Eq. (1) using $D = 2.5$ g/cm^3 and $\phi = 0.9$.

fractional solubilization shows a general increase with depth in the sediments but also exhibits a local maximum between 12 and 15 cm. These observations indicate that the rate of release of iron and arsenic from the sediments is not uniform and is not controlled by their sediment concentrations. Other factors, such as the rate of supply of organic matter, may be influencing the rate of arsenic release from the sediments.

The goal of the interim arsenic management plan is accomplished by coagulant addition to the LAA water supply at Cottonwood. Arsenic is removed from the water supply and deposited with the iron floc in North Haiwee Reservoir. Although the deposited floc becomes unstable under the reducing conditions in the deeper sediments, only a small fraction of the deposited arsenic is released to the porewaters. The uppermost (presumably oxic) layer of the sediment provides a barrier to the diffusive flux of arsenic from the deep porewater into the overlying water. However, physical disturbance of the sediment would allow the elevated concentrations of dissolved arsenic in the porewater to be dispersed into the overlying water. Diffusive flux of arsenic into the overlying water could occur if the oxygen demand in the surficial sediments increased sufficiently that the oxic–anoxic boundary coincided with the sediment–water interface. The deposition of arsenic-rich floc and consequent supply of arsenic to the porewater may also pose a threat to groundwater if shallow groundwater is recharged from the reservoir.

V. IMPLICATIONS FOR WATER SUPPLY AND TREATMENT

The occurrence of arsenic in source waters (and particularly groundwater) can be rationalized by examining the mechanisms of arsenic release from the aquifer solids to the contacting water. Equilibrium calculations provide a constraint on arsenic concentrations that may be expected for water in contact with various solids but slow kinetics (e.g., of dissolution) may result in significant discrepancies between predicted and observed concentrations of dissolved arsenic.

Only a small fraction of U.S. source waters would require treatment to comply with a drinking water standard of 10 µg/L. Although most of the systems that would be out of compliance with this standard without treatment would be small groundwater systems, some large groundwater systems (e.g., Albuquerque, NM) and large surface water systems (e.g., Los Angeles, CA) would also be affected by a standard at this level.

Because of the spatial variability in arsenic occurrence in groundwater, selective siting or screening of wells may provide an acceptable water supply. This approach was successfully applied in the City of Hanford, CA (located in the San Joaquin Valley) to comply with the (current) MCL of 50 µg/L. Arsenic concentrations in shallow wells, constructed prior to 1981, routinely exceeded this value. These wells were constructed with maximum depths of approximately 203 m and long, gravel-packed perforated intervals extending within 42 m of the surface. After 1981, deeper wells were constructed (to depths of 460 m) with short perforated intervals located only in coarse-grained sediments not shallower than 216 m and sealed above and below the perforated intervals (127). This strategy is, however, unlikely to be sufficient to comply with a standard of 10 μg/L (128).

When treatment systems must be installed to achieve arsenic concentrations below 10 μg/L, constituents other than arsenic must also be considered in choosing a treatment technology. Common groundwater constituents, such as iron, can interfere with treatment technologies for arsenic removal, such as ion-exchange (see Chap. 9). The redox speciation of arsenic in the source water must also be considered in designing treatment systems. Since As(V) is more effectively removed that As(III) by most treatment technologies, preoxidation will generally be required except for technologies, such as manganese greensand filtration, that incorporate an oxidant (7). Simple aeration is insufficient to oxidize As(III) to As(V) (128,129) but many conventional oxidants are suitable for this purpose (130).

A subsurface treatment strategy for groundwaters has been recently field tested in Germany (131). At a municipal well field, extraction of groundwater was alternated, on a 24-hr cycle, with aeration and reinjection of a portion of the extracted water. Introduction of oxygen into the subsurface resulted in the oxidation of ambient (reduced) iron and manganese and precipitation of the corresponding oxyhydroxides in the aquifer. No change in permeability was observed. Arsenic was immobilized in the subsurface; with ambient As(III), however, a "ripening" period, attributed to biofilm formation and onset of biological oxidation of As(III) to As(V), was required for efficient arsenic removal.

This treatment strategy relies on the ambient iron to form a subsurface treatment zone. In iron-poor groundwaters, it is possible that iron could be injected into the subsurface (as a solution of ferric or ferrous salts) and precipitated as Fe(III) oxyhydroxides by subsequent injection of a neutralizing solution (i.e., either base or an oxidant). In calcite-bearing aquifers, precipitation may occur as the ferric salt solution contacts aquifer minerals as has been demonstrated in laboratory experiments by Morrison et al. (132). Principal concerns in implementing such an in situ treatment strategy would be to avoid clogging the porous medium in the subsurface treatment zone, exceeding the capacity for arsenic retention within the treatment zone, or remobilizing arsenic from the treatment zone.

Although groundwaters generally contain higher concentrations of trace constituents (derived from water–rock interactions) than do surface waters, they also often provide superior protection against pathogens and greater reliability of supply. A thorough assessment of groundwater quality is particularly important when major changes in water resource allocation are made (e.g., the large-scale shifts from surface water to groundwater use in West Bengal, India, and Bangladesh) to avoid exposure to naturally occurring groundwater constituents, such as arsenic.

ACKNOWLEDGMENTS

The work on Haiwee Reservoir was supported by the U.S. Environmental Protection Agency (R826202-01-0). Assistance from the staff of the Water Quality and Distribution Division of the Los Angeles Department of Water and Power is gratefully acknowledged.

REFERENCES

1. National Research Council. Arsenic in Drinking Water. Washington, DC: National Academy Press, 1999.
2. BK Mandal, TR Chowdhury, G Samanta, GK Basu, PP Chowdhury, CR Chanda, D Lodh, NK Karan, RK Dhar, DK Tamili, D Das, KC Saha, D Chakraborti. Arsenic in groundwater in seven districts of West Bengal, India. The biggest arsenic calamity in the world. Curr Sci India 70:976–986, 1996.
3. BK Biswas, RK Dhar, G Samanta, BK Mandal, D Chakraborti, I Faruk, KS Islam, MM Chowdhury, A Islam, S Roy. Detailed study report of Samta, one of the arsenic-affected villages of Jessore District, Bangladesh. Curr Sci India 74:134–145, 1998.
4. United States Environmental Protection Agency. 40 CFR parts 9, 141, and 142, National Primary Drinking Water Regulations; Arsenic and Clarifications to Compliance and New Source Contaminant Monitoring; Proposed Rule. Washington, DC: Fed Regist 66:20579–20584, 2001.
5. United States Environmental Protection Agency. 40 CFR parts 141 and 142, National Primary Drinking Water Regulations; Arsenic and Clarifications to Compliance and New Source Contaminant Monitoring; Final Rule. Washington, DC: Fed Regist 66:6976–7066, 2001.
6. D Langmuir. Aqueous Environmental Geochemistry. Upper Saddle River, NJ: Prentice-Hall, 1997.
7. United States Environmental Protection Agency. 40 CFR parts 141 and 142, National Primary Drinking Water Regulations; Arsenic and Clarifications to Compliance and New Source Contaminant Monitoring; Proposed Rule. Washington, DC: Fed Regist 65:38888–38983, 2000.
8. HW Chen, MM Frey, D Clifford, LS McNeill, M Edwards. Arsenic treatment considerations. J Am Water Works Assoc 91(3):74–85, 1999.

9. AH Welch, DB Westjohn, DR Helsel, RB Wanty. Arsenic in groundwater of the United States: Occurrence and geochemistry. Ground Water 38:589–604, 2000.
10. C Choprapawon, A Rodcline. Chronic arsenic poisoning in Ronpibool Nakhon Sri Thammarat, the Southern Province of Thailand. In: CO Abernathy, RL Calderon, WR Chappell, eds. Arsenic Exposure and Health Effects. London: Chapman & Hall, 1997, pp 69–77.
11. AH Welch, MS Lico, JL Hughes. Arsenic in groundwater of the western United States. Ground Water 26:333–347, 1988.
12. WR Cullen, KJ Reimer. Arsenic speciation in the environment. Chem Rev 89:713–764, 1989.
13. JF Ferguson, J Gavis. A review of the arsenic cycle in natural waters. Water Res 6:1259–1274, 1972.
14. NE Korte, Q Fernando. A review of arsenic(III) in groundwater. Crit Rev Environ Control 21:1–39, 1991.
15. JW Moore. Inorganic Contaminants of Surface Water. New York: Springer-Verlag, 1991.
16. JO Nriagu, ed. Arsenic in the Environment. Part I: Cycling and Characterization. New York: Wiley, 1994.
17. MO Andreae. Arsenic in rain and the atmospheric balance of arsenic. J Geophys Res 85:4512–4518, 1980.
18. G Kiely. Environmental Engineering. London: McGraw-Hill, 1997.
19. L Eccles. Sources of Arsenic in Streams Tributary to Lake Crowley, CA. U.S. Geological Survey, Water Resources Investigations 76-36, 1976.
20. KH Johannesson, WB Lyons, S Huey, GA Doyle, EE Swanson, E Hackett. Oxyanion concentrations in eastern Sierra Nevada rivers. 2. Arsenic and phosphate. Aquat Geochem 3:61–97, 1997.
21. DM Imboden, A Wuest. Mixing mechanisms in lakes. In: A Lerman, DM Imboden, JR Gat, eds. Physics and Chemistry of Lakes. Berlin: Springer-Verlag, 1995, pp 83–138.
22. RA Freeze, JA Cherry. Groundwater. Upper Saddle River, NJ: Prentice-Hall, 1979.
23. GA Cutter. Kinetic controls on metalloid speciation in seawater. Marine Chem 40: 65–80, 1992.
24. A Kuhn, L Sigg. Arsenic cycling in eutrophic Lake Greifen, Switzerland: Influence of seasonal redox processes. Limnol Oceanogr 38:1052–1059, 1993.
25. LCD Anderson, KW Bruland. Biogeochemistry of arsenic in natural waters: The importance of methylated species. Environ Sci Technol 25:420–427, 1991.
26. AC Aurilio, RP Mason, HF Hemond. Speciation and fate of arsenic in three lakes of the Abejona Watershed. Environ Sci Technol 28:577–585, 1994.
27. PE Kneebone, JG Hering. Behavior of arsenic and other redox sensitive elements in Crowley Lake, CA: A reservoir in the Los Angeles aqueduct system. Environ Sci Technol 34:4307–4312, 2000.
28. P Seyler, JM Martin. Biogeochemical processes affecting arsenic species distribution in a permanently stratified lake. Environ Sci Technol 23:1258–1263, 1989.
29. Y Sohrin, M Matsui, M Kawashima, M Hojo, H Hasegawa. Arsenic biogeochemistry affected by eutrophication in Lake Biwa, Japan. Environ Sci Technol 31:2712–2720, 1997.

30. MK Davis, KD Reich, M Tikkanen. Nationwide and California arsenic occurrence studies. In: WR Chappell, CÒ Abernathy, CR Cothern, eds. Arsenic Exposure and Health. Northwood, UK: Science Technology Letters, 1994, pp 31–40.

31. MM Frey, M Edwards. Surveying arsenic occurrence. J Am Water Works Assoc 89(3):105–117, 1997.

32. J Reid. Arsenic occurrence: USEPA seeks a clearer picture. J Am Water Works Assoc 86(9):44–51, 1994.

33. AH Welch, MS Lico. Aqueous geochemistry of ground-water with high concentrations of arsenic and uranium, Carson River basin, Nevada. Chem Geol 70:19–19, 1988.

34. DA Nimick. Arsenic hydrogeochemistry in an irrigated river valley: A reevaluation. Ground Water 36:743–753, 1998.

35. TC Winter, JW Harvey, OL Franke, WM Alley. Ground Water and Surface Water: A Single Resource. Denver, CO: U.S. Geological Survey, Circular 1139, 1999.

36. JW Harvey, CC Fuller. Effect of enhanced manganese oxidation in the hyporheic zone on basin-scale geochemical mass balance. Water Res Res 34:623–636, 1998.

37. RJ Swartz, GD Thyne, JM Gillespie. Dissolved arsenic in the Kern Fan, San Joaquin Valley, California: Naturally occurring or anthropogenic? Environ Geosci 3:143–153, 1996.

38. H Yan-Chu. Arsenic distribution in soils. In: JO Nriagu, ed. Arsenic in the Environment, Part 1: Cycling and Characterization. New York: Wiley, 1994, pp 17–49.

39. DK Bhumbla, RF Keefer. Arsenic mobilization and bioavailability in soils. In: JO Nriagu, ed. Arsenic in the Environment, Part 1: Cycling and Characterization. New York: Wiley, 1994. p 51–82.

40. RJ Bowell. Sorption of arsenic by iron oxides and oxyhydroxides in soils. Appl Geochem 9:279–286, 1994.

41. D Ahmann, AL Roberts, LR Krumholz, FM Morel. Microbe grows by reducing arsenic. Nature 371:750, 1994.

42. DK Newman, EK Kennedy, JD Coates, D Ahmann, DJ Ellis, DR Lovley, FM Morel. Dissimilatory arsenate and sulfate reduction in desulfotomaculum auripigmentum sp. Nov. Arch Microbiol 168:380–388, 1997.

43. N Korte. Naturally occurring arsenic in groundwaters of the midwestern United States. Environ Geol Water Sci 18:137–141, 1991.

44. SH Stow. The occurrence of arsenic and the color-causing components in Florida land-pebble phosphate rock. Econ Geol 64:667–671, 1969.

45. SC Peters, JD Blum, B Klaue, MR Karagas. Arsenic occurrence in New Hampshire drinking water. Environ Sci Technol 33:1328–1333, 1999.

46. TR Holm, CD Curtiss. A comparison of oxidation-reduction potentials calculated from the As(V)/As(III) and Fe(III)/Fe(II) couples with measured platinum-electrode potentials in groundwater. J Contam Hydrol 5:67–81, 1989.

47. I Grenthe, W Stumm, M Laaksuharju, AC Nilsson, P Wikberg. Redox potentials and redox reactions in deep groundwater systems. Chem Geol 98:131–150, 1992.

48. YF Wang, P Van Cappellen. A multicomponent reactive transport model of early diagenesis: Application to redox cycling in coastal marine sediments. Geochim Cosmochim Acta 60:2993–3014, 1996.

49. WM Mok, CM Wai. Distribution and mobilization of arsenic and antimony species in the Coeur d'Alene River, Idaho. Environ Sci Technol 24:102–108, 1990.

50. LE Eary, JA Schramke. Rates of inorganic oxidation reactions involving dissolved oxygen. In: DC Melchior, RL Bassett, eds. Chemical Modeling of Aqueous Systems, II. Washington, DC: American Chemical Society, 1990, pp 379–396.

51. BA Manning, S Goldberg. Adsorption and stability of arsenic(III) at the clay mineral-water interface. Environ Sci Technol 31:2005–2011, 1997a.

52. EA Rochette, BC Bostick, GC Li, S Fendorf. Kinetics of arsenate reduction by dissolved sulfide. Environ Sci Technol 34:4714–4720, 2000.

53. CA Jones, HW Langner, K Anderson, TR McDermott, WP Inskeep. Rates of microbially mediated arsenate reduction and solubilization. Soil Sci Soc Am J 64:600–608, 2000.

54. J Zobrist, PR Dowdle, JA Davis, RS Oremland. Mobilization of arsenite by dissimilatory reduction of adsorbed arsenate. Environ Sci Technol 34:4747–4753, 2000.

55. FH Osborne, HL Ehrlich. Oxidation of arsenite by a soil isolate of Alcaligenes. J Appl Bacteriol 41:295–305, 1976.

56. SE Phillips, ML Taylor. Oxidation of arsenite to arsenate by Alcaligenes faecalis. Appl Environ Microbiol 32:392–399, 1976.

57. TM Salmassi, K Venkateswaren, M Satomi, KH Nealson, DK Newman, JG Hering. Oxidation of arsenite by Agrobacterium albertimagni, AOL15 sp. nov., isolated from Hot Creek, CA. Geomicrobiol J 19:xxx, 2002.

58. JM Santini, LI Sly, RD Schnagl, JM Macy. A new chemolithoautotrophic arsenite-oxidizing bacterium isolated from a gold mine: Phylogenetic, physiological, and preliminary biochemical studies. Appl Environ Microbiol 66:92–97, 2000.

59. VQ Chiu, JG Hering. Arsenic adsorption and oxidation at manganite surfaces. 1. Method for simultaneous determination of adsorbed and dissolved arsenic species. Environ Sci Technol 34:2029–2034, 2000.

60. W Driehaus, R Seith, M Jekel. Oxidation of arsenate (III) with manganese oxides in water treatment. Water Res 29:297–305, 1995.

61. JN Moore, JR Walker, TH Hayes. Reaction scheme for the oxidation of As(III) to As(V) by birnessite. Clays Clay Miner 38:549–555, 1990.

62. DW Oscarson, PM Huang, WK Liaw. Role of manganese in the oxidation of arsenite by freshwater lake sediments. Clays Clay Miner 29:219–225, 1981.

63. DW Oscarson, PM Huang, WK Liaw, UT Hammer. Kinetics of oxidation of arsenite by various manganese dioxides. Soil Sci Soc Am J 47:644–648, 1983.

64. MJ Scott, JJ Morgan. Reactions at oxide surfaces. 1. Oxidation of As(III) by synthetic birnessite. Environ Sci Technol 29:1898–1905, 1995.

65. DW Oscarson, PM Huang, WK Liaw. The oxidation of arsenite by aquatic sediments. J Environ Qual 9:700–703, 1980.

66. BA Manning, DL Suarez. Modeling arsenic(III) adsorption and heterogeneous oxidation kinetics in soils. Soil Sci Soc Am J 64:128–137, 2000.

67. EA Crecelius. The geochemical cycle of arsenic in Lake Washington and its relation to other elements. Limnol Oceanogr 20:441–451, 1975.

68. HM Edenborn, N Belzile, A Mucci, J Lebel, N Silverberg. Observations on the diagenetic behavior of arsenic in a deep coastal sediment. Biogeochem 2:359–376, 1986.

69. JN Moore, WH Ficklin, C Johns. Partitioning of arsenic and metals in reducing sulfidic sediments. Environ Sci Technol 22:432–437, 1988.

70. HM Spliethoff, RP Mason, HF Hemond. Interannual variability in the speciation and mobility of arsenic in a dimictic lake. Environ Sci Technol 29:2157–2161, 1995.

71. A Widerlund, J Ingri. Early diagenesis of arsenic in sediments of the Kalix River estuary, northern Sweden. Chem Geol 125:185–196, 1995.

72. JG Reynolds, DV Naylor, SE Fendorf. Arsenic sorption in phosphate-amended soils during flooding and subsequent aeration. Soil Sci Soc Am J 63:1149–1156, 1999.

73. J Aggett, GA O'Brian. Detailed model for the mobility of arsenic in lacustrine sediments based on measurements in Lake Ohakuri. Environ Sci Technol 19:231–238, 1985.

74. JM Azcue, JO Nriagu, S Schiff. Role of sediment porewater in the cycling of arsenic in a mine-polluted lake. Environ Int 20:517–527, 1994.

75. WM Mok, CM Wai. Distribution and mobilization of arsenic species in the creeks around the Blackbird mining district, Idaho. Water Res 23:7–13, 1989.

76. MA Huerta-Diaz, A Tessier, R Carignan. Geochemistry of trace metals associated with reduced sulfur in freshwater sediments. Appl Geochem 13:213–233, 1998.

77. KA Sullivan, RC Aller. Diagenetic cycling of arsenic in Amazon shelf sediments. Geochim Cosmochim Acta 60:1465–1477, 1996.

78. PH Masscheleyn, RDJ Delaune, PW H. Arsenic and selenium chemistry as affected by sediment redox potential and pH. J Environ Qual 20:522–527, 1991.

79. F Juillot, P Ildefonse, G Morin, G Calas, AM de Kersabiec, M Benedetti. Remobilization of arsenic from buried wastes at an industrial site: Mineralogical and geochemical control. Appl Geochem 14:1031–1048, 1999.

80. R Wagemann. Some theoretical aspects of stability and solubility of inorganic arsenic in the freshwater environment. Water Res 12:139–145, 1978.

81. ME Essington. Solubility of barium arsenate. Soil Sci Soc Am J 52:1566–1570, 1988.

82. JG Webster. The solubility of As_2S_3 and speciation of As in dilute and sulfide-bearing fluids at 25 and 90°C. Geochim Cosmochim Acta 54:1009–1017, 1990.

83. GR Helz, JA Tossell, JM Charnock, RAD Pattrick, DJ Vaughan, CD Garner. Oligomerization in As(III) sulfide solutions: Theoretical constraints and spectroscopic evidence. Geochim Cosmochim Acta 59:4591–4604, 1995.

84. MJ Kim, J Nriagu, S Haack. Carbonate ions and arsenic dissolution by groundwater. Environ Sci Technol 34:3094–3100, 2000.

85. MB Clarke, GR Helz. Metal-thiometalate transport of biologically active trace elements in sulfidic environments. 1. Experimental evidence for copper thioarsenite complexing. Environ Sci Technol 34:1477–1482, 2000.

86. ME Schreiber, JA Simo, PG Freiberg. Stratigraphic and geochemical controls on naturally occurring arsenic in groundwater, Eastern Wisconsin, USA. Hydrogeol J 8:161–176, 2000.

87. HW Nesbitt, LJ Muir, AR Pratt. Oxidation of arsenopyrite by air and air-saturated, distilled water, and implications for mechanism of oxidation. Geochim Cosmochim Acta 59:1773–1786, 1995.

88. DK Nordstrom, G Southam. Geomicrobiology of sulfide mineral oxidation. In: JF

Banfield, KH Nealson, eds. Geomicrobiology: Interactions Between Microbes and Minerals. Washington, DC: Mineralogical Society of America, 1997, pp 361–390.

89. N Belzile, A Tessier. Interactions between arsenic and iron oxyhydroxides in lacustrine sediments. Geochim Cosmochim Acta 54:103–109, 1990.

90. T Takamatsu, M Kawashima, M Koyama. The role of Mn^{2+}-rich hydrous manganese oxide in the accumulation of arsenic in lake sediments. Water Res 19:1029–1032, 1985.

91. CC Fuller, JA Davis. Influence of coupling of sorption and photosynthetic processes on trace element cycles in natural waters. Nature 340:52–54, 1989.

92. RA Feely, JH Trefry, GJ Massoth, S Metz. A comparison of the scavenging of phosphorus and arsenic from seawater by hydrothermal iron oxyhydroxides in the Atlantic and Pacific oceans. Deep-Sea Res 38:617–623, 1991.

93. T Pichler, J Veizer, GEM Hall. Natural input of arsenic into a coral reef ecosystem by hydrothermal fluids and its removal by Fe(III) oxyhydroxides. Environ Sci Technol 33:1373–1378, 1999.

94. R Nickson, J McArthur, W Burgess, KM Ahmed, P Ravenscroft, M Rahman. Arsenic poisoning of Bangladesh groundwater. Nature 395:338, 1998.

95. RT Nickson, JM McArthur, P Ravenscroft, WG Burgess, KM Ahmed. Mechanism of arsenic release to groundwater, Bangladesh and West Bengal. Appl Geochem 15:403–413, 2000.

96. FN Robertson. Arsenic in ground-water under oxidizing conditions, South-west United States. Environ Geochem Health 11:171–185, 1989.

97. MA Anderson, JF Ferguson, J Gavis. Arsenate adsorption on amorphous aluminum hydroxide. J Coll Int Sci 54:391–399, 1976.

98. MM Ghosh, JR Yuan. Adsorption of inorganic arsenic and organoarsenicals on hydrous oxides. Environ Prog 6:150–157, 1987.

99. CC Fuller, JA Davis, GA Waychunas. Surface chemistry of ferrihydrite: Part 2. Kinetics of arsenic adsorption and coprecipitation. Geochim Cosmochim Acta 57:2271–2282, 1993.

100. TS Hsia, SL Lo, CF Lin, DY Lee. Characterization of arsenate adsorption on hydrous iron oxide using chemical and physical methods. Coll Surf A 85:1–7, 1994.

101. JO Leckie, MM Benjamin, K Hayes, G Kaufman, S Altmann. Adsorption/Coprecipitation of Trace Elements from Water with Iron Oxyhydroxide. Rep 910-1. Palo Alto, CA: Electric Power Resource Institute, 1980.

102. BA Manning, S Goldberg. Modeling arsenate competitive adsorption on kaolinite, montmorillonite and illite. Clays Clay Miner 44:609–623, 1996.

103. BA Manning, S Goldberg. Modeling competitive adsorption of arsenate with phosphate and molybdate on oxide minerals. Soil Sci Soc Am J 60:121–131, 1996.

104. BA Manning, S Goldberg. Arsenic (III) and arsenic (V) adsorption on three California soils. Soil Sci 162:886–895, 1997.

105. ML Pierce, CB Moore. Adsorption of arsenite on amorphous iron hydroxide from dilute aqueous solution. Environ Sci Technol 14:214–216, 1980.

106. ML Pierce, CB Moore. Adsorption of arsenite and arsenate on amorphous iron hydroxide. Water Res 16:1247–1253, 1982.

107. JA Wilkie, JG Hering. Adsorption of arsenic onto hydrous ferric oxide: Effects

of adsorbate/adsorbent ratios and co-occurring solutes. Coll Surf A 107:97–110, 1996.

108. NT Livesey, PM Huang. Adsorption of arsenate by soils and its relation to selected chemical properties and anions. Soil Sci 131:88–94, 1981.

109. KP Raven, A Jain, RH Loeppert. Arsenite and arsenate adsorption on ferrihydrite: Kinetics, equilibrium, and adsorption envelopes. Environ Sci Technol 32:344–349, 1998.

110. JG Hering, PY Chen, JA Wilkie, M Elimelech. Arsenic removal from drinking water during coagulation. J Environ Eng ASCE 123:800–807, 1997.

111. TJ Sorg, GS Logsdon. Treatment technology to meet the interim primary drinking water regulations for inorganics: Part 2. J Am Water Works Assoc 70(7):379–393, 1978.

112. JA Wilkie. Processes controlling arsenic mobility in natural and engineered systems. PhD dissertation. University of California, Los Angeles, Los Angeles, CA, 1997.

113. FJ Peryea. Phosphate-induced release of arsenic from soils contaminated with lead arsenate. Soil Sci Soc Am J 55:1301–1306, 1991.

114. HCB Hansen, TP Wetche, K Raulundrasmussen, OK Borggaard. Stability-constants for silicate adsorbed to ferrihydrite. Clay Min 29:341–350, 1994.

115. XG Meng, S Bang, GP Korfiatis. Effects of silicate, sulfate, and carbonate on arsenic removal by ferric chloride. Water Res 34(4):1255–1261, 2000.

116. GP Miller. Prediction of the environmental mobility of arsenic: Evaluation of a Mechanistic Approach to Modeling Water-Rock Partitioning. Rep 1000547. Palo Alto, CA: Electric Power Research Institute, 2000.

117. J Aggett, LS Roberts. Insight into the mechanism of accumulation of arsenate and phosphate in hydro lake sediments by measuring the rate of dissolution with ethylenediaminetetraacetic acid. Environ Sci Technol 20:183–186, 1986.

118. DE Cummings, F Caccavo, S Fendorf, RF Rosenzweig. Arsenic mobilization by the dissimilatory Fe(III)-reducing bacterium Shewanella alga BrY. Environ Sci Technol 33:723–729, 1999.

119. JA Wilkie, JG Hering. Rapid oxidation of geothermal arsenic (III) in streamwaters of the eastern Sierra Nevada. Environ Sci Technol 32(5):657–662, 1998.

120. ML Sorey, MD Clark. Changes in the Discharge Characteristics of Thermal Springs and Fumaroles in the Long Valley Caldera, California, Resulting from Earthquakes on May 25–27, 1980. Open-File Rep 81-203. Denver, CO: U.S. Geological Survey, 1981.

121. ML Sorey, EM Colvard, SE Ingebritsen. Measurements of Thermal-Water Discharge Outside Lassen Volcanic National Park, California, 1983–94. Water Resources Investigations Rep 94-4180-B. Denver, CO: U.S. Geological Survey, 1994.

122. GF Stolarik, JD Christie. Interim arsenic management plan for Los Angeles. Proceedings of American Water Works Association Annual Conference, 1999, Chicago, IL, June 20–24.

123. PE Kneebone. Arsenic geochemistry in a geothermally impacted system: The Los Angeles Aqueduct. PhD disseration, California Institute of Technology, Pasadena, CA, 2000.

124. PE Kneebone, PA O'Day, N Jones, JG Hering. Rapid evolution of arsenic specia-

tion in arsenic- and iron-enriched reservoir sediments. Environ Sci Technol, in press.

125. MD Krom, P Davison, H Zhang, W Davison. High-resolution pore-water sampling with a gel sampler. Limnol Oceanogr 39:1967–1972, 1994.

126. DA Cornwell. Water treatment plant residuals management. In: American Water Works Association Water Quality and Treatment. 5th ed. New York: McGraw-Hill, 1999, pp 16.1–16.51.

127. CS Johnson. The occurrence of arsenic in ground water. Hanford, CA: Possible geologic and geochemical explanations. MS thesis, California State University, Fresno, Frensno, CA, 1990.

128. JG Hering, VQ Chiu. Arsenic occurrence and speciation in municipal ground-water-based supply system. J Environ Eng ASCE 126:471–474, 2000.

129. JD Lowry, SB Lowry. Oxidation of As(III) by aeration and storage. U.S. Environmental Protection Agency, in press.

130. G Ghurye, D Clifford. Laboratory study of the oxidation of As III to As V. U.S. Environmental Protection Agency, EPA/600/R-01/021. Washington, DC, 2001.

131. U Rott., M Friedle. Subterranean removal of arsenic from groundwater. In: WR Chappell, CO Abernathy, RL Calderon, eds. Arsenic Exposure and Health Effects. Oxford: Elsevier, 1999, pp 389–396.

132. SJ Morrison, RR Spangler, SA Morris. Subsurface injection of dissolved ferric chloride to form a chemical barrier: Laboratory investigations. Ground Water 34: 75–83, 1996.

133. WD Schecher, DC McAvoy. MINEQL$^+$: A Chemical Equilibrium Modeling System. Hallowell: Environmental Research Software, 1998.

134. DA Dzombak, FM Morel. Surface Complexation Modeling: Hydrous Ferric Oxide. New York: Wiley, 1990.

8
Arsenic (V)/(III) Cycling in Soils and Natural Waters: Chemical and Microbiological Processes

William P. Inskeep and Timothy R. McDermott
Montana State University, Bozeman, Montana

Scott Fendorf
Stanford University, Stanford, California

I. INTRODUCTION

Arsenic (As) cycling among different valence states and chemical species in soils and natural waters is influenced by both abiotic and biotic processes. Consequently, like all biogeochemical processes, a thorough understanding of factors responsible for As transformation requires appreciation of chemical and microbiological contributions, and in most instances these contributions are inextricably linked. Chemists and microbiologists often work within paradigms that are too simplistic for understanding and predicting electron transfer (redox) reactions in natural systems. For example, geochemists have too often relied on the notion that the thermodynamic stability of oxidized and reduced species as determined using the conceptual electron activity (pe $= -\log \{e-\}$) is a meaningful predictor of the concentrations of oxidized and reduced species (1). Although misleading, the conceptual electron activity is often referred to as a "master variable" along with pH in defining the ratio of oxidized to reduced species (1). However, it is well known that a single measure of the *system* redox potential (pe) useful for defining the ratio of oxidized to reduced species for all redox couples simultaneously does not exist (2,3).

Several studies (4–8) have documented the *nonequilibrium* behavior of the As(V)/As(III) couple relative to other indicators of redox status (i.e. dissolved O_2, Pt electrode measurement), where As(III) is often observed in oxic environments and As(V) persists in anoxic systems. In such cases, slow kinetics and/or biological phenomena are usually invoked to explain the apparent lack of thermodynamic equilibrium. Indeed, more accurate prediction of the speciation and fate of As in natural waters will require further information on the rates of individual reactions, which contribute either indirectly or directly to As oxidation and reduction. This is true regardless of whether the reactions may be biological or abiological; in actuality, processes such as surface complexation, mineral dissolution, dissimilatory reduction, or detoxification are linked in real systems to define As cycling. Regarding biological pathways of oxidation and reduction, microbiologists too often rely on cultivation and simple characterization of pure culture isolates as the primary tool for understanding microbially driven electron transfer reactions in natural systems. Such simplistic approaches effectively disregard the physical–chemical context that defines actual microbial habitats in soils and waters, and ignore the questions regarding how the cultivated organisms actually relate to the measured oxidation–reduction processes observed in the systems under study.

We believe that progress toward an improved understanding and prediction of redox transformations important in As cycling will be realized with greater emphasis on comprehensive approaches that (1) define rates of specific abiotic and biotic pathways in appropriate and realistic environmental contexts, (2) explore patterns in microbial ecology that associate microbial communities with their chemical–physical environments, and (3) elucidate functional pathways of microbially driven redox transformations in situ. That said, we are limited in part by the scale at which oxidation–reduction processes occur and by the heterogeneity and complexity of microenvironments typical of soil and water systems. Our measurement tools are often not appropriate for resolving spatial heterogeneity and the complexity of microbial interactions with aqueous and surface phases under realistic in situ conditions. Nevertheless, where possible, future studies intended to elucidate mechanisms of As transformation should employ a compliment of measurement tools that provide more accurate insights to the chemical and microbiological processes important in a natural setting. In this chapter, we focus primarily on chemical and microbiological processes that mediate transformations between As(III) and As(V). Our goals are to (1) provide an updated review of the important chemical processes that control the distribution and speciation of As(III) and As(V) in soils and natural waters, (2) compare and contrast abiotic and biotic processes that may contribute to oxidation of As(III) or reduction of As(V), and (3) provide several examples of microbially mediated As transformations that can be linked to microbial populations using cultivation-independent molecular methods.

II. CHEMISTRY OF INORGANIC ARSENIC (III) AND ARSENIC (V) IN SOILS AND NATURAL WATERS

A. Chemical Equilibria

A comprehensive thermodynamic model including all chemical equilibria important for predicting the distribution of As species in natural waters is not available at this time; however, we review here some pertinent details regarding As speciation important for understanding the distribution of various As species in different environments. As mentioned above, pe–pH diagrams are often used for predicting the stability fields of various oxidized and reduced species. For arsenic, such diagrams have been published using a variety of assumptions (9–12), and the general conclusion is that As(V) is predicted to be the thermodynamically stable valence at E_h values greater than approximately -100 mV at pH 8 and greater than 300 mV at pH 4 (Fig. 1). Below these redox potentials, As(III) is the thermodynamically stable valence, present either as the $H_3AsO_3^\circ$ species, As-S complexes (e.g. $H_2As_3S_6^-$), or As(III) solid phases such as As_2S_3 (Fig. 1). We have not shown the stability regions for more reduced forms of As [e.g., As(0), As($-$III)] due to limited knowledge of their occurrence in soils and natural waters, and uncertainties regarding appropriate thermodynamic data.

The pe values at which the activities of the As(III) and As(V) species are equal can be defined using the equilibria expressions given in Table 1 [E_h values

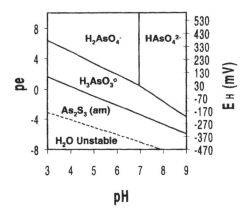

Figure 1   Simplified pe–pH diagram for the As-S-H_2O system at 25°C constructed based on the following assumptions: $SO_4 = 10^{-3}M$, $H_3AsO_3^\circ = 10^{-5}$ M, and $As_2S_3 =$ amorphous orpiment (see Table 1). The diagram does not consider the importance of As(V)–Fe(III)-hydroxide solid phases likely to be predominant under oxidized conditions, nor does it consider more reduced forms of arsenic potentially found in solid phases, such as realgar (AsS) and arsenopyrite (FeAsS), or as gaseous forms such as arsine (AsH$_3$).

Table 1 Partial List of Chemical Equilibria and Thermodynamic Constants Important to Speciation and Cycling of As(V) and As(III) in Soils and Natural Waters

Description	Reaction	log K	Ref.
As(V)/As(III) Couple	$H_3AsO_3^\circ + H_2O = HAsO_4^{2-} + 4H^+ + 2e$	-28.63	14
As(III) Dissociation	$H_3AsO_3^\circ = H_2AsO_3 + H^+$	-9.29	14
As(V) Dissociation	$H_3AsO_4^\circ = H_2AsO_4 + H^+$	-2.24	14
	$H_2AsO_4 = HAsO_4^{2-} + H^+$	-6.94	14
	$HAsO_4^{2-} = AsO_4^{3-} + H^+$	-12.19	14
As(III)-S Solubility			
• amorph As_2S_3	$0.5As_2S_3\,(s) + 3H_2O = H_3AsO_3^\circ + 1.5\,H_2S(aq)$	-11.9	15
	$0.5As_2S_3\,(s) + H_2O + 0.5H_2S = AsO(SH)_2 + H^+$	-7.9	16
	$1.5\,As_2S_3\,(s) + 1.5H_2S = H_2As_3S_6 + H^+$	-5.5	16
• orpiment	$0.5As_2S_3\,(c) + 3H_2O = H_3AsO_3^\circ + 1.5\,H_2S(aq)$	-12.6	18
As(III)-S	$H_3AsO_3^\circ + 2H_2S(aq) = 0.33H_2As_3S_6 + 0.33H^+ + 3H_2O$	10.3	a
Complexation	$H_3AsO_3^\circ + 2H_2S(aq) = AsO(SH)_2 + H^+ + 2H_2O$	~4	16

a log K values: 10.17 (16), 10.26 (15), 10.34 (18) and 10.39 (17).

can be used alternatively to pe, where at 25°C, E_h (mV) = 59 pe (1)]. For example, the pe–pH relationships defining the equivalence of predominant As(III) and As(V) species are as follows:

$$H_3AsO_3^\circ\text{-}H_2AsO_4^-: \qquad pe = 10.9 - 1.5\ pH$$
$$H_3AsO_3^\circ\text{-}HAsO_4^{2-}: \qquad pe = 14.3 - 2pH$$

where $H_2AsO_4^-$ is the predominant form of As(V) between pH 2.5 and 7, $HAsO_4^{2-}$ is the predominant form of As(V) between pH 7 and 12, and $H_3AsO_3^\circ$ is the predominant form of As(III) below pH 9.3 (see pK_a values in Table 1). From these thermodynamic stability relationships, the standard state electrode potentials (i.e., pe^0 values) at pH 7 for the As(V)/As(III) redox couple can be shown to fall roughly between the $NO_3^-/N_2(g)$ and the $Fe(OH)_3(s)/Fe^{2+}$ redox couples (1). Of course, these calculations rely on assumptions regarding the activities of the oxidized and reduced species, and readers are reminded that the relative position of these redox couples depends dramatically on assumptions used to define the activity of the oxidized and reduced species, and on concentration gradients. Stumm and Morgan (1) show an excellent example of the former using the Fe(III)/Fe(II) redox couple where, depending on the complexing ligands present or the types of solid phases controlling the activity of Fe species, the calculated pe for the Fe(III)/Fe(II) redox couple can range anywhere from 1000 mV to less than -400 mV! Nevertheless, the electrode potential of the As(V)/As(III) couple is positioned such that oxidation–reduction reactions involving As(III) or As(V) species can be mediated either abiotically via a variety of potential oxidants and reductants important in soils and natural waters, and/or biotically via the biochemical pathways of both prokaryotic and eukaryotic organisms. More specifically, the electrode potential of the As(V)/As(III) couple falls within the dynamic range of redox potential values commonly observed in sediments, aquifers, lakes, rivers, oceans, and in soils with fluctuating water contents and/or water tables. Consequently, it is to be expected that the cycling of As(V)/As(III) is as common to terrestrial and aquatic systems as is the cycling of Fe(III)/Fe(II), N(V)/N(0)/N(−III), Mn(IV)/Mn(II), or S(VI)/S(IV)/S(0)/S(−II).

B. Solid Phases and Surface Complexation Reactions

Solid phases of As(V) or As(III) that may be important in natural system include the Fe, Mn, and Ca arsenates (13,14), and the arsenic (III) sulfides such as orpiment (As_2S_3), amorphous As_2S_3 and perhaps realgar (AsS) (15–18). In addition, arsenopyrite (FeAsS) is an important primary mineral source of As, and in some instances, a possible diagenetic or pedogenic mineral phase formed under reduced conditions (19–21). The arsenate salts of Fe, Ca, Mn, Mg, and Al are generally too soluble to control activities of As(V) in soil solutions and natural waters (14), although scorodite ($FeAsO_4 \cdot 2H_2O$) may form during the early stages of the oxidation of As-rich pyrite or FeAsS (22). More commonly, the activity of arsenate is controlled by surface complexation (sorption) reactions on (hydro)oxide minerals of Al, Mn, and especially Fe (23–29).

Arsenate demonstrates a strong affinity for most metal (hydr)oxides and clay minerals common to soils and waters, forming surface complexes analogous to phosphate. Arsenite, in contrast, is selective, exhibiting a strong preference for (hydr)oxides of iron. Both species do, however, show typical trends of oxyanions having either increased sorption with decreasing pH or a sorption maximum centered around the pK_a of the conjugate base of the oxyanion (24). Although arsenite is often touted to have a low sorption capacity for soil and water particulates relative to arsenate, in fact arsenite has a greater sorption capacity on ferrihydrite and goethite than arsenate except at very low solution concentrations (28,30,31). The selectivity of arsenite for iron oxides is exemplified by the data of Manning and Goldberg (30) and Manning et al. (31) (Fig. 2a). On iron oxides, the sorption capacity of arsenite compares or exceeds that of arsenate, the former showing only slight pH dependence with an adsorption envelope centered at pH 8 while the latter increases continually with decreasing pH. However, a reverse trend is noted for sorption on amorphous $Al(OH)_3$ (30). The sorption of arsenite is highly pH dependent, with the envelope centered at pH 9, and never completely diminishes the dissolved phase As even at low surface coverages (Fig. 2b). In contrast, arsenate sorption to $Al(OH)_3$ has limited pH dependence and is completely removed from solution across the pH range of 4–10.

Despite their different trends in sorption with pH, both arsenate and arsenite form similar surface complexes on goethite. Surface complexes of arsenate and

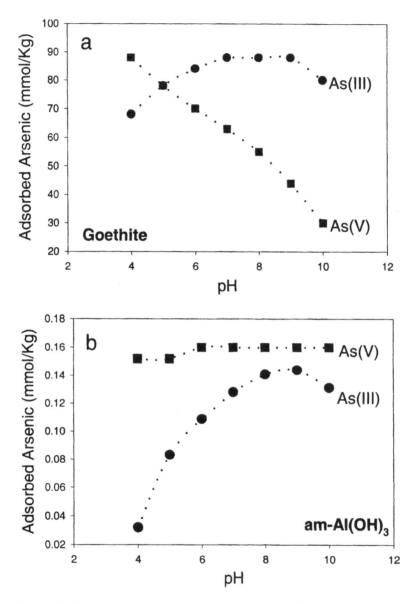

Figure 2 Sorption of arsenate and arsenite on (a) goethite ([As] = 0.267 mM, {goethite} = 2.5 g/L) and (b) amorphous $Al(OH)_3$ ([As] = 0.4 μM, {$Al(OH)_3$} = 2.5 g/L). (Data from Refs. 30 and 31).

arsenite on iron oxides have been examined using both infrared (32) and extended x-ray absorption fine structure (EXAFS) spectroscopy (33–37). Although the subtleties of these studies differ slightly, the general consensus of all but one (34) of the EXAFS studies is that the bidentate, binuclear complex formed by arsenate complexing with two adjacent iron octahedral corner sites predominates (Fig. 3)—a finding consistent with infrared studies of phosphate adsorption on iron (hydr)oxides (38) Interestingly, arsenite also forms a bidentate, binuclear complex on goethite comparable to arsenate with a slightly longer d(As-Fe) (31) (Fig. 3). Thus, based on the similar surface configuration of arsenate and arsenite on goethite, it is not surprising both ions form strong complexes with this class of solids.

In summary, the sorption of arsenate and arsenite by Fe(III)-oxide solid phases is perhaps one of the most important sinks for As in aquatic and terrestrial environments. It has been shown that Fe(III)-oxide phases can moderate the discharge of high concentrations of arsenic from hydrothermal fluids (39) and acid mine waters (40) by accumulating significant quantities of As. Analysis of the fate and transport of As through individual soils, aquifers, lakes, or watersheds (5,8,19,41–43) must consider the reactions of As(III) and As(V) species with Fe(III)-oxide phases. For example, the pH dependence of arsenate sorption by Fe(III) oxides has important implications for the management of acidic mine tailings remaining from mining activities. Although it is widely known and ac-

Surface Complexes on α-FeOOH
(dominant)

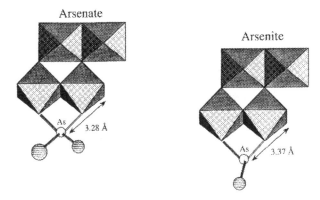

Figure 3 Dominant surface structure of arsenate and arsenite on goethite, showing Fe(III)-As bond lengths determined using extended x-ray absorption fine structure spectroscopy (EXAFS).

cepted that liming (and subsequent increases in soil pH) is desirable for enhancing revegetation and for minimizing the mobility of trace metals such as Cu, Zn, Cd, and Pb, it can have the opposite effect on As mobility where increases in soil pH values >8 have been shown to increase mobilization of arsenate out of the soil profile (42,44). Furthermore, as discussed below, the reductive dissolution of Fe(III) oxides containing sorbed As can represent an important source of As in natural waters. Given the propensity of both arsenate and arsenite to sorb to Fe(III) oxides, it is imperative to recognize that microorganisms capable of utilizing As(V) or As(III) for metabolic purposes often function under conditions where Fe(III)-oxide phases control As activities and perhaps As bioavailability.

III. REDUCTION OF ARSENIC (V) TO ARSENIC (III) IN SOILS AND NATURAL WATERS

A. Solubilization of As From Fe(III)-Oxide Phases

The reduction of As(V) to As(III) is commonly observed under microaerobic to anoxic conditions such as found in sediments, flooded soils, and aquifers (12,45–47). Under these conditions, Pt electrode measurements generally confirm the presence of *reduced* environments with Pt redox potentials commonly ranging from 100 to -200 mV. In the absence of dissolved sulfide, it is common to observe significant increases in concentrations of total soluble As (As_{TS}), predominantly as As(III) (12,45,48). We will consider the effect of sulfide species on the behavior of As(V) and As(III) in separate sections below. Given that arsenite is often more mobile and considered more toxic to microbiota and plants than arsenate, the increased concentration and mobility of As(III) in such systems can become a significant environmental health concern (43,47). One of the most plausible mechanisms explaining the increases in As_{TS} and the mobilization of As under reduced conditions involves the reductive dissolution of Fe(III)-oxide sorbing phases (48–51) allowing release of As(V) into the aqueous phase (Fig. 4), followed by the rapid reduction of aqueous As(V) via either abiotic or biotic pathways (discussed below). Alternatively, As(V) may be reduced to As(III) on the surface, then released upon reductive dissolution of the Fe(III)-oxide phase.

The enhanced dissolution rates of Fe(III)-containing minerals by microorganisms utilizing Fe(III) as a terminal electron acceptor (e.g., during respiration on lactate or acetate) is well established. In the case of Fe(III) oxides, the rates of reductive dissolution can depend markedly on the crystallinity and surface area of the solid (48,50); rates are generally much faster for high-surface-area, amorphous ferrihydrite compared with well-crystalline goethite. Jones et al. (48) showed that a mixed microbial culture can induce the reductive dissolution of an amorphous ferrihydrite phase resulting in significant solubilization of sorbed As(V) (at rates of 38 μM As hr^{-1}). Additionally, it has been shown that the

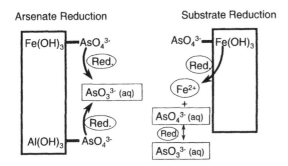

Figure 4   Schematic representation of potential pathways for the reductive dissolution of sorbed arsenic. Arsenic may be released from the solid either through reduction to arsenite (left) or, more likely, through degradation of the substrates (right) via reductive dissolution. As noted by Zobrist et al. (51), transformation to arsenite alone does not induce desorption from ferric-(hydr)oxide–rich environments, but may occur if Al oxides are the dominant substrate for retention.

Fe(III)-reducing *Shewanella alga* (strain BrY) can release As(V) at rates approaching 35 μM hr^{-1} during the reductive dissolution of scorodite (FeAsO$_4$ · 2H$_2$O) in the presence of 10 mM lactate as a C source (49). Although this particular microorganism is not capable of reducing As(V) to As(III) after reducing the Fe(III) phase, the reduction of aqueous As(V) to As(III) can occur quite rapidly in soil and water environments via other microorganisms. For example, Zobrist et al. (51) have recently reported on an anaerobic organism (*Sulfurospirillum barnesii*) capable of both reductive dissolution of Fe(III) oxides and reduction of As(V) to As(III). These mechanisms may be important to the release of As(III) from anaerobic sediments containing As sorbed to Fe(III) oxides. For example, the solubilization of As from aquifer sediments to groundwater in Bangladesh is thought to be due to the reductive dissolution of Fe(III)-oxide phases containing sorbed As (43). The highest groundwater concentrations of As are found at aquifer depths corresponding to zones with high organic matter contents where greater microbial activity would likely result in higher rates of reductive dissolution of the Fe(III)-oxide phase.

B.   Microbial Pathways for the Reduction of As(V)

Microbial reduction of arsenate to arsenite may occur by at least two principal mechanisms: dissimilatory reduction where As(V) is utilized as a terminal electron acceptor during anaerobic respiration (52–57), and detoxification activity which involves an arsenate reductase and an arsenite efflux pump (58–65). Given that both dissimilatory and detoxification processes are discussed elsewhere in this volume (see Chaps. 10 and 11), they are not reviewed in detail here. However, we do draw attention the fact that apparent first-order half-lives for microbially mediated As(V) reduction estimated from studies using pure bacterial cultures show that rapid rates of As(V) reduction are possible for either the dissimilatory reduction pathway or the detoxification process (Table 2). Further, it is important to note that As(V) reduction via detoxification will occur under both anaerobic and aerobic conditions and that, in soils and natural waters, the role of the detoxification pathway relative to dissimilatory reduction may be underestimated for several reasons. First, the concentration of As(V) in many As-contaminated soils is generally not high enough to represent a dominant electron acceptor capable of supporting significant growth of As(V)-respiring organisms, yet these concentrations may be sufficient to induce the *ars* genes, which appear to be widely distributed phylogenetically (60,65). Second, dissimilatory reduction of As(V) may require strict anaerobic conditions where organic acids such as lactate are the primary electron donors. However, As(III) has been repeatedly observed in highly oxidized environments (6,8,58) suggesting that reduction processes other than anaerobic respiration are responsible for the reduction of As(V), or that the reoxidation of As(III) is kinetically limited. We provide examples from our current research that demonstrate the reduction of As(V) to As(III) via microbial

Table 2 Approximate Time Scales of Microbiologically Mediated Reduction of Arsenate to Arsenite Performed by Bacteria Capable of Utilizing As(V) as a Terminal Electron Acceptor During Anaerobic Respiration (Dissimilatory Reduction) or by Organisms Either Containing, or Presumed to Contain, *ars* Genes that Code for an Arsenate Reductase and an Arsenite Efflux Pump

Microbial process	Organism/experimental conditions	Half-life (hr)[a]	Ref.
Respiration	• SES-3, *Geospirillum*		
	C source = lactate, initial As(V) = 5–10 mM	6–30	54
	• *Desulfotomaculum*		
	C source = lactate, initial As(V) = 5–10 mM	20–45	56
	• *Desulfomicrobium*		
	C source = lactate, initial As(V) = 8.2 mM	30–40	55
Detoxification	Diverse genera (e.g.)		
	• *Desulfovibrio*; while using SO₄ and lactate	80–100	55
	• *Clostridium*; while fermenting glucose	6–12	48
	• *Psuedomonas*, other aerobic heterotrophs	6–50	58

[a] Half-lives were estimated assuming first-order kinetics and are dependent on the experimental conditions, some of which are provided.

detoxification, similar to that described in microorganisms containing the well-characterized *ars* genes.

In one recent study by Macur et al. (58), the solubilization and speciation of aqueous As from acidic (pH 3) mine tailings (containing 20% Fe and 0.3% As) was studied after liming amendments (lime autoclaved prior to addition) under column transport conditions. In sterile (killed) columns, increases in mine tailing pH to 7.8 after liming resulted in significant solubilization of As primarily as As(V) (Fig. 5), consistent with the pH dependence of As(V) sorption reactions on Fe(III)-oxide minerals (see Fig. 2a). These results verified those obtained by

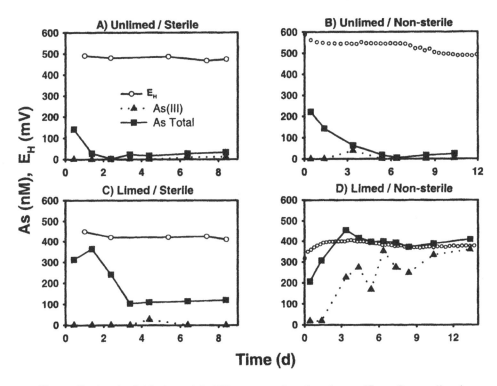

Time (d)

Figure 5   Total soluble As and As(III) concentrations in column effluent from unlimed (pH = 3.9) and limed (pH = 7.8) reprocessed mine tailings (RT) containing 0.3% (w/w) total As, under sterile and nonsterile conditions. Influent was not supplemented with C. Steady-state As concentrations in unlimed treatments (A and B) were less than ~20 nM, but increased to 99 (±14) nM As in sterile limed treatments (C) and to 300 (±61) nM As in nonsterile treatments (D). The increase in total soluble As in the nonsterile treatments was predominantly As(III), despite oxic column conditions as measured with a Pt electrode positioned in the center of the columns. (From Ref. 58.)

Jones et al. (44) who used similar tailings materials and demonstrated the potential for As mobilization after increasing the pH of acidic mine tailings. However, in nonsterile columns, the solubilization of As after liming was threefold greater than that observed under sterile conditions (steady-state concentrations of As increase from 0.1–0.3 µM). Moreover, the speciation of As in the column effluent showed that the majority of As_{TS} in the nonsterile columns was As(III) despite the fact that the columns were maintained under aerobic conditions, the Pt electrode readings remained above 400 mV, and there was no evidence of reductive dissolution of the Fe solid phase. Tailings samples from these columns were used for molecular analysis of 16S rDNA genes contained in the total microbial population and also for cultivation of As(V)-reducing microorganisms. Despite attempts to cultivate anaerobic As(V)-respiring organisms on lactate, the only As(V)-reducing microorganisms that could be cultivated were obtained under aerobic conditions. Experiments with these isolates (*Caulobacter*, *Sphingomonas*, and *Rhizobium*-like isolates) demonstrated that all could reduce As(V) rapidly under well-aerated conditions (Fig. 6). Unequivocal verification that each isolate was a component of the column microbial community came from comparing the denaturing gradient gel electrophoresis (DGGE) (66) profile of the community DNA with the band signatures derived from each isolate. Sequencing of cloned bands from the column environment showed perfect matches with similarly migrating, individual bands obtained from each isolate (Fig. 7). In other column studies, Macur et al. (58) isolated three additional bacteria, including *Pseudomonas fluorescens*, *Pseudomonas aeruginosa*, and *Sphingomonas echinoids*, all of which could reduce As(V) under aerobic conditions. For all six of these heterotrophs, reduction of As(V) was coincident with microbial growth under aerobic conditions, and maximum culture optical densities were not significantly different in the absence of As(V). These results suggest that growth was not coupled to reduction of As(V) and that growth was not inhibited by the presence of As(V) at these levels. Cai et al. (60) demonstrated that strains of *P. aeruginosa* and *P. fluorescens*, which are closely related to two of the isolates obtained in Macur et al. (58), carry *ars* operon homologues that confer increased resistance to As. The results from Macur et al. (58) are some of the first to verify that the rapid reduction of As(V) to As(III) in soils and natural waters can occur via microbial processes under aerobic conditions, and by mechanisms that appear consistent with detoxification encoded by *ars*-like genes rather than dissimilatory reduction of As(V) via anaerobic respiration. In summary, we suggest that the ability to reduce As(V) to As(III) via the detoxification pathway may be a widely distributed trait in microbes inhabiting soil and aquatic environments; if so, this pathway must be considered to fully appreciate As cycling in natural systems, especially if the reduction of As(V) occurs rapidly under well-aerated conditions.

It is also possible that obligate anaerobic microorganisms reduce As(V) via detoxification pathways. Jones et al. (48) isolated a glucose-fermenting *Clostrid-*

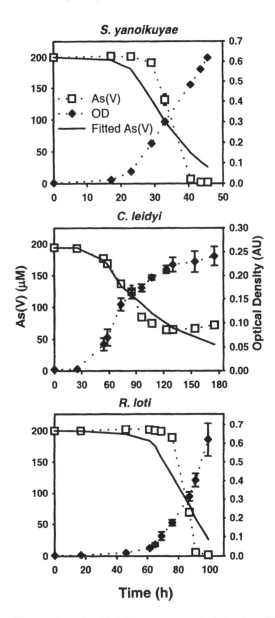

Figure 6 Microbial biomass (as optical density, OD) and As(V) concentrations as a function of time for As(V)-reducing isolates obtained from reprocessed tailings (see Figs. 3 and 4) incubated under aerated conditions. Serum bottles inoculated with *S. yanoikuyae*, *C. leidyi*, and *R. loti* contained 50 μM P and 200 μM As(V). Error bars represent standard errors of three replicate serum bottle experiments. (From Ref. 58.)

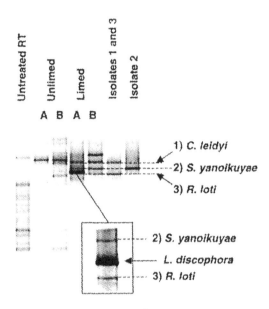

Figure 7 Denaturing gradient gel (35–70%) of PCR-amplified 16S rDNA fragments from untreated reprocessed mine tailings (RT), unlimed RT columns (replicates A and B), limed RT columns (replicates A and B), and *C. leidyi, S. yanoikuyae,* and *R. loti*– like isolates cultivated from the tailings after column treatments. To resolve the *S. yanoiku-yae* and *R. loti* bands in the limed RT column (replication A), the sample was run with a narrower (40–60%) denaturing gradient (inset). Comigration of isolate bands with bands from limed RT columns is indicated with dashed lines (From Ref. 58.)

ium from an As-contaminated soil in the Madison River basin, Montana, that reduces As(V) during growth but does not require or benefit from the presence of As(V). Additional studies with this organism showed that although it is capable of rapidly reducing aqueous As(V) to As(III), it does not appear to reduce As(V) sorbed to ferrihydrite, or to result in the solubilization of As via reductive dissolution of the Fe(III)-oxide phase (67). In summary, given the number of diverse microorganisms [including a chemolithotrophic *Thiobacillus* sp. (59) and an anaerobic sulfate reducer, *Desulfovibrio* sp. (55)] that have been reported to reduce As(V) to As(III) via detoxification pathways, we need not invoke anaerobic respiration where As(V) is utilized as a terminal electron acceptor to observe rapid rates of As(V) reduction in soils and waters. In fact, it may be more common that solubilization of As from soils and sediments under anoxic conditions is first initiated by the reductive dissolution of As(V)-sorbing Fe(III)-oxide phases, followed by the reduction of As(V) to As(III) by microorganisms containing *ars* or *ars*-like inducible genes.

C. Abiotic Pathways Responsible for the Reduction of As(V) to As(III)

At this juncture it is important to clarify the potential role of chemical reductants (electron donors) on As(V) reduction rates commonly observed in soils and natural waters. One of the chemical species that might contribute to As(V) reduction rates is dissolved sulfide (H_2S, HS^-), especially at low pH. Studies by Cherry et al. (9), Newman et al. (56), and Rochette et al. (68) suggest that the reduction of arsenate by dissolved sulfide is very slow at circumneutral pH values. However, at pH values less than 5, the reduction rates of arsenate due to sulfide may be significant in natural systems, where half-lives as short as 21 hr have been reported (68) for this abiotic pathway (Table 3). Rochette et al. (68) also revealed the potential importance of intermediate As-O-S species in electron transfer reactions between sulfide and arsenate, such as $H_2As^VO_3S^-$, $H_2As^{III}O_2S^-$, and $H_2As^{III}OS_2^-$. It is not known whether these chemical species may also serve as important redox active species for microbial metabolism. These authors have also compared the rates of As(V) reduction in the presence of sulfide versus those rates expected via dissimilatory reduction by an arsenate-respiring organism (strain SES-3) (54) and for those measured in lake sediments (69); at pH values less than 5, reduction rates due to dissolved sulfide can become more significant than reduction rates due to anaerobic respiration where As(V) is used as the terminal electron acceptor (Fig. 8).

Table 3 Time Scales of Abiotic Reduction Rates of Arsenate to Arsenite in the Presence of Dissolved Sulfide, at pH Values Ranging from 4–7

Reductant	Experimental conditions		Half-life (hr)[a]	Ref.
H_2S/HS^-	pH = 4	133 µM initial As(V)		
		• 25 µM H_2S	346	68
		• 270 µM H_2S	21	
	pH = 6.8	5.5 mM initial As(V)		
		• 1 mM H_2S	500	56
		• 3 mM H_2S	900	
	pH = 7	0.7 µM initial As(V)		
		• 10 mM H_2S	63	9

[a] Half-lives were estimated assuming first-order kinetics and are dependent on the experimental conditions, some of which are provided.

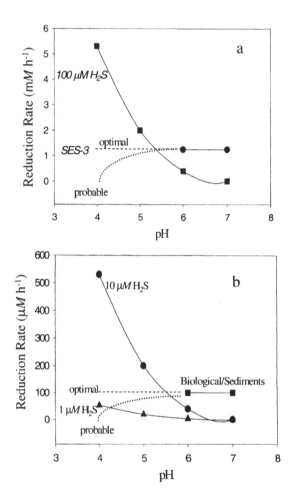

Figure 8 Predicted contributions, based on comparison of rates, for biological and sulfide reduction of arsenate in (a) pure cultures of arsenate respiring bacteria and (b) for natural settings. (Rate data for biological reduction by strain SES3 are from Ref. 54, and for sediments from Ref. 69; data from Refs. 56 and 68 used to calculate sulfide reduction rates.)

D. Reactions of As(III) with Dissolved Sulfide

The speciation and solubility of As(III) in natural systems can be further complicated in the presence of dissolved sulfide. It has been noted that in environments where As(V) and SO_4^{2-} are reduced respectively to As(III) and dissolved sulfide (either H_2S or HS^-, depending on pH), concentrations of As_{TS} actually decrease as a result of formation of As(III)-sulfide phases (19,20,56,70). For example, the arsenate-respiring, anaerobic organisms described by Macy et al. (55) and Newman et al. (56) can utilize both arsenate and sulfate as electron acceptors, and may subsequently precipitate As_2S_3 in the surrounding media. The As-sulfide phases have been suggested as important sinks for As(III) in reduced environments (19,21). In one specific example, Langner et al. (70) performed a greenhouse study using controlled wetland chambers in the presence and absence of wetland plants and observed rapid reduction of As(V) and S(VI) to As(III) and S(-II) species, respectively, followed by a slower but significant decrease in As_{TS} (Fig. 9A). The gravel substrates utilized in this wetland environment were transferred to serum bottles for further analysis; additional inputs of As(V) and S(VI) resulted in the formation of an amorphous As_2S_3 phase (Fig. 9B), similar to that reported in Newman et al. (56). Consequently, under conditions that result in sulfate reduction and subsequent formation of As_2S_3 or FeAsS-like phases, the

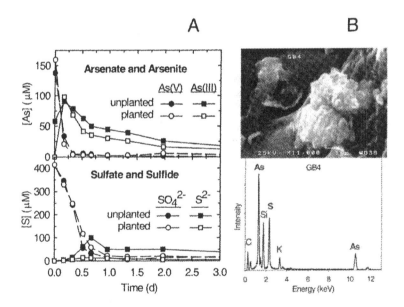

Figure 9 Reduction of As(V) to As(III) and reduction of SO_4^{2-} to H_2S (A) in constructed wetland chambers designed to treat wastewater C and N. The formation of As(III) and H_2S results in the subsequent precipitation of amorphous As_2S_3 phases as indicated by decreasing As(III) concentrations, and as shown (B) in the scanning electron micrograph (SEM) with corresponding elemental analysis (energy dispersive analysis of x-rays, EDX). Differences between unplanted and planted chambers are not discussed here; however, patterns in the reduction of As(V) followed by precipitation of As-S phases were similar. (From Ref. 70.)

concentrations of As_{TS} [primarily $H_3AsO_3°$ and soluble As(III)-sulfide complexes] may be controlled by equilibrium with As(III)-sulfide phases. The solubility of orpiment (As_2S_3) and amorphous As_2S_3 has been the subject of considerable uncertainty (15,16,18); furthermore, debate continues regarding the nature and number of possible As(III)-sulfide complexes present under different solution conditions (16,68). However, the fact that $H_3AsO_3°$ forms significant aqueous complexes with sulfide (see Table 1) results in a complicated As(III)-solubility behavior (18) as a function of pH and sulfide concentration. For example, the solubility of amorphous As_2S_3 reaches a minimum at activities of HS^- ranging from $10^{-6}–10^{-3}$ M over the pH range 4–9 (Fig. 10). Higher concentrations of HS^- result in increases in As_2S_3 solubility as a result of the formation of $H_2As_3S_6^-$ soluble complexes. Consequently, in some reduced environments, the mobilization of As(III) may actually increase if HS^- levels increase above 10^{-3} M. It should also be noted that the beneficial decrease in As(III) concentrations observed as a result of As_2S_3 formation is extremely sensitive to reoxidation and solubilization of As_2S_3 solid phases. Nevertheless, the formation of As(III)-sulfide phases under wetland environments may represent one possible immobili-

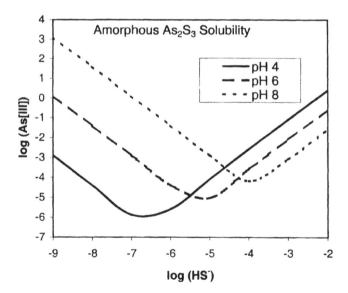

Figure 10 Solubility of amorphous As_2S_3 as a function of pH and HS^- activity (see Table 1). The total concentration of As(III) shown on the y-axis was taken as the sum of $H_3AsO_3°$ and $H_2As_3S_6^-$. The increase in As_2S_3 solubility that occurs at HS^- activities greater than $10^{-6}–10^{-3}$M (depending on pH) is due to the formation of the $H_2As_3S_6^-$ complex.

zation strategy for minimizing As transport to surface water or groundwater (19,70).

IV. OXIDATION PATHWAYS OF ARSENIC (III) TO ARSENIC (V) IN SOILS AND NATURAL WATERS

A. Chemical Processes

There are numerous potential oxidants that may contribute to the oxidation of As(III) in soils and natural waters and these are discussed briefly here along with the pathways mediated by microorganisms (Tables 4 and 5). Solutions of arsenite [As(III)] salts are generally fairly stable under room conditions and do not show rapid oxidation to As(V) (71,72). In fact, the apparent half-life reported for As(III) oxidation via $O_2(g)$ has been reported to be approximately 1 yr (71); however, other recent work suggests that homogeneous oxidation of As(III) can take place, but only at pH values >9 (30). Needless to say, although thermodynamically favorable, it is not considered an important mechanism of As(III) oxidation in most soil and water environments. Aqueous Fe(III) may contribute to

Table 4 Approximate Time Scales of Abiotic Pathways that May Contribute to Oxidation of As(III) to As(V) in Soils and Natural Waters

Oxidant	Experimental conditions/comments	Half-life (hr)[a]	Ref.
$O_2(g)$	Very slow, not considered significant	8760	71
Fe(III)	pH = 5, 10µM Fe	336	
	pH = 2, 5,000 µM Fe	29	9
	Initial As(III) = 0.65 µM		
H_2O_2	• pH = 8.1 H_2O_2 = 227 µM	14	
	• pH = 10.3 H_2O_2 = 227 µM	0.02	75
	• pH = 8.1 H_2O_2 = 50 µM[b]	60	
$Mn^{IV}O_2(s)$	Birnessite, 1.4 g/L, 1.3 mM As(III), pH = 7	2.6	
	Pyrolusite, 1.4 g/L, 1.3 mM As(III), pH = 7	1580	73
	Birnessite, 0.2 g/L, 0.1 mM As(III), pH = 4	0.15	
	Birnessite, 0.2 g/L, 0.1 mM As(III), pH = 6.8	0.3	74
Fe(III)-oxalate + UV light	Initial As(III) ~ Fe(III) = 18 µM, Oxalate = 1 mM		
• OH^- radical	• pH = 7	1	
	• pH = 5	0.1	76
	• pH = 3	0.05	

[a] Apparent half-lives were estimated assuming first-order kinetics and are dependent on the experimental conditions, some of which are provided.
[b] More realistic upper limit of H_2O_2 concentration in soils and natural waters.

Table 5 Approximate Time Scales of Microbiologically Mediated Oxidation of Arsenite to Arsenate, Using Examples of Pure Culture Isolates or Microbial Populations Present in Hot-Spring Ecosystems

Microbial process	Organism/experimental conditions	Half-life (hr)[a]	Ref.
Chemolithoautotroph	NT-26 (novel species belonging to *Agrobacterium/Rhizobium* branch)		
	• Periplasmic arsenite oxidase	1.8	82
	• 5 mM As, log growth		
Detoxification	*Alcaligenes faecalis*		
	• Cytoplasmic arsenite oxidase	—	83
	• Fe or Fe-S protein, e⁻ acceptor for detoxification of As(III), energy yield not clear		
Mechanism unknown	Microorganisms associated with submerged macrophytes: thermal stream, Hot Creek, CA, pH 8.3		
	• in situ rates	0.3	87
Mechanism unknown	Microbial mats: thermal spring, Norris Basin, Yellowstone National Park, WY, pH 3.2		
	• in situ rates	0.017	85, 86
	• ex situ rates	0.25	

[a] Apparent half-lives were determined assuming first-order kinetics and are dependent on the experimental conditions, some of which are provided.

As(III) oxidation rates at low pH (e.g., pH <3) and high Fe(III) concentration (9), but does not likely play a significant role in most natural waters. Further, although oxidation of sorbed As(III) to As(V) on Fe(III)-oxide surfaces may occur to a limited (~20%) extent (32) other recent studies do not suggest significant oxidation of As(III) by Fe(III)-oxide surfaces (31). Conversely, $Mn^{IV}O_2$ solid phases such as birnessite, are effective oxidants of As(III). Under controlled laboratory conditions at high MnO_2 suspension densities, the apparent half-lives for oxidation of As(III) are less than 1 hr (73,74). Based on such rapid rates of As(III) oxidation, these authors suggested that the oxidation of As(III) in marine sediments is controlled by MnO_2 phases. At a minimum, it appears that the Mn(IV)-oxide surfaces are indeed an important pathway that may contribute to As(III) oxidation rates in natural systems.

Recent studies have also evaluated the oxidation of As(III) in the presence of H_2O_2 (75) and irradiated solutions of Fe(III) oxalate (76). Under appropriate conditions, H_2O_2 can be shown to be an effective oxidant for As(III) (Table 4).

Pettine et al. (75) have shown that at pH values of 10 [where $H_2AsO_3^-$ is the dominant As(III) solution species] and at high H_2O_2:As(III) ratios (350:1), half-lives of As(III) oxidation are as short as 0.02 hr. The oxidation pathway is pH dependent and rates decline significantly with decreasing pH. At pH values <8, and at more realistic concentrations of H_2O_2, this pathway is unlikely to contribute significantly to oxidation rates of As(III) observed in most natural systems. Conversely, in natural waters exposed to UV light, the oxidation rates of As(III) via free-radical–generating reactions such as the ferrioxalate system may be significant (76). For example, at pH values ranging from 3–7, rapid oxidation of 18 μM As(III) can be achieved in the presence of irradiated 18 μM Fe(III) and 1 mM oxalate solutions (Fig. 11), with apparent half-lives <1 hr (Table 4). The oxidation of As(III) was shown to correlate with the production of the ·OH$^-$ free radical species produced during the decomposition of H_2O_2 in the presence of Fe(II) (77). Consequently, although the production of H_2O_2 is required to generate the ·OH$^-$ free-radical species, H_2O_2 is not a significant oxidant of As(III) under these conditions. This particular As(III) oxidation pathway may only play a limited role in soils and sediments due to the logical constraints of UV-light penetration; however, this may be a very important process in surface waters where it would not be uncommon to expect a significant fraction of dissolved Fe(III) to be complexed with UV-absorbing organic chromophores. Further, this

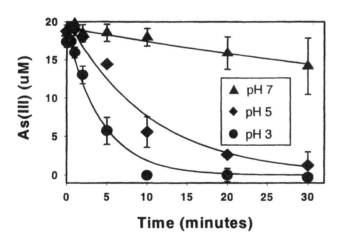

Figure 11 Oxidation of As(III) to As(V) in irradiated ferrioxalate solutions containing 1 mM oxalate, and 18 μM Fe(III) and As(III). Solutions were irradiated using a halogen light source (97 μE cm^{-2}hr^{-1} between 300 and 500 nm) at pH values of 3, 5, and 7. The rapid oxidation of As(III), especially at pH 3–5, was shown to be due to the production of ·OH$^-$ free-radical species during irradiation of ferrioxalate (From Ref. 76.)

particular oxidation pathway is a reminder that the role of free-radical species in mediating oxidation–reduction reactions is probably under looked in many environments.

B. Microbial Oxidation of As(III) to As(V)

Although several of the abiotic processes discussed above may contribute to As(III) oxidation rates in natural systems, it is clear that microbial pathways are also important in the cycling of As(III) to As(V) in the environment (Table 5). The oxidation of As(III) by microorganisms has been studied for many years (78–81), but only recently have definitive mechanisms of oxidation been elucidated (see Chaps. 12–14). For example, the work by Santini et al. (82) clearly documents an organism capable of chemolithoautotrophic growth on As(III). The half-life of As(III) oxidation under optimal laboratory growth conditions was reported to be approximately 1.8 hr (Table 5). Another potential mechanism of As(III) oxidation is also referred to as "As detoxification," wherein little to no energy appears to be acquired by the cell and often the oxidation does not occur until cultures have reached stationary phase (78–80). As discussed in more detail elsewhere in this volume (see Chap. 15), the oxidation mechanism characterized in *Alcaligenes* is due to an inducible arsenite oxidase found on the outer surface of the cytoplasmic membrane (83). The oxidation of As(III) has also been documented in cases where microorganisms excrete fatty acids into the surrounding media conferring As(III)-oxidative capability to the filtered spent culture fluids (81,84); however, it is not entirely clear what, if any, regulatory mechanisms may be operative, and whether these extracellular oxidants are actually inducible with As(III). Certainly, additional work is necessary to fully appreciate the diversity of bacterial species in soils and waters that possess the capability to oxidize As(III) to As(V), and the various mechanisms by which As(III) oxidation may actually occur in the environment.

A recent research thrust by Inskeep and McDermott (85,86) has focused on examining structure–function relationships within an As(III)-oxidizing thermal, acidic spring in Yellowstone National Park (Norris basin, WY). Biogeochemical analyses of aqueous samples have documented rapid in situ As(III) oxidation rates with half-lives as low as 0.02 hr (Fig. 12). These represent As(III) oxidation rates that are roughly an order of magnitude greater than previously documented for the oxidation of As(III) observed in Hot Creek, CA, mediated by microorganims associated with submerged macrophytes (87). The rapid oxidation of As(III) observed in our study is biological, and is associated with dense microbial mats covering the bottom of the spring, which ranges in temperature from approximately 62°C at the source to 48°C in the outflow channel. Specifically, the oxidation of As(III) is most highly correlated with a brown microbial mat comprised of filamentous bacteria in association with amorphous Fe(III) hydroxides exhib-

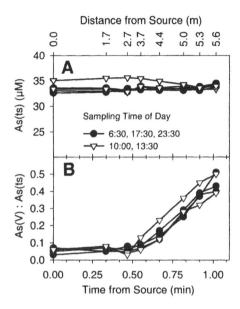

Figure 12 (A) Concentrations of total soluble As and (B) the ratio of As(V)/As$_{TS}$ in an acidic thermal (62°C) spring (Norris basin, Yellowstone National Park) as a function of distance (or travel time) from the point of discharge. The rapid oxidation of As(III) to As(V) occurs from approximately 4–6 m, over the course of 0.5 min in an area dominated by brown microbial mats on the spring bottom, and is independent of daylight intensity. The rapid oxidation is due to microbial activity and the populations responsible are under investigation (From Refs. 85 and 86.)

iting a high As/Fe ratio of nearly 0.7. The primary organisms responsible for As(III) oxidation are still under study; however, extraction of community DNA and phylogenetic analysis of near-full-length PCR-amplified 16S rRNA genes suggest the presence of *Hydrogenobacter*, a novel *Desulfurella* sp., *Meithermus*, *Acidomicrobium*, and several novel archael 16S rDNA sequences of both *Crenarchaeota* and *Euryarchaeota* (86). The majority of archaeal sequences were most similar to sequences obtained from marine hydrothermal vents and other acidic hot springs, although the level of similarity was typically just 90%. The oxidation of As(III) in this thermal spring may be mediated by these unknown archaeal taxa, and/or bacteria listed above. Current efforts are underway to enrich for the As(III)-oxidizing populations responsible for the rapid oxidation observed in situ, and to determine the biological basis for As(III) oxidation.

In other recent collaborative work, investigators have been studying the oxidation of As(III) to As(V) in an As-contaminated soil (Van Dyke) from the Madison River basin in Montana (88). This soil comes from a site downriver from the geothermal sources of As in Yellowstone National Park (Wyoming) and contains elevated levels of total and soluble As (89). Macur et al. (88) conducted experiments utilizing carefully designed, aseptic transport columns as enrichment vessels to characterize rates of As transformation that potentially can be linked to specific microbial populations. In one treatment, the column received influent containing 75 µM As(III) and the effluent was monitored for several weeks under unsaturated flow conditions. Characterization of the effluent showed that As(III) oxidation occurred rapidly and continued for 15 days (Fig. 13A). Other column treatments included As(V) in the influent under either saturated or unsaturated conditions. Reduction of influent As(V) to As(III) was complete in the column conducted under saturated conditions, whereas little net reduction of As(V) was observed under unsaturated flow (data on As not shown). Soil samples collected after termination of column experiments were subjected to cultivation efforts on various media and to molecular analyses that coupled polymerase chain reaction (PCR) and DGGE to qualitatively assess the presence of specific microbial populations based on variation in the community 16S rRNA gene sequences. The DGGE banding patterns of 16S rDNA fragments showed that the various treatments enriched for several consistent populations and several different microorganisms (Fig. 14). In addition, as was shown for the mine tailings experiments, DGGE fingerprints of several of the pure culture isolates could be matched exactly (based on sequence analysis) with similarly migrating bands cloned from

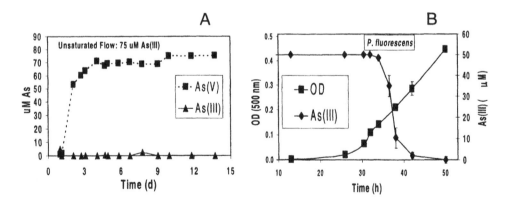

Figure 13 Oxidation of As(III) in column studies using an As-contaminated soil from the Madison River basin (A). The influent contained 75 µM As(III), and after 3 days, the effluent was dominated by As(V). Under sterile conditions, effluent As remained as As(III) (not shown). The oxidation of As(III) under serum bottle conditions is shown (B) for one of the bacterial isolates (*Pseudomonas fluorescens*) cultivated from this soil (see Table 6) (From Ref. 88.)

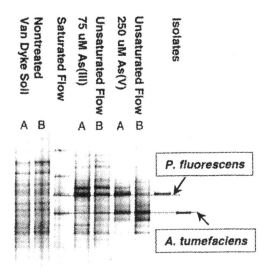

Figure 14 Denaturing gradient gel (35–70%) of PCR-amplified 16S rDNA fragments from nontreated Van Dyke soil (Madison River basin) and from column experiments using this soil under saturated flow [As(V) influent], unsaturated flow [As(III) influent], and unsaturated flow [As(V) influent] conditions. In addition, the banding patterns for two of the isolates cultivated from these treatments (*Psuedomonas fluorescens* and *Agrobacterium tumefaciens*) are shown to comigrate with bands in the environmental samples as indicated with dashed lines. (From Ref. 88.)

the community DGGE profile (e.g., *Psuedomonas fluorescens* and *Agrobacterium tumefaciens*).

In total, seven unique isolates from the Madison River soil were characterized regarding their ability to either oxidize As(III) or reduce As(V) (Table 6); four were found to reduce As(V) under aerobic conditions and three were found to oxidize As(III). For example, the *P. fluorescens*–like isolate was found to oxidize As(III) rapidly in separate serum bottle experiments (Fig. 13B). Intriguingly, two isolates sharing 99% 16S rRNA gene sequence similarity to *Agrobacterium*, but identical to each other, responded in opposite fashion to As treatment: one strain oxidized As(III) and one strain reduced As(V) (Table 6). We are currently investigating the possibility that plasmid-borne *ars* genes are absent from the *Agrobacterium* isolate capable of oxidizing As(III). More importantly, this study documents the occurrence of both As-oxidizing and As-reducing organisms in the same soil, establishing definitive linkages among specific microbial popula-

Table 6 Phylogenetic Affiliation and As(III) Oxidation or As(V) Reduction Rates of Isolates Cultivated from Unsaturated Columns Containing Van Dyke Soil (VD). Maximum Rates (Max rate) Are Transformation Rates of As Observed Under Aerated Serum Bottle Conditions

	Nearest GenBank relative[a]				As transformations	
Isolate	Species	Phylogenetic group	Percent similarity	GenBank accession no.	Transformation	Max rate (mM day^{-1})
31	Pseudomonas fluorescens	γ Proteobacteria	99.1	AJ278813	Oxidation	0.33
34	Variovorax paradoxus	β Proteobacteria	99.3	D30793	Oxidation	0.09
52	Agrobacterium tumefaciens	α Proteobacteria	99.9	M11223	Oxidation	0.26
36	Flavobacterium heparinum	Sphingobacteria	94.9	M11657	Reduction	0.31
42	Agrobacterium tumefaciens	α Proteobacteria	99.9	M11223	Reduction	0.93
46	Microbacterium sp.	Actinobacteria	98.1	Y14699	Reduction	0.70
51	Arthrobacter aurescens	Actinobacteria	99.6	X83405	Reduction	1.16
53	Arthrobacter sp.	Actinobacteria	97.8	AF197050	Reduction	0.41

[a] Phylogenetic affiliation was determined by comparing near-full-length 16S rDNA sequences of these isolates to sequences in the GenBank database (90).
Source: Ref. 88.

tions and As(V)/As(III) cycling. The capabilities to oxidize or reduce arsenic appear widely distributed among soil microorganisms. Consequently, it is easy to conceive how transient environmental conditions may select or favor specific members of the microbial community, with resulting consequences on the "equilibrium" chemistry of As. This interrelationship among abiotic and biotic factors is in essence what controls As(V)/As(III) cycling, and in turn, the solubility, mobility, and subsequent bioavailability of As within watersheds.

V. SUMMARY AND CONCLUSIONS

We have provided an overview of possible abiotic and microbiological processes that may contribute to transformations between As(V) and As(III) in soils and natural waters. Clearly, predictions regarding the fate of As in real systems require an integration of physical, chemical, and biological processes. A greater understanding of factors contributing to As cycling will allow for the development of suitable models for predicting As speciation, bioavailability, and subsequent risk exposure. Specifically, the identification of key elementary processes (whether abiotic or biotic) and characterization of rate constants under realistic conditions will provide the foundation for a more comprehensive conceptual model for predicting transformation rates between As(V) and As(III), and subsequent partitioning between solid and aqueous phases, which may be necessary for transport, or for risk exposure models.

Several of the more important processes controlling the oxidation and reduction of inorganic As species (i.e., arsenate and arsenite) are discussed in this chapter, including: surface complexation of As on Fe(III) (hydr)oxides, precipitation of As solid phases such as As_2S_3; reductive dissolution of Fe(III) phases containing sorbed As; microbiological reduction of As(V) via dissimilatory or detoxification processes; abiotic reduction of As(V) via dissolved sulfide: abiotic oxidation of As(III) via either $MnO_2(s)$, H_2O_2, or light-induced free-radical reactions; and microbiological oxidation of As(III) not coupled to microbial growth or by chemolithoautotrophs. In several examples, we have demonstrated the promise of linking observed transformation rates of As(V)/As(III) with specific members of the microbial community using both cultivation and cultivation-independent molecular methods and subsequent phylogenetic analysis (16S rDNA). A greater understanding of the functional pathways responsible for biologically mediated As transformation rates in real systems would also be beneficial, especially if these pathways can be linked to characteristic rates of transformation in specific chemical and physical contexts. In this regard, patterns in microbial population dynamics across environments, taken together with hydrological and chemical controls, may be quite useful for guiding predictions of the environmental fate of As in soils and natural waters.

ACKNOWLEDGMENTS

The authors appreciate support from the U.S. Environmental Protection Agency (Projects R827457-01-0 and R825403-01-0), NASA (Project NAG5-8807), and the Montana Agricultural Experiment Station (Projects 914398 and 911310). Although supported in part by the U.S. Environmental Protection Agency, it has not been subjected to the Agency's required peer and policy review and therefore does not necessarily reflect the views of the Agency, and no official endorsement should be inferred.

REFERENCES

1. W Stumm, JJ Morgan. Aquatic Chemistry. 3rd ed. New York: Wiley, 1996, pp 425–515.
2. W Stumm. Interpretation and measurement of redox intensity in natural waters. Schweiz Z Hydrol 46:291–296, 1984.
3. TR Holm, CD Curtiss. A comparison of oxidation-reduction potentials calculated from the As(V)/As(III) and Fe(III)/Fe(II) couples with measured platinum-electrode potentials in groundwater. J Contam Hydrol 5:67–81, 1989.
4. MI Abdullah, Z Shiyu, K Mosgren. Arsenic and selenium species in the oxic and anoxic waters of the Oslofjord, Norway. Marine Pollut Bull 31:116–126, 1995.
5. AC Aurillo, RP Mason, HF Hemond. Speciation and fate of arsenic in three lakes of the Aberjona watershed. Environ Sci Technol 28:577–585, 1994.
6. P Seyler, J-M Martin. Biogeochemical processes affecting arsenic species distribution in a permanently stratified lake. Environ Sci Technol 23:1258–1263, 1989.
7. Y Sohrin, M Matsui, M Kawashima, M Hojo, H Hasegawa. Arsenic biogeochemistry affected by eutrophication in Lake Biwa, Japan. Environ Sci Technol 31:2712–2720, 1997.
8. HM Spliethoff, RP Mason, HF Hemond. Interannual variability in the speciation and mobility of arsenic in a dimictic lake. Environ Sci Technol 29:2157–2161, 1995.
9. JA Cherry, AU Shaikh, DE Tallman, RV Nicholson. Arsenic species as an indicator of redox conditions in groundwater. J Hydrol 43:373–392, 1979.
10. WR Cullen, KJ Reimer. Arsenic speciation in the environment. Chem Rev 89:713–764, 1989.
11. JF Ferguson, J Gavis. A review of the arsenic cycle in natural waters. Water Res 6:1259–1274, 1972.
12. PH Masscheleyn, RD Delaune, WH Patrick Jr. Effect of redox potential and pH on arsenic speciation and solubility in a contaminated soil. Environ Sci Technol 25:1414–1419, 1991.
13. JV Bothe Jr, PW Brown. The stabilities of calcium arsenates. J Haz Mater B69:197–207, 1999.
14. M Sadiq. Arsenic chemistry in soils: An overview of thermodynamic predictions and field observations. Water Air Soil Pollut 93:117–136, 1997.

15. LE Eary. The solubility of amorphous As_2S_3 from 25 to 90°C. Geochim Cosmochim Acta 56:2267–2280, 1992.

16. GR Helz, JA Tossell, JM Charnock, RAD Pattrick, DJ Vaughan, CD Garner. Oligomerization in As(III) sulfide solutions: Theoretical constraints and spectroscopic evidence. Geochim Cosmochim Acta 59:4591–4604, 1995.

17. NF Spycher, MH Reed. As(III) and Sb(III) sulfide complexes: An evaluation of stoichiometry and stability from existing experimental data. Geochim Cosmochim Acta 53:2185–2194, 1989.

18. JG Webster. The solubility of As_2S_3 and speciation of As in dilute and sulphide-bearing fluids at 25 and 90°C. Geochim Cosmochim Acta 54:1009–1017, 1990.

19. H McCreadie, DW Blowes, CJ Ptacek, JL Jambor. Influence of reduction reactions and solid-phase composition on porewater concentrations of arsenic. Environ Sci Technol 34:3159–3166, 2000.

20. JN Moore, WH Ficklin, C Johns. Partitioning of arsenic and metals in reducing sulfidic sediments. Environ Sci Technol 22:432–437, 1988.

21. KA Rittle, JI Drever, PJS Colberg. Precipitation of arsenic during bacterial sulfate reduction. Geomicrobiol J 13:1–11, 1995.

22. PM Dove, JD Rimstidt. The solubility and stability of scorodite, $FeAsO_4 \cdot 2H_2O$. Am Mineral 70:838–844, 1985.

23. S Goldberg. Chemical modeling of arsenate adsorption on aluminum and iron oxide minerals. Soil Sci Soc Am J 50:1154–1157, 1986.

24. FJ Hingston, AM Posner, JP Quirk. Anion adsorption by goethite and gibbsite. I. The role of the proton in determining adsorption envelopes. J Soil Sci 23:177–192, 1972.

25. FJ Hingston, AM Posner, JP Quirk. Competitive adsorption of negatively charged ligands on oxide surfaces. In: FC Tompkins, ed. Surface Chemistry of Oxides. Discussions of the Faraday Society, No. 52. London: The Faraday Society, 1971, pp 334–343.

26. HW Nesbitt, IJ Muir and AR Pratt. Oxidation of arsenopyrite by air and air-saturated, distilled water and implications for mechanisms of oxidation. Geochim Cosmochim Acta 59:1773–1786, 1995.

27. ML Pierce, CB Moore. Adsorption of arsenite and arsenate on amorphous iron hydroxide. Water Res 16:1247–1253, 1982.

28. KP Raven, A Jain, RH Loeppert. Arsenite and arsenate adsorption on ferrihydrite: Kinetics, equilibrium, and adsorption envelopes. Environ Sci Tech 32:344–349, 1998.

29. RG Robins. Solubility and stability of scorodite, $FeAsO_4 \cdot 2H_2O$: Discussion. Am Mineral 72:842–844, 1987.

30. BA Manning, S Goldberg. Adsorption and stability of arsenic (III) at the clay mineral-water interface. Environ Sci Technol 31:2005–2011, 1997.

31. BA Manning, SE Fendorf and S Goldberg. Surface structures and stability of arsenic (III) on goethite: Spectroscopic evidence for inner-sphere complexes. Environ Sci Tech 32:2383–2388, 1998.

32. X Sun, HE Doner. Adsorption and oxidation of arsenite on goethite. Soil Sci 163:278–287, 1998.

33. SE Fendorf, MJ Eick, PR Grossl, DL Sparks. Arsenate and chromate retention mechanisms on goethite. I. Surface structure. Environ Sci Technol 31:315–320, 1997.
34. AA Manceau. The mechanism of anion adsorption on iron oxides: Evidence for the bonding of arsenate tetrahedra on free $Fe(O, OH)_6$ edges. Geochim Cosmochim Acta 59:3647–3653, 1995.
35. GA Waychunas, BA Rea, CC Fuller, JA Davis. Surface chemistry of ferrihydrite. Part I. EXAFS studies of the geometry of coprecipitated and adsorbed arsenate. Geochim Cosmochim Acta 57:2251–2270, 1993.
36. GA Waychunas, JA Davis, CC Fuller. Geometry of sorbed arsenate on ferrihydrite and crystalline FeOOH: Re-evaluation of EXAFS results and topological factors in predicting sorbate geometry, and evidence for monodentate complexes. Geochim Cosmochim Acta 59:3655–3661, 1995.
37. GA Waychunas, CC Fuller, BA Rea, JA Davis. Wide angle X-ray scattering (WAXS) study of "two-line" ferrihydrite structure: Effect of arsenate sorption and counterion variation and comparison with EXAFS results. Geochim Cosmochim Acta 60:1765–1781, 1996.
38. RL Parfitt, RJ Atkinson, RStC Smart. The mechanism of phosphate fixation by iron oxides. Soil Sci Soc Am Proc 39:837–842, 1975.
39. T Pilcher, J Veizer, GEM Hall. Natural input of arsenic into a coral-reef ecosystem by hydrothermal fluids and its removal by Fe(III) oxyhydroxides. Environ Sci Technol 33:1373–1378, 1999.
40. M Leblanc, B Achard, DB Othman, JM Luck. Accumulation of arsenic from acidic mine waters by ferruginous bacterial acretions (stromatolites). Appl Geochem 11:541–554, 1996.
41. JE Darland, WP Inskeep. Effects of pore water velocity on the transport of arsenic. Environ Sci Technol 31:704–709, 1997.
42. JE Darland, WP Inskeep. Effects of pH and phosphate competition on arsenic transport. J Environ Qual 26:1133–1139, 1997.
43. JM McArthur, P Ravenscroft, S Safiulla, MF Thirwall. Arsenic in groundwater: Testing pollution mechanisms for sedimentary aquifers in Bangladesh. Water Res Res 37:109–117, 2001.
44. CA Jones, WP Inskeep and DR Neuman. Arsenic transport in contaminated mine tailings following liming. J Environ Qual 26:433–439, 1997.
45. PH Masscheleyn, RD Delaune, WH Patrick Jr. Arsenic and selenium chemistry as affected by sediment redox potential and pH. J Environ Qual 20:522–527, 1991.
46. SL McGeehan, DV Naylor. Sorption and redox transformation of arsenite and arsenate in two flooded soils. Soil Sci Soc Am J 58:337–342, 1994.
47. DA Nimick. Arsenic hydrogeochemistry in an irrigated river valley: A reevaluation. Groundwater 36:743–753, 1998.
48. CA Jones, HW Langner, K Anderson, TR McDermott, WP Inskeep. Rates of microbially mediated arsenate reduction and solubilization. Soil Sci Soc Am J 64:600–608, 2000.
49. DE Cummings, F Caccavo Jr., S Fendorf, RF Rosenzweig. Arsenic mobilization by the dissimilatory Fe(III)-reducing bacterium *Shewanella alga* BrY. Environ Sci Technol 33:723–729, 1999.
50. EE Roden, JM Zachara. Microbial reduction of crystalline iron (III) oxides: Influence

of oxide surface area and potential for cell growth. Environ Sci Technol 30:1618–1628, 1996.

51. J Zobrist, PR Dowdle, JA Davis, RS Oremland. Mobilization of arsenite by dissimilatory reduction of adsorbed arsenate. Environ Sci Technol 34:4747–4753, 2000.

52. D Ahmann, AL Roberts, LR Krumholz, FMM Morel. Microbe grows by reducing arsenic. Nature 371:750, 1994.

53. PR Dowdle, AM Laverman, RS Oremland. Bacterial reduction of arsenic (V) to arsenic (III) in anoxic sediments. Appl Environ Microbiol 62:1664–1669, 1996.

54. AM Laverman, JS Blum, JK Schaefer, EJP Phillips, DR Lovley, RS Oremland. Growth of strain SES-3 with arsenate and other diverse electron acceptors. Appl Environ Microbiol 61:3556–3561, 1995.

55. JM Macy, JM Santini, BV Pauling, AH O'Neill, LI Sly. Two new arsenate/sulfate-reducing bacteria: Mechanisms of arsenate reduction. Arch Microbiol 173:49–57, 2000.

56. DK Newman, EK Kennedy, JD Coates, D Ahmann, DJ Ellis, DR Lovely, FMM Morel. Dissimilatory arsenate and sulfate reduction in *Desulfotomaculum auripigmentum* sp. nov. Arch Microbiol 168:380–388, 1997.

57. JF Stolz, RS Oremland. Bacterial respiration of arsenic and selenium. FEMS Microbiol Rev 23:615–627, 1999.

58. RE Macur, JT Wheeler, TR McDermott, WP Inskeep. Microbial populations associated with the reduction and enhanced mobilization of arsenic in mine tailings. Environ Sci Technol 35: in press, 2001.

59. BG Butcher, SM Deane, DE Rawlings. The chromosomal arsenic resistance genes of *Thiobacillus ferrooxidans* have an unusual arrangement and confer increased arsenic and antimony resistance to *Escherichia coli*. Appl Environ Microbiol 66:1826–1833, 2000.

60. J Cai, K Salmon, MS DuBow. A chromosomal ars operon homologue of *Psuedomonas aeruginosa* confers increased resistance to arsenic and antimony in *Escherichia coli*. Microbiol 144:2705–2713, 1998.

61. C Cervantes, G Ji, JL Ramirez, S Silver. Resistance to arsenic compounds in microorganisms. FEMS Microbiol Rev 15:355–367, 1994.

62. C Diorio, J Cai, J Marmor, R Shinder, MS DuBow. An *Escherichia coli* chromosomal *ars* operon homolog is functional in arsenic detoxification and is conserved in gram-negative bacteria. J Bacteriol 177:2050–2056, 1995.

63. G Ji, S Silver. Regulation and expression of the arsenic resistance operon from *Staphylococcus aureus* plasmid pI258. J Bacteriol 174:3684–3694, 1992.

64. G Ji, S Silver. Reduction of arsenate to arsenite by the ArsC protein of the arsenic resistance operon of *Staphylococcus aureus* plasmid pI258. Proc Natl Acad Sci USA 89:7974–7978, 1992.

65. S Silver, G Ji, S Broer, S Dey, D Dou, BP Rosen. Orphan enzyme or patriarch of a new tribe: The arsenic resistance ATPase of bacterial plasmids. Mol Microbiol 8:637–642, 1993.

66. G Muyzer, EC DeWaal, AG Uitterlinden. Profiling of complex microbial populations by denaturing gradient gel electrophoresis analysis of polymerase chain reaction-amplified genes coding for 16S rRNA. Appl Environ Microbiol 59:695–700, 1993.

67. HW Langner, WP Inskeep. Microbial reduction of arsenate in the presence of ferrihydrite. Environ Sci Technol 34:3131–3136, 2000.
68. EA Rochette, BC Bostick, G Li, S Fendorf. Kinetics of arsenate reduction by dissolved sulfide. Environ Sci Technol 34:4714–4720, 2000.
69. JM Harrington, SE Fendorf, RF Rosenzwieg. Biotic generation of arsenic (III) in metal contaminated lake sediments. Environ Sci Technol 32:2425–2430, 1998.
70. HW Langner, WC Allen, WP Inskeep. Reduction and solid phase control of dissolved arsenic in constructed wetlands. In: Annual Meeting Abstracts. Madison, WI: Soil Science Society of America 1999, p 349.
71. LE Eary, JA Schramke. In: DC Melchoir, RL Bassett, eds. Chemical Modeling of Aqueous Systems II. American Chemical Society Symposium Series 416. Washington DC: American Chemical Society, 1990, pp 379–396.
72. DE Tallman, AU Shaikh. Redox stability of inorganic arsenic (III) and arsenic (V) in aqueous solution. Anal Chem 52:196–199, 1980.
73. DW Oscarson, PM Huang, WK Liaw and UT Hammer. Kinetics of oxidation of arsenite by various manganese dioxides. Soil Sci Soc Am J 47:644–648, 1983.
74. MJ Scott, JJ Morgan. Reactions of oxide surfaces. 1. Oxidation of As(III) by synthetic birnessite. Environ Sci Technol 29:1898–1905, 1995.
75. M Pettine, L Campanella, FJ Millero. Arsenite oxidation by H_2O_2 in aqueous solutions. Geochim Cosmochim Acta 63:2727–2735, 1999.
76. BD Kocar, RE Macur, WP Inskeep. Photochemically induced oxidation of arsenite in the presence of Fe(III) and oxalate. In: Annual Meeting Abstracts. Madison, WI: Soil Science Society of America, 2000, p 235.
77. KA Hislop, JR Bolton. The photochemical generation of hydroxyl radicals in the UV-vis/ferrioxalate/H_2O_2 system. Environ Sci Technol 33:3119–3126, 1999.
78. FH Osborne, HL Ehrlich. Oxidation of arsenite by a soil isolate of Alcaligenes. J Appl Bacteriol 41:295–305, 1976.
79. SE Phillips, ML Taylor. Oxidation of arsenite to arsenate by Alcaligenes faecalis. Appl Environ Microbiol 32:392–399, 1976.
80. N Wakao, H Koyatsu, Y Komai, H Shimokawara, Y Sakurai, H Shiota. Microbial oxidation of arsenite and occurrence of arsenite-oxidizing bacteria in acid mine water from a sulfur-pyrite mine. Geomicrobiol 6:11–24, 1988.
81. AN Ilyaletdinov, SA Abdrashitova. Autotrophic oxidation of arsenic by a culture of Pseudomonas arsenitoxidans. Mikrobiologya 50:197–204 (translation), 1979.
82. JE Santini, LI Sly, RD Schnagl, JM Macy. A new chemolithoautotrophic arsenite-oxidizing bacterium isolated from a gold mine: Phylogenetic, physiological, and preliminary biochemical studies. Appl Environ Microbiol 66:92–97, 2000.
83. CL Anderson, J Williams, R Hille. The purification and characterization of arsenite oxidase from Alcaligenes faecalis, a molybdenum-containing hydroxylase. J Biol Chem 267:23674–23682, 1992.
84. SA Abdrashitova, GG Abdullina, AN Ilyaletdinov, VK Orlov. Influence of arsenite on peroxide oxidation of lipids and fatty acid composition of Alcaligenes eutrophus cells. Mikrobiologya 57:231–235 (translation), 1988.
85. HW Langner, CR Jackson, TR McDermott, WP Inskeep. Rapid oxidation of arsenite in a hot spring ecosystem in Yellowstone National Park. Environ Sci Technol 35: in press, 2001.

86. CR Jackson, H Langner, J Christiansen, WP Inskeep, TR McDermott. Molecular analysis of the microbial community involved in arsenite oxidation in an acidic, thermal spring. Microbiol Ecol, in press, 2001.

87. JA Wilkie, JG Hering. Rapid oxidation of geothermal arsenic (III) in streamwaters of the eastern Sierra Nevada. Environ Sci Technol 32:657–662, 1998.

88. RE Macur, TR McDermott, WP Inskeep. Microbially mediated arsenic cycling in a contaminated soil. In: Annual Meeting Abstracts. Madison, WI: Soil Science Society of America, 2000, p 235.

89. CA Jones, WP Inskeep, JW Bauder, KE Keith. Arsenic solubility and attenuation in soils of the Madison River basin, Montana: Impacts of long-term irrigation. J Environ Qual 28:1314–1320, 1999.

90. BL Maidak, N Larson, MJ McCaughey, R Overbeek, GJ Olsen, K Fogel, J Blandy, CR Woese. The ribosomal database project. Nucleic Acids Res 22:3485–3487, 1994.

9

Metal-Oxide Adsorption, Ion Exchange, and Coagulation– Microfiltration for Arsenic Removal from Water

Dennis A. Clifford and Ganesh L. Ghurye
University of Houston, Houston, Texas

I. INTRODUCTION

As with other toxic inorganic contaminants, arsenic (As) is almost exclusively a groundwater problem. Although it can exist in both organic and inorganic forms, only inorganic arsenic in the $+III$ or $+V$ oxidation state has been found to be significant where potable water supplies are concerned (1). Redox potential governs the As(III)/(V) distribution. Arsenite [As(III)] species are present in anoxic and reducing waters while arsenate [As(V)] species are found in oxidizing waters.

For many years, the maximum contaminant level (MCL) for total arsenic in drinking water worldwide was 50 µg/L (ppb). More recently, however, most countries, including the United States, have adopted an MCL of 10 µg/L following a reassessment of the health effects of higher arsenic levels. In the United States, more than 3000 community water supplies serving about 11 million persons contain arsenic above the 10 ppb MCL, but nearly all of these contain less than 50 ppb. However, in Bangladesh and West Bengal (India), an estimated 65 million persons are served by shallow tube wells containing more than 50 ppb arsenic with many wells containing more than the 500 ppb.

The pH of the water is very important in determining the arsenic speciation. The primary As(V) species found in groundwater in the pH range of 6–9 are monovalent $H_2AsO_4^-$ and divalent $HAs_2O_4^{2-}$. These anions result from the dissociation of arsenic acid (H_3AsO_4), which exhibits pK_a values of 2.2, 7.0, and 11.5. Uncharged arsenious acid (H_3AsO_3) is the predominant species of trivalent arsenic found in natural water. Only at pH values above its pK_a of 9.2 does the monovalent arsenite anion ($H_2AsO_3^-$) predominate.

Arsenic in drinking water supplies can be removed by a variety of treatment processes including those cited in Table 1, which also lists the typical applications of each process. All these processes do a much better job of removing As(V) compared with As(III). Thus, before using these processes, it will often be necessary to oxidize As(III) to As(V) using chlorine or an alternative oxidant. This chapter focuses arsenic treatment by metal-oxide adsorption (MOA), ion exchange (IX), and iron (III) coagulation–microfiltration (C-MF), because these processes have proven to be the most efficient and cost effective in bench- and pilot-scale studies, especially for point-of-use (POU), point-of-entry (POE), well-head, and small community treatment systems.

In addition to the water source, location, and system size, background water quality will also affect the selection of a treatment process. For example, ion exchange is generally not an option for waters containing high total dissolved solids (TDS >500 mg/L) and/or high sulfate (SO_4^{2-} >120 mg/L), because these contaminants lead to rapid exhaustion of the columns (run length less than 400 bed volumes). Similarly, metal oxide adsorbents may not be reasonable alternatives for treating supplies containing relatively high concentrations of competing

Table 1 Processes for Arsenic Removal

Process	Potential application[a]
Metal-oxide adsorption using packed beds of activated alumina, modified activated alumina, granular ferric hydroxide, iron-oxide coated sands, and other specialty adsorbents	GW, Large Systems, Small Systems, POU, POE, Wellhead
Ion exchange using packed beds of chloride-form anion-exchange resins	GW, Small Systems, Wellhead
Iron(II) coagulation–microfiltration without flocculation and presedimentation	GW, SW, Large Systems, Small Systems
Lime softening	GW, SW, Large Systems, Small Systems
Reverse osmosis and nanofiltration	GW, SW, POE, POU, Large Systems, Small Systems
Enhanced coagulation	SW, Large Systems

[a] GW, groundwater; SW, surface water; POU, point of use; POE, point of entry.

ions, such as fluoride, phosphate, and silicate, which can dramatically shorten the run length of an alumina column.

Process complexity, hazardous materials handling, and waste disposal considerations can also have a major influence on the choice of a process. For example, when removing As(V) by anion exchange, pH adjustment is not required, and a small excess of ordinary sodium chloride can be used to achieve essentially complete elution of arsenic from the spent resin. Activated alumina, however, may require feed and effluent pH adjustment and will require a large excess of both sodium hydroxide and sulfuric or hydrochloric acid to elute the adsorbed arsenic and reacidify the alumina. A significant potential disadvantage of ion exchange is that chromatographic peaking or dumping of arsenic into the process effluent is possible if the run is not terminated before arsenic breakthrough. Careful monitoring of the flow or chemical quality of the effluent, or operating several columns in parallel at different stages of exhaustion can solve this potential problem, but this adds complexity to the process.

The simplest version of the iron (III) coagulation-microfiltration process is by nature more complex and costly than the simplest versions of the ion-exchange and metal-oxide adsorbent processes. Furthermore, as is the case with the MOA processes, high feed pH, and the presence of significant silica, phosphate, and fluoride concentrations will decrease the arsenic-removal efficiency of the C-MF process. Nevertheless, C-MF is attractive for arsenic removal because of lower chemical consumption and less waste production.

The primary objectives of this chapter are to (1) explain the theory of the processes, (2) present the influence of water quality on arsenic removal efficiency, (3) identify reasonable process design parameters, and (4) compare the processes.

II. GRANULAR METAL-OXIDE ADSORBENTS

The granular metal-oxide adsorbents discussed in this chapter are activated alumina (AAI), modified activated aluminas, iron-oxide–coated sand (IOCS), granular ferric hydroxide (GFH or GEH), and proprietary filter media such as ADI. Although they differ in physical appearance, they all involve hydrous oxides of iron or aluminum that remove arsenic by a process of ligand exchange. Generally, they are employed in packed beds or small filters containing 28 × 48 mesh (0.6–0.3 mm diam) particles. In the simplest process, arsenic-contaminated raw water is passed through a bed of granular media without pH adjustment or oxidation of As(III) to As(V). Figure 1 is an example of an activated alumina adsorption process for arsenic. Any of the MOAs may be used in this way to remove arsenic, but the arsenic removal efficiency can usually be improved dramatically by oxi-

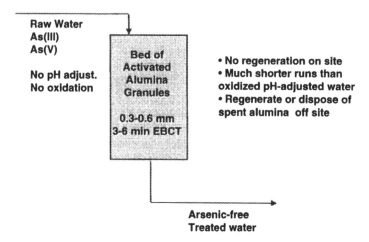

Figure 1 Schematic for a simple activated alumina process for arsenic removal from groundwater without pretreatment.

dizing As(III) to As(V) and lowering the feed water pH to about 6. Why pH reduction and arsenic oxidation are necessary to improve process performance is explained in the following section.

A. Activated Alumina Adsorption Theory

Packed beds of activated alumina can be used to remove arsenic, fluoride, selenium, silica, and natural organic material (NOM) anions from water. The mechanism, which is one of exchange of contaminant anions for surface hydroxides on the alumina, is generally called *adsorption*, although *ligand exchange* is a more appropriate term for the highly specific surface reactions involved (2).

The typical activated aluminas used in water treatment are 28 × 48 mesh (0.6–0.3 mm diam) mixtures of amorphous and gamma aluminum oxide (γ-Al_2O_3) prepared by low-temperature (300–600°C) dehydration of precipitated $Al(OH)_3$. These highly porous materials have surface areas of 50–300 m^2/g. Using the model of hydroxylated alumina surface subject to protonation and deprotonation, the following ligand exchange reaction, Eq. (1), can be written for arsenate adsorption in acid solution (alumina exhaustion) in which $\equiv Al$ represents the alumina surface and an overbar denotes the solid phase:

$$\overline{\equiv Al-OH} + H^+ + H_2AsO_4^- \rightarrow \overline{\equiv Al-H_2AsO_4} + HOH \qquad (1)$$

The equation for arsenate desorption by hydroxide (alumina regeneration) is presented in Eq. (2).

$$\overline{\equiv Al - H_2A_sO_4} + OH^- \rightarrow \overline{\equiv Al - OH} + H_2AsO_4^- \tag{2}$$

Activated alumina processes are sensitive to pH, and anions are best adsorbed below pH 8.2, a typical zero point of charge (ZPC), below which the alumina surface has a net positive charge and excess protons are available to fuel Eq. (1). Above the ZPC, alumina is predominantly a cation exchanger, but its use for cation exchange is relatively rare in water treatment.

Ligand exchange as indicated in Eqs. (1) and (2) occurs chemically at the internal and external surfaces of activated alumina. A more useful model for process design, however, is one that assumes that the adsorption of arsenate onto alumina at the optimum pH of 5.5–6 is analogous to weak-base anion exchange. For example, the uptake of $H_2AsO_4^-$ or F^-, requires the protonation of the alumina surface, and that is accomplished by preacidification with HCl or H_2SO_4, and reducing the feed water pH into the 5.5–6.0 region. The positive charge caused by excess surface protons may then be viewed as being balanced by exchanging anions, i.e., ligands such as arsenate, hydroxide, and fluoride. To reverse the adsorption process and remove the adsorbed arsenate, an excess of strong base, e.g., NaOH, must be applied. The following series of reactions [Eqs. (3) to (7)] is presented as a model of the adsorption–regeneration cycle that is useful for design purposes.

The first step in the cycle is acidification in which neutral (water-washed) alumina (alumina · HOH) is treated with acid, e.g., HCl, and protonated (acidic) alumina is formed as follows:

$$\overline{Alumina \cdot HOH} + HCl \rightarrow \overline{Alumina \cdot HCl} + HOH \tag{3}$$

When HCl-acidified alumina is contacted with arsenate ions, they strongly displace the chloride ions, especially when the alumina surface remains acidic (pH 5.5–6) This displacement of chloride by arsenate, analogous to weak-base ion exchange, can be written as

$$\overline{Alumina \cdot HCl} + H_2AsO_4^- \rightarrow \overline{Alumina \cdot HH_2AsO_4} + Cl^- \tag{4}$$

To regenerate the arsenate-contaminated adsorbent, a 0.25–1.0 N NaOH solution is used. Because alumina is both a cation and an anion exchanger, Na^+ is exchanged for H^+, which immediately combines with OH^- to form HOH in the alkaline regenerant solution. The regeneration reaction of arsenate-spent alumina is

$$\begin{aligned} \overline{Alumina \cdot HH_2AsO_4} + 2NaOH \rightarrow \overline{Alumina \cdot NaOH} \\ + NaH_2AsO_4 + HOH \end{aligned} \tag{5}$$

Recent experiments have suggested that Eq. (5) can be carried out using fresh or recycled NaOH from a previous regeneration. This suggestion is based on arsenic-removal field studies in which arsenic-spent alumina was regenerated with equally good results using fresh or once-used 1.0 N NaOH (3,4). Probably, the spent regenerant fortified with NaOH to maintain its hydroxide concentration at 1.0 N could have been used many times, but the optimum number of spent-regenerant reuse cycles was not determined in the field study.

To restore the arsenate removal capacity, the basic alumina is acidified by contacting it with an excess of dilute acid, typically 0.1–0.5 N HCl or H_2SO_4:

$$\overline{Alumina \cdot NaOH} + 2HCl \rightarrow \overline{Alumina \cdot HCl} + NaCl + HOH \qquad (6)$$

The acidic alumina, alumina · HCl, is now ready for another arsenate ligand-exchange cycle as summarized by Eq. (4). Alternatively, the feed water may be acidified prior to contact with the basic alumina, thereby combining acidification and adsorption into one step as summarized by Eq. (7):

$$\overline{Alumina \cdot NaOH} + NaH_2AsO_4 + 2HCl \rightarrow \overline{Alumina \cdot HH_2AsO_4}$$
$$+ 2NaCl + HOH \qquad (7)$$

B. Other Metal-Oxide Adsorbents

Many proprietary adsorbents are being marketed for arsenic removal from drinking water. Most are based on the above ligand-exchange reactions in which arsenate is exchanged for hydroxide on a hydroxylated aluminum- or iron-oxyhydroxide surface that coats a porous solid. Only by testing the adsorption capacity of a media sample can the capacity for arsenic adsorption be evaluated, because it is strongly dependent on the physical–chemical properties of the media, including particle size, crystal structure, surface area, porosity, and degree of hydration. As with alumina, one can expect that the sensitivity of the proprietary metal oxide adsorbent to pH, and competing ions, will depend on its zero point of charge (ZPC). Hydrated iron (III) oxide, e.g., FeOOH, has slightly higher ZPC (8.6) than AlOOH (8.2), thus the solid iron oxide is somewhat less sensitive to increasing pH than is aluminum oxide (2). The decreased sensitivity to increasing pH and its intrinsically higher affinity for arsenate are reasons why ferric rather than alum salts are used for arsenic removal by coagulation (5).

Three of the recently introduced arsenic adsorbents, iron-doped alumina, iron-oxide coated sand, and granular ferric hydroxide, merit a brief discussion here because of their demonstrated effectiveness in removing arsenic. Because ferric hydroxide has a higher capacity for arsenic than does an equivalent surface area of aluminum hydroxide, "iron-doped" aluminas have been designed for the purpose of improving their arsenic capacity. One such adsorbent is Alcan AAFS-50, a brown-colored "promoted" alumina that is advertised to have five times the arsenic capacity and less pH sensitivity than conventional activated aluminas (6). Unlike conventional aluminas, AAFS-50 cannot be regenerated, but it reportedly can be landfilled without special treatment. Our recent research (7) showed that AAFS-50 had about twice the As(V) capacity of a conventional alumina when adsorption isotherms (20–25°C) were compared in the pH 6–8.5 range in waters containing 20–30 mg/L silica.

Iron-oxide–coated sand is another recently introduced arsenic adsorbent that has been shown to have promise for arsenic removal (8,9). However, because the effective adsorption area is only on the surface of the particle, minimal capacity should be expected compared with adsorbents that are pure hydrated iron oxide and are truly porous. An example of the latter type of adsorbent is granular ferric hydroxide (5).

Granular ferric hydroxide (GFH) is prepared from aqueous solutions of ferric chloride by alkaline precipitation of ferric hydroxide and conversion of the precipitate to a hydrated granular form by centrifugation and high-pressure dewatering. Granular ferric hydroxide is reported to be a poorly crystallized β-FeOOH that has chloride incorporated into the tunnel structure. The commercial product has a water content of 45%, particle porosity of 75%, a surface area of 250–300 m^2/g, and a water-saturated particle density of 1.6 g/cm^3. Although the grain size is large, ranging from 0.3–2.0 mm, adsorption capacity is reportedly very high, and the arsenic uptake rate is good, because the media has never been dried, and the pores are completely filled with water (5,10). One would expect the adsorption kinetics of such a media to be faster than activated alumina, which is made by low-temperature dehydration of aluminum hydroxide precipitate at 300–600°C.

We recently compared the performance of GFH with conventional alumina using rapid, small-scale column tests (RSSCTs) with 4 mL of 60 × 100 mesh adsorbents (7). The column feed water was Three Forks, Montana, well water at pH 7.5 containing 70 µg/L As(V). Granular ferric hydroxide outperformed the conventional alumina with a run length of 6750 BV to the 10 µg/L MCL compared with 2250 BV for the conventional alumina. The competing ions—phosphate, silicate, and fluoride—were not measured. Granular ferric hydroxide has shown a similarly large advantage over conventional aluminas in other comparison tests, such as the pilot-scale field test in the United Kingdom conducted by the Severn–Trent Water Authority on a well water containing 22 µg/L arsenic (11).

According to the manufacturer, before use on a drinking water supply, GFH must be disinfected upflow using free chlorine at a dosage of about 600 g free Cl_2/m^3 GFH. Water fed to a GFH adsorber should have the following characteristics: pH 5.5–9; dissolved oxygen >0.5 mg/L; Fe(II) <0.2 mg/L; Mn(II) <0.05 mg/L; and Al <0.2 mg/L. Granular ferric hydroxide is widely used in Europe for arsenic and phosphate removal applications. The spent GFH is not regenerated, but rather landfilled without further treatment. Because run lengths are so long, GFH should be backwashed once or twice per month. Although GFH capacity for arsenic is much greater than alumina, the overall arsenic-removal treatment costs including arsenic waste disposal, are reportedly comparable to conventional alumina processes.

C. Effect of Oxidation State on Arsenic Capacity and Run Length

To achieve effective removal of arsenic from groundwater by means of activated alumina, As(III) must be oxidized to As(V). The following are some examples of the arsenic capacity of alumina observed during some University of Houston laboratory and field studies (12–15). These examples illustrate the importance of oxidative pretreatment ahead of activated alumina columns operated at the optimum pH of 6.0. A bench-scale minicolumn experiment with a feed water containing 100 μg As(III)/L, reached 50 μg As/L after only 300 BV. However, the same water oxidized with chlorine so that it contained 100 μg As(V)/L, did not reach 50 μg As/L until 23,400 BV. This was an 80-fold improvement in performance due simply to converting As(III) to As(V). It is noted, however, that at relatively high pH (8.6), it did not matter whether As(III) was oxidized to As(V), because about the same mediocre performance was obtained, i.e., 800–900 BV. At pH 8.6, the adsorption or ligand-exchange capacity of alumina was severely reduced by competition from hydroxide ions.

D. Effect of pH, Sulfate, and Hardness on Arsenic Capacity

Although the subsequent discussion refers primarily to activated alumina, the observations are generally valid for iron-doped aluminas and GFH. Activated alumina adsorption of arsenate is highly dependent on pH; at alkaline pHs where hydroxide competition is significant, arsenate adsorption is poor. As the pH is lowered, arsenate adsorption increases dramatically until about pH 6. Based on a limited number of full-scale and pilot-scale column studies, the recommended pH for operation ranges from 5.5 (16) to 6.0 (3,4,7,11–15). Figure 2 shows the arsenic capacity declining rapidly as pH increases above about pH 6.0. Simms and Azizian (17) found a similar trend in arsenic-capacity reduction with increasing pH in their Severn–Trent Water Authority pilot studies in the United Kingdom where the ground water contained about 22 μg/L As(V).

Figure 3 compares equilibrium isotherms for adsorption of As(V) onto activated alumina in the presence of very high concentrations of chloride and sulfate. The isotherms indicate that the effect of sulfate is much more pronounced than chloride. For example, at an equilibrium As(V) concentration of 1 mg/L, arsenic loading was reduced by 16% in the presence of 532 mg Cl^-/L (15 meq/L) compared with 50% in the presence of 720 mg SO_4^{2-}/L (15 meq/L). This suggests

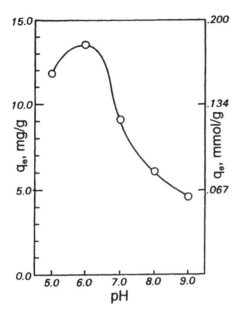

Figure 2 Effect of pH on equilibrium As(V) capacity of alumina in soft water without silica. C_e = 1 mg/L, TDS = 1000 mg/L, SO_4^{2-} = 240–270 mg/L, hardness = 0. (From Ref. 12.)

that HCl rather than H_2SO_4 would be preferred for pH adjustment. A similar influence of sulfate on arsenic capacity was found with GFH (5), but only at pH less than 7.0. Above pH 7.0, sulfate did not reduce the arsenic capacity of GFH.

In actual field operation (Table 2), however, the alumina column capacity at an arsenic MCL of 50 μg/L was far less than the batch-equilibrium values shown in Figure 3. The four column capacities shown in Table 2 were obtained from single columns operated at pH 5.5–6.0 until the breakthrough of 50 As(V)/L. The effects of competing anions, nonequilibrium mass transfer limitations, and fouling by particulate matter are reasons for the low column capacities observed. The exceptionally long run length (110,500 BV) and high As(V) capacity (2.6 mg/g AAl) in the Severn–Trent U.K. study are difficult to explain in light of the performance of the same alumina (Alcan AA-400G) in Albuquerque, NM (3,4), where only 15,600 BV and 0.41 mg/g AAl capacity were observed. Based on the higher sulfate concentration and shorter (3 min) empty bed contact time (EBCT), the U.K. water should have produced a much shorter run length and lower capacity. Neither waters contained significant iron, which might have aided arsenic removal. The Albuquerque water contained 51 mg/L silica, which was not reported for the U.K. water. The high silica might explain a large part of the

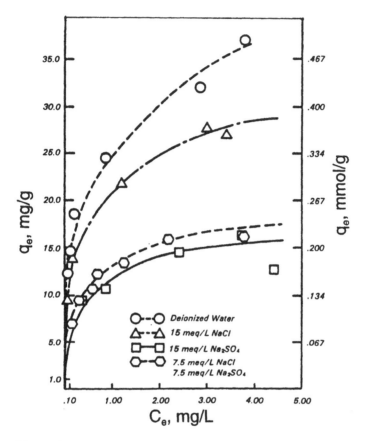

Figure 3 Effect of competition by chloride and sulfate on adsorption of As(V) onto activated alumina. T = 25°C, C_0 = 5 mg As V/L, pH = 6, t_{eq} = 7 days. (From Ref. 12.)

difference in alumina performance between the two locations. Phosphate, a known competitor with arsenate for adsorption sites (5) on the alumina was not measured in either water. The only known significant difference was the sixfold higher hardness level in the U.K. water, which suggested that calcium and/or magnesium may be involved in the uptake of arsenate by alumina. Our recent research has shown that the arsenic capacity of alumina at pH 7.3 was 30–50% higher in a hard water (250 mg/L hardness as $CaCO_3$) compared with a soft water (5 mg/L hardness as $CaCO_3$) at the same TDS level. Curiously, the influence of hardness was reversed at pH 6.0 where it produced a 10–30% reduction in arsenic capacity compared with soft water (7).

Table 2 As(V) Capacities (mg As(V)/g alumina) for Activated Aluminas Used in Field Studies at pH 5.5–6.0

	Location (ref.)[a]			
	Fallon, NV (16)	Hanford, CA (15)	Albuquerque, NM (3,4)	Severn–Trent, UK (17)
Feed As(V), μg/L	110	98	22	23
Sulfate, mg/L	96	5	70	117
Hardness, mg/L as $CaCO_3$	5	10	53	331
Bed volumes	14,450	16,000	15,600	110,000
As(V) capacity, mg As(V)/g alumina	1.5	1.7	0.4	2.6

[a] Refs. 15 and 16 used F-1 activated alumina from Alcoa (no longer available commercially) whereas Refs. 3, 4, and 17 used Alcan AA-400G. EBCT was 5 min except for Ref. 17 where it was 3.0 min. Arsenic capacities are based on old MCL of 50 μg/L.

E. Effect of Silica on Arsenic Capacity

Dissolved silica, SiO_2, is present in water as silicic acid, H_4SiO_4, or as the silicate anion, $H_3SiO_4^-$. As pH increases, the relative concentration of silicate increases because of the dissociation of H_4SiO_4, which is a weak acid with $pK_1 = 9.77$. Silicate anions are strong ligands that compete well with arsenate for adsorption sites on the alumina. In fact, activated alumina is sometimes used to remove silica from cooling water (18). With this background it is not difficult to understand why the silica concentration of a water supply is an important determinant of arsenic removal capacity. Figure 4, compares the arsenic adsorption isotherms for a conventional alumina (Alcan AA-400G) using hard synthetic groundwater with (15 mg/L) and without (0 mg/L) silica (7). The effect of silica on arsenic capacity was dramatic even at pH 7.5, which is far below pH 9.77, the pK_1 of silicic acid. The equilibrium As(V) capacity of the alumina dropped nearly 75% from 0.55 down to 0.15 mg/g as a result of adding 15 mg/L silica to the raw water. The influence of silica decreases as pH decreases, because of decreasing ionization of silicic acid. Nevertheless, the negative effect of silica was seen even at pH 6.0 where very little dissociated silica exists.

The expected influence of silica on other metal oxide adsorbents, such as granular ferric hydroxide, is similar to its negative effect on alumina. Phosphate and fluoride are other strong ligands commonly found in groundwater that exhibit a negative influence similar to that of silicate on the arsenic capacity of alumina and GFH (5,12). Thus, the concentrations of these ions must be known before a reasonable estimate of arsenic capacity can be made.

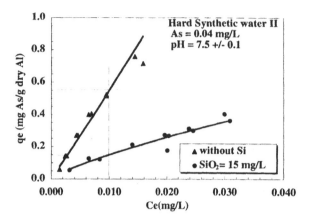

Figure 4 Effect of silica on As(V) adsorption onto Alcan AA-400G alumina with hard synthetic water at pH 7.5. T = 22 – 25°C, t_{eq} = 2 days. Initial As(V) conc. = 0.04 mg/ L. (From Ref. 7.)

F. Process Design Considerations

The process variables known to influence arsenic capacity and column performance of alumina, GFH, and other adsorbents are as follows: adsorbent, adsorbent particle size, flow rate, EBCT, and water quality parameters including arsenic concentration, As(III)/(V) speciation, pH, silica, phosphate, fluoride, hardness, and sulfate concentrations. Even with a complete water analysis, it is prudent to perform pilot studies with competitive adsorbents on the water to be treated because of the numerous factors that influence arsenic adsorption. Equilibrium isotherms and rapid small-scale column tests (RSSCTs) are typically run prior to the pilot study (7).

Alumina particle size and empty bed contact time (EBCT) can significantly influence arsenic removal by alumina. Clifford and Lin (14,15) and Simms and Azizian (17) reported that finer particles of alumina (28 × 48 mesh, 0.6–3 mm) have higher arsenic capacity, lower arsenic leakage, and longer run length than larger alumina particles (14 × 28, 1.18–0.6 mm). During the same study, Simms and Azizian (17) found that arsenic run length was linearly proportional to EBCT in the range of 3 min (9000 BV) to 12 min (14,000 BV) when operating with 14 × 28 mesh Alcan AA-400G alumina at pH 7.5. To minimize bed size and alumina inventory, however, they preferred to operate in the 3–6 min EBCT range. In the recent Albuquerque arsenic studies (3,4), a similar relationship between EBCT and run length was observed when using the finer 28 × 48 mesh alumina and operating at pH 6.0. At EBCTs of 5 and 10 min, the run lengths were 6400 and 8800 BV, respectively.

Equilibrium isotherms are generally done by the "bottle-point" method in which each bottle represents an equilibrium data point. Predetermined amounts of ground-up (60 × 100 mesh) adsorbent are added to bottles of raw water such that the final equilibrium arsenic concentration is within a specified range, which is generally between the influent concentration and one-half of the desired effluent concentration of arsenic. The bottles are capped, and then shaken or tumbled for 24–48 hr to achieve adsorption equilibrium. The final arsenic concentration in a filtered sample is determined and the loading on the adsorbent is calculated as the difference between the initial and final arsenic concentration divided by the mass of adsorbent. All points are replicated. Finally, a graph of arsenic loading q_e, mg As/g adsorbent, is plotted vs. C_e, the mg As/L water. Isotherms are a rapid way to screen adsorbents on the basis of arsenic capacity under a variety of water-quality conditions.

A rapid small-scale column test protocol for bench-scale evaluation of alumina adsorbents was recently developed at the University of Houston (7). The RSSCT allows for faster evaluation of arsenic adsorbents using bench-scale column tests with actual or synthetic groundwaters. The RSSCT protocol uses low-pressure (<5 psig) glass minicolumns containing 4.0 mL of 60 × 100 mesh adsorbent. Compared with the standard (28 × 48 mesh) adsorbents currently used in pilot-scale and minicolumn tests, the RSSCT allows the use of faster flow rates so that the column runs can be completed in one-fourth to one-tenth the time required for the standard tests.

G. Alumina Regeneration

Arsenic is much more difficult to elute from alumina compared with adsorbed fluoride, which is eluted with 1% NaOH. For this reason, higher concentrations, typically 1–4% (0.25–1 N) NaOH, are used for the base-regeneration step. Acid concentrations in the range of 0.1–0.5 N HCl or H_2SO_4 are typically used for the acid-neutralization step. These recommendations are based on field regeneration research completed in San Ysidro, NM, where it was found that the lower NaOH concentrations took proportionately longer to elute the arsenic (14). [Note: The alumina supplier, Alcan Chemicals, recommends 0.5% NaOH (0.125 N) and 0.1% HCl (0.03 N) (6).] However, even with excess caustic, less than 70% of the adsorbed arsenic was eluted from the column during the first regeneration in San Ysidro (14). In spite of incomplete removal of arsenic from spent alumina during regeneration and greater arsenic leakage from regenerated alumina, the process appears to be feasible based on more recent field studies in Albuquerque, NM (3,4), and in the United Kingdom (17). After four regenerations, the column capacity for arsenic at feed pH 7.5 and 6 min EBCT stabilized at about 80% of the virgin run capacity during the U.K. studies (17).

Due to the difficulties associated with regenerating alumina and disposing of the arsenic-contaminated residue, the use of point-of-use (POU) or point-of-entry (POE) treatment without pH adjustment should be considered. Point-of-use systems without preoxidation or pH reduction, used intermittently, should achieve about 1000-BV throughput prior to exhaustion. Exhausted medium would simply be thrown away, not regenerated. Although not verified, the spent medium would probably pass the standard toxicity characteristic leaching procedure (TCLP) or extraction procedure (EP) toxicity tests as a nonhazardous waste. The reason is that the arsenic loading is very low, and toxicity tests are performed at pH 5, which is near the optimum pH for arsenic adsorption onto alumina. Furthermore, arsenic-laden $Al(OH)_3(s)$ sludges from alumina spent-regenerant treatment passed the EP toxicity test (14) and these sludges have very similar chemistry to that of activated alumina containing adsorbed arsenic.

H.   Alumina Waste Disposal

During normal regeneration and acidification of spent alumina, enough aluminum dissolves to make precipitation of $Al(OH)_3(s)$ a feasible treatment step for the removal of arsenic from the spent-regenerant wastewater. When lowering the pH to approximately 6.5 with HCl or H_2SO_4, the As(V) quantitatively coprecipitates with the resulting $Al(OH)_3(s)$. Following dewatering, the dried arsenic-contaminated sludge should easily pass the TCLP test if 5.0 mg As/L is allowed in the leachate. Hathaway and Rubel (16) and Clifford and Lin (14) used this procedure to treat spent alumina regenerant and produced leachates containing 0.036 and 0.6 mg As/L, respectively. The latter sludge contained some As(III), which caused the higher arsenic concentration in the leachate. The leachate arsenic concentration can be minimized by oxidizing the sludge, e.g., with chlorine, to ensure the presence of As(V) as opposed to As(III).

I.   Activated Alumina Process with Regenerant Reuse

Figure 5 describes an activated alumina process for arsenic removal that maximizes run length and minimizes regenerant consumption and disposal problems. Raw water containing As(III) and As(V) would be oxidized with chlorine or alternative oxidant prior to pH adjustment to 5.5–6. The oxidized, pH-adjusted water would then be fed to the 28 × 48 mesh alumina column with an EBCT of 4–6 min. The column effluent pH would be raised to 7.5–8 using lime or caustic. The media would be regenerated with 0.25–1.0 N NaOH and then acid rinsed with 0.2–0.5 N H_2SO_4 or HCl. Spent regenerant, with NaOH added as necessary to maintain its initial concentration, would be reused 5–10 times or until it became ineffective. Prior to disposal of the spent regenerant, its pH would be adjusted to 6.5 or lower with H_2SO_4 or HCl to coprecipitate the arsenic with

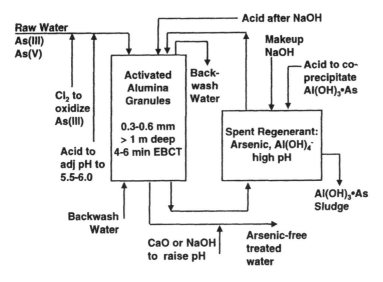

Figure 5 Process schematic for optimized AAl process

Al(OH)$_3$. The main disadvantages of the process are complexity, the need to handle hazardous chemicals, and a large increase in effluent total dissolved solids (TDS) due to the added acid and caustic. Because of process complexity and high operational costs, USEPA favored the use of throw-away AAl in its estimate of the national compliance with the new arsenic MCL.

III. ION EXCHANGE

Anion exchange for arsenic removal is one of the BAT (best available technology) recommended by the U.S. Environmental Protection Agency (EPA). Extensive studies, both at the bench and pilot scale have shown that for a source water containing <120 mg/L sulfate and <500 mg/L TDS, ion exchange may be the arsenic-removal process of choice (3,4,19–21).

A. Arsenic Oxidation

It should be noted that As(V) is more efficiently removed by anion exchange than is As(III) (13–15). Oxidation of As(III) to As(V) is easily achieved by commonly used oxidants such as chlorine, ozone, or permanganate (13,22). Solid oxidizing media like Filox™ can also be used for As(III) oxidation (22).

B. Reactions and Process Schematic

Typically, the chlorinated and filtered raw water, without pH adjustment, is passed downflow through a 2.5- to 5-ft-deep bed of chloride-form strong-base anion-exchange resin and the chloride-arsenate ion-exchange reaction [Eq. (8)] takes place in the near-neutral pH range. Regeneration, according to Eq. (9) is not difficult because a divalent ion (arsenate) is being replaced by a monovalent ion (chloride) in high ionic strength solution where electroselectivity reversal favors monovalent ion uptake by the resin. Regeneration returns the resin to the chloride form, ready for another exhaustion cycle:

$$2\ \overline{RCl} + HAsO_4^{2-} = \overline{R_2HAsO_4} + 2\ Cl^- \qquad (8)$$

$$\overline{R_2HAsO_4} + 2NaCl = \overline{RCl} + Na_2HAsO_4 \qquad (9)$$

The spent brine may either be wasted or reused by adding enough salt to bring the chloride concentration back to its initial value. Regeneration and brine reuse are discussed in detail later.

C. Resins and Selectivity

Strong-base anion-exchange resins used for arsenate removal are typically polystyrene divinylbenzene polymers with quaternary trimethyl amine (type 1) or dimethylethanolamine (type 2) functional groups. Polyacrylic divinylbenzene resins with triethylamine functional groups can also be used when more resistance to organic fouling is desired. All of these resins prefer sulfate to divalent arsenate, and exhibit the following selectivity sequence in dilute (<0.010 M) solution for the common background ions found in arsenic-contaminated ground water:

$$SO_4^{2-} > HAsO_4^{2-} > CO_3^{2-},\ NO_3^- > Cl^- > H_2AsO_4^-,\ HCO_3^-, \ggg H_3AsO_3$$

   In the above sequence, sulfate is the most-preferred anion for strong-base-anion (SBA) resins and it will eventually displace ions with lower selectivity from the resin. Monovalent arsenate, $H_2AsO_4^-$, and bicarbonate have low affinity for the resin. Arsenious acid, H_3AsO_3, is the least preferred; it simply passes through the resin because it's not ionic. This resin selectivity sequence, which had been suggested by Horng (23) based on his As(V)-removal studies was confirmed by other studies (3,4,19–21).

   During exhaustion of the resin bed, the selectivity sequence manifests itself as zones of sulfate-rich, arsenate-rich, bicarbonate-rich, etc. bands with the sulfate-rich zone located at the beginning of the column and the next-preferred ion after it and so on. In the beginning of an exhaustion cycle, the resin-bed profile is similar to Figure 6a in composition. As the exhaustion progresses, the respective anion bands are displaced in the direction of the outlet where they "break through" into the effluent as wave fronts. The first wave front to break through

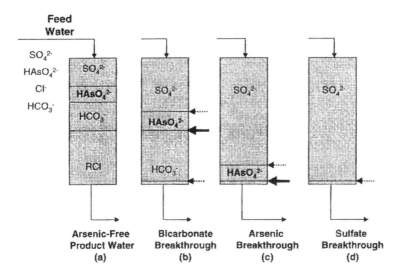

Figure 6   Resin concentration profiles at various stages of SBA resin exhaustion during arsenic removal by ion exchange: (a) beginning of an exhaustion run, (b) at bicarbonate breakthrough, (c) at arsenate breakthrough, and (d) at sulfate breakthrough.

is bicarbonate, the least preferred ion (Fig. 6b). Then the arsenic wave front breaks through (Fig. 6c). Finally, if the column is run beyond arsenic breakthrough, the most-preferred ion, sulfate, breaks through (Figure 6d). A typical series of breakthrough curves is shown in Figure 7. The first ion to show up in the effluent is bicarbonate at 70 BV, then after a relatively long run (16 hrs @ EBCT = 1.5 min with a feed sulfate concentration of 80 mg/L and feed As(V) concentration of 22 µg/L), arsenic breaks through and reaches the 10 µg/L MCL at 680 BV. After its breakthrough, arsenic peaks at twice its influent concentration because it is driven by sulfate, which breaks through at about 750 BV. Unlike the very gradual breakthrough observed with alumina, arsenic breakthrough on ion-exchange resin is rapid. It takes less than 100 BV for arsenic to rise from near zero to above its influent concentration. Note that in normal operation, the run would have been stopped at a predetermined volume of throughput, before the arsenic peak occurred.

D.   Effect of Sulfate on Run Length

Of the common anions present in groundwater, sulfate has the greatest adverse effect on arsenic run length (21) as seen in Figure 8. Based on this figure, it is recommended that anion exchange for arsenic removal be considered only when

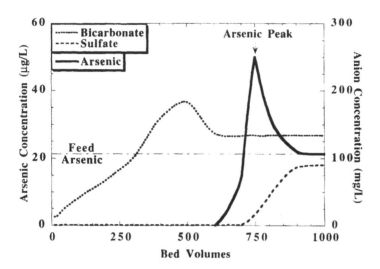

Figure 7   Typical ion-exchange exhaustion run showing bicarbonate, arsenate, and sulfate breakthrough curves from chloride-form anion-exchange column.

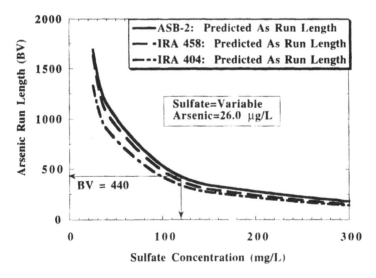

Figure 8   Predicted arsenic run lengths for SBA resins Ionac ASB-2 (PS Type 2), IRA 458 (Acrylic Type 1), and IRA 404 (PS Type 1). (From Ref. 21.)

the feed sulfate concentration is less than 120 mg/L, which should provide a typical run length of 400 BV or more.

E. Chromatographic Peaking

The phenomenon of a less-preferred ion appearing in the effluent at concentrations much greater than its influent concentration is referred to as chromatographic peaking (see Fig. 7). Chromatographic peaking is one of the major drawbacks of ion exchange for arsenic removal. Chromatographic peaking can be circumvented by terminating an exhaustion cycle at about 80% of the arsenic run length. Peaking can also be minimized or avoided entirely by operating two or more columns in parallel at different stages of exhaustion. If one column is run beyond arsenic breakthrough, its effluent concentration is diminished by averaging with the other effluents that essentially contain no arsenic.

F. Regeneration and Brine Reuse

During the original (1984) Hanford studies (14), arsenic recoveries upon downflow (co-current) regeneration were essentially complete. Three BV of 1.0 N NaCl (11 lb NaCl/ft^3 resin) was more than adequate to elute all the adsorbed arsenic, which was even easier to elute than bicarbonate, a very nonpreferred ion. One reason why arsenic elutes so readily is that it is a divalent ion ($HAsO_4^{2-}$) and is thus subject to a selectivity reversal in the pH 9 high-ionic-strength (>1 M) environment of the regenerant solution. This ease of regeneration is a strong point in favor of ion exchange as compared with alumina for arsenic removal in low-TDS, low-sulfate waters.

The 1984 Hanford studies (14) also showed that dilute regenerants (0.25–0.5 N) were more efficient than concentrated ones for eluting arsenic. The greater efficiency of dilute regenerants was further verified in the Albuquerque field studies (3,4,19–21) where 0.5 N NaCl outperformed 1.0 and 2.0 N regenerants in downflow regeneration experiments on similar aliquots of exhausted ASB-2 resin. It is also advisable to perform regeneration at a superficial linear velocity (SLV) ≥ 2 cm/min to avoid regenerant channeling, and consequently poor arsenic recoveries.

G. Reuse of Spent Regenerant

The major finding of the Albuquerque arsenic study (3,4) was that spent regenerant could be reused several times without treatment to remove arsenic. In a series of exhaustion–regeneration cycles, a type-2 resin column (1-in. i.d.) was exhausted and regenerated 18 times using recycled regenerant that had been compensated each cycle with NaCl to maintain the chloride concentration at 1 N. These results were verified in the scaled-up experiments shown in Figure 9. Despite the fact that arsenic concentration in the reuse brine rose to 19,000 μg/L, sulfate reached 151 g/L (3.1 N), and bicarbonate reached 24.4 g/L (0.39 N), the arsenic leakage during exhaustion was typically <0.4 μg/L and run lengths were consistently in the 400–450 BV range. In order to avoid the inconvenience of having to make up the chloride concentration after each regeneration, a series of exhaustion–regeneration cycles can be started with a much higher brine volume and concentration.

The Albuquerque runs showed that spent arsenic brine could be reused at least 20 times, and possibly more. Eventually, the spent brine will have to be wasted, but it should be treated to remove arsenic before disposal. This can be accomplished by iron (III) or aluminum (III) precipitation as discussed in the waste disposal section.

H. Waste Disposal

The spent brine can be safely disposed as a nonhazardous waste after removing the accumulated arsenic and reducing the arsenic concentration in the spent brine to acceptable levels. This can be accomplished by precipitating the arsenic using either ferric or aluminum salts such as $FeCl_3 \cdot 6H_2O$ or $Al_2(SO_4)_3 \cdot 18H_2O$.

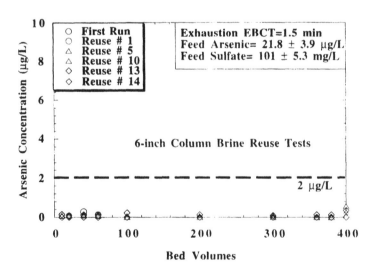

Figure 9 Results of scale-up tests with 6-in. column of ASB-2 resin, 14 cycles of spent brine reuse with chloride make-up, downflow cocurrent regeneration. (From Ref. 3.)

Experiments to remove arsenic from both low- (3.45 mg/L) and high-arsenic (11.3 mg/L) brines were performed in the Albuquerque field study (3,4). The low-arsenic brine was extensively tested to determine an optimum Fe(III)/As molar ratio in order to obtain ≥99.5% arsenic removal. As expected, arsenic removals achieved were found to be strongly pH dependent. A lower Fe/As dose of 20 could be used if the pH was lowered to at least 5.5. However, a much higher molar dose of 50 was required when operating at a higher pH of 6.2. Using the low-pH–low-dose approach, arsenic was precipitated from the high-arsenic brine. The results of 17 replicate tests yielded a treated brine containing a residual arsenic concentration of only 37 ± 15 µg/L (99.7% removal). TCLP tests on the sludge obtained from the high-arsenic brine precipitations produced a TCLP extract of 270 µg/L arsenic. At the current limit of 5,000 µg/L, the Fe-As sludge will easily pass the TCLP test requirements.

I. Process Design Considerations

Ion exchange is not attractive for single-column point-of-use or point-of-entry arsenic treatment because of the potential arsenic peaking problem. (When nitrate is in the water it will peak before arsenic breakthrough, and this must also be considered.) Effluent pH as low as 5–6.0 can also be a problem during the first 100 BV of a single-column run (3,4,23). However, whenever several columns can be operated in parallel, e.g., in a community water supply application, peaking and low-initial-pH problems can be eliminated and ion exchange should be considered (3,4,19–21). This assumes, too, that spent regenerant disposal is not an insurmountable problem. At least three columns operating in parallel should be employed. Our research has shown that both polystyrene and polyacrylic type 1 and type 2 resins are acceptable for arsenic removal. Bench- or pilot-scale column tests are recommended before final design. Finally, when using strong-base-anion resins in water supply applications, one must also consider the potential for nitrosodimethylamine (NDMA) formation. This carcinogen has been recently associated with use of SBA resins in water supply (24).

IV. COAGULATION–MICROFILTRATION

Arsenic removal via adsorption onto hydrolyzing metal salts has been shown to be an effective technology (3,4,25–30). Typically, ferric chloride (or alum) is added to the arsenic-containing water. Ferric chloride hydrolyzes and precipitates as insoluble ferric hydroxide, which is then filtered. In general, contaminant removal by coagulant addition occurs via the following mechanisms: (1) adsorption to effect surface charge neutralization, (2) enmeshment in precipitate, (3) adsorption to permit interparticle bridging, (4) surface precipitation, (5) compression of the double layer, (6) ligand exchange–surface complexation, and (7) hydrogen bonding. Hering and Elimilech (29) reported that adsorption was the dominant mechanism for arsenic removal by ferric hydroxide. The adsorption of As(V) on ferric hydroxide has been described in terms of a ligand exchange reaction (surface complex reaction) of arsenate for surface hydroxyls (31–34). The formation of As(V) surface complexes on oxides is supported by infrared spectroscopy (32,34).

During typical surface complex formation, or ligand exchange, the surface hydroxyl group on the hydrous oxide exchanges with a similar Lewis base electron pair donor in the solution. Adsorption of either protons or hydroxide ions is interpreted in terms of an acid–base reaction at the oxide surface, i.e., the surface hydroxyl group is either protonated or deprotonated. The adsorption of ligands (anions and weak acids) on a metal-oxide surface can also be compared with complex formation reactions in solution, e.g.:

$$Fe(OH)^{2+} + H_2AsO_4^- = Fe(H_2AsO_4)^{2+} + OH^- \tag{10}$$

$$S-OH + H_2AsO_4^- = S-H_2AsO_4 + OH^-, \tag{11}$$

where $S-OH$ corresponds to $\equiv Fe-OH$. In the above reactions, the central atom of the metal surface acts as a Lewis acid and exchanges its structural OH^- for other ligands (ligand exchange).

A. Reactions and Competing Ions

Of the common anions encountered in arsenic-contaminated waters, the one found to have the most significant effect on arsenic removal was silicate (3,4,28,35–39). Phosphate (0.13–0.5 μM) is also a significant competitor for arsenate adsorption, although it is not adsorbed as strongly as arsenate (5). Sulfate and bicarbonate were found to have no significant effect on arsenic removal (3,4,28,38). Cations such as Ca^{2+} and Mg^{2+} have been reported to have a slightly beneficial effect on arsenic adsorption onto ferric hydroxide (29,38). The beneficial effect of Ca^{2+} and Mg^{2+} ions is likely due to the neutralization of negative surface charges that result from silica adsorption on iron hydroxide. Natural organic matter (NOM) generally had no effect on As(V) adsorption at pH values less than 7 although the adsorption was significantly decreased at pH 9 (29). This study also determined that the effect of NOM was more significant for As(III) adsorption with arsenic removals lower in the presence of NOM in the pH range of 4–9.

As mentioned before, the effect of silica on arsenic adsorption by ferric hydroxide has been reported by a number of researchers. Swedlund and Webster (39) determined that the adsorption of silica onto ferric hydroxide was the predominant factor in the inhibition of arsenic adsorption. In addition to adsorption, silica may also interact with Fe(III) to form soluble polymers and highly dispersed colloids that are not removed by filtration (35,36). Gregory and Duan (37) found that silica levels above 50 mg/L strongly inhibited flocculation.

Tong (28) reported that the adverse effect of silica on As(V) adsorption was almost completely attenuated at a pH of 6.5. Similar attenuation has been reported by other researchers at pH 6.8 (38). The deleterious effect of silica on arsenic removal can be clearly seen in Figure 10. As the pH and concentration of silica increased, the adsorption density of arsenic on ferric hydroxide was sharply reduced.

B. Effect of Oxidation State and pH on Arsenic Removal

Arsenic removal has been observed to be much more efficient for As(V) than As(III) (27–29). Figure 11 shows As(V) and As(III) removals as a function of pH. As(V) removal was strongly dependent on the pH. For a dose of 2 mg Fe(III)/L, approximately 100% removal was observed at pH 5.5, while no removal occurred at pH 9.6. Figure 11 also shows that more As(V) was removed at lower pH than at higher pH when the coagulant dose was the same, i.e., the lower the pH, the greater the As(V) removal.

As expected, As(III) removal was extremely low at pH 5.5 and changed very little in the pH range of 7.5–9.6. This is because As(III) exists predominately as the neutral species H_3AsO_3 when the pH of water is below 9.23 and is a poor

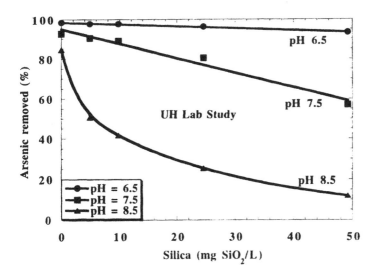

Figure 10 Effect of silica on As(V) removal. (From Ref. 28.)

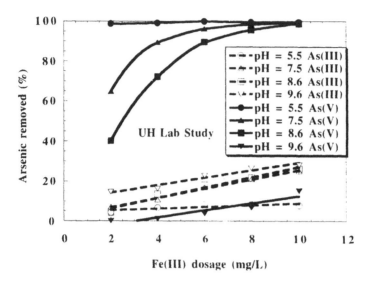

Figure 11   Effect of pH, oxidation state, and Fe(III) dose on As(III) and As(V) removal. (From Ref. 28.)

ligand when compared with As(V). However, as the pH increases, the greater concentration of hydroxide, which is an excellent ligand, outcompetes the $H_2AsO_3^-$ anion for complexation sites. As(V) occurs in the anionic form over the pH range examined, and thus was removed far more efficiently than As(III). However, at pH 9.6, hydroxide competition was significant, and both As(V) and As(III) were adversely affected.

C. Effect of Filter Pore Size on Arsenic Removal

The microfilter pore size was found to be an important parameter for effective filtration of the ferric hydroxide floc particles. Tong (28) studied three pore sizes, 0.22, 0.45, and 0.8 μm; and both synthetic water (without silica) and natural Albuquerque water (containing approx. 51 mg/L silica) were tested. For both the waters tested, better arsenic removals were obtained with smaller filter pore size. Moreover, for the same pore size, higher removals were obtained for the synthetic water when compared with the natural Albuquerque water. The lower arsenic removal with the Albuquerque water was attributed to its silica content (52 mg/L) compared with the synthetic water, which had no silica. By comparison, Hering and Elimilech (29) reported only minor effects of filter pore size in the range of 0.1–1.0 μm for both As(III) and As(V) removal. It is, however, not clear if the test waters in their study had significant concentrations of silica.

D. Effect of Fe(III) Dosage on As Removal

Numerous studies have shown that both As(III) and As(V) removals increase with increasing coagulant dose (3,4,27–30). The increased removal is consistent with the expected linear increase in surface area and the concentration of surface sites for adsorption (2,29). However, coagulant dose is secondary in importance to pH in the range of 5.0–8.0 (3,4,27). For example, during pilot studies performed in Albuquerque at pH 7.4 with a feed As(V) concentration of 34 μg/L, an increase in the coagulant dose from 1.9 to 5.8 mg Fe(III)/L resulted in only a slight decrease in filtrate As(V) concentrations from 3.9 to 3.0 μg/L, respectively. Under similar conditions at pH 6.4, filtrate As(V) concentrations were virtually unchanged.

E. Pilot-Scale Studies on Fe(III) Coagulation–Microfiltration

Usually, in a conventional coagulation–filtration process for arsenic removal, the addition of the coagulant is followed by a short rapid-mix step followed by a slow-mix step for flocculation. Flocculation is usually followed by sedimentation and filtration. To do away with the multitude of steps required in the conventional process, the authors designed and tested a simplified iron coagulation–direct microfiltration process, as described below (3,4).

The feasibility of iron coagulation–microfiltration was first demonstrated by Chang et al. (26). This study used a static mixer (for the rapid-mix step), a flocculation step (20 min) followed by microfiltration. Fe(III) doses as high as 6.9 mg/L (20 mg/L as $FeCl_3$) were used without any fouling problems. Another study (30) was also successful in designing a microfiltration scheme that easily met a target MCL of 5 μg/L using a Fe(III) dose of only 2.4 mg/L (7 mg/L as $FeCl_3$). Once again, a static mixer was used for the rapid-mix step followed by a flocculation step and then microfiltration. However, they reported fouling (reversible when using a shorter backwash interval) when using a higher coagulant dose of 3.4 mg/L as Fe(III) (10 mg/L as $FeCl_3$) to achieve a target filtrate arsenic concentration of 2 μg/L.

The authors' pilot-plant studies in Albuquerque (3,4) aimed to build on these previous studies. Several important improvements were made. A simpler but more demanding microfiltration scheme was used (Fig. 12) where the coagulant was added to the feed water in a *single* rapid-mix step followed by *direct* microfiltration, thereby eliminating a flocculation step. The rapid-mix step used

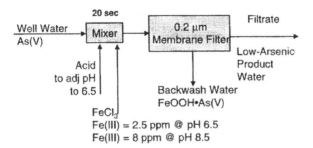

Figure 12 Process schematic for Fe(III)-coagulation–direct microfiltration process tested in Albuquerque, NM, showing two possible Fe(III) doses at pH 6.5 and 8.5. (From Ref. 3.)

a very short contact time of 16–20 sec. Furthermore, the Albuquerque study was designed with a more stringent MCL of 2.0 µg/L and also aimed to validate the findings of the University of Houston laboratory studies of Tong (28).

The design and operation of the custom-built rapid mixer are described elsewhere (3,4). An effective rapid-mix step rendered a flocculation step unnecessary and enabled the use of Fe(III) doses as high as 20.6 mg/L (60 mg/L as $FeCl_3$) without any fouling. Moreover, backwash intervals as high as 29 min were

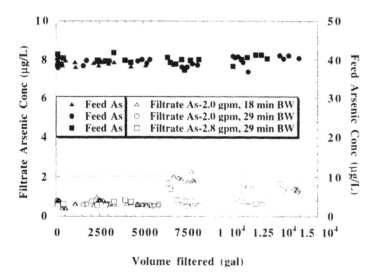

Figure 13 Results of extended iron coagulation–microfiltration tests on a groundwater in Albuquerque, NM. (From Ref. 3.)

employed with no adverse effect on filtrate flux. Figure 13 shows the results of extended testing with feed arsenic concentrations in the range of 37–40 µg/L and at a pH of 6.3 and an iron dose of 2.4 mg Fe(III)/L.

F. Waste Disposal

According to the TCLP Method 1311 (EPA SW 846) (40), for a liquid waste containing less than 0.5% solids, the liquid portion of the waste after filtration is defined as the TCLP extract. For a microfiltration system operated at a 2.5 mg Fe(III)/L dose, 2.8 gpm flow rate (1.4 gpm/m^2), and 29-min backwash interval, and assuming that all the solids are removed from the filter upon backwash, the backwash water (assuming a backwash volume of 1.8 gal/m^2) will have a solids content (calculated) of 0.01% (by wt). Arsenic concentration in such a filtered backwash water (average of 20 filtered samples) was determined to be 2.6 ± 2.4 µg/L. The backwash water can thus be directly disposed as a nonhazardous waste assuming that the arsenic TCLP limit stays at its current value of 5000 µg/L.

REFERENCES

1. MO Andreae. Distribution and speciation of arsenic in natural waters and some marine algae. Deep-Sea Res 25:391–402, 1978.
2. W Stumm, JJ Morgan. Aquatic Chemistry: Chemical Equilibria and Rates in Natural Waters. 3rd edition, New York: Wiley, 1996.
3. DA Clifford, GL Ghurye, AR Tripp, T Jian. Final Report: Phases 1 and 2, City of Albuquerque Arsenic Study, Field Studies on Arsenic Removal in Albuquerque, NM, Using the University of Houston/USEPA Mobile Drinking Water Treatment Research Facility. Houston, TX: University of Houston, 1997.
4. DA Clifford, GL Ghurye, AR Tripp, T Jian. Final Report: Phase 3, City of Albuquerque Arsenic Study, Field Studies on Arsenic Removal in Albuquerque, NM, Using the University of Houston/USEPA Mobile Drinking Water Treatment Research Facility. Houston, TX: University of Houston, 1998.
5. W Driehaus, M Jekel, U Hildebrandt. Granular ferric hydroxide: A novel adsorbent for the removal of arsenic from natural water. J Water SRT, 47:30–35, 1998.
6. F Azizian. Removal of Arsenic from Potable Water Using Activated Alumina. Oxon, England: Alcan Chemicals, 2000.
7. DA Clifford and M Wu. Arsenic treatment technology demonstration: Predicting the effect of water quality on arsenic adsorption by activated alumina. Report to the Montana State University Water Resources Center, 2001.
8. MM Benjamin, RS Sletten, RP Bailey, T Bennett. Sorption and filtration of metals using iron-oxide coated sand. Water Res 30:2609–2620, 1996.
9. A Joshi, M Chaudhuri. Removal of arsenic from ground water by iron oxide-coated sand. J Environ Eng 122:769–771, 1996.

10. GEH Wasserchemie. GEH-granular ferric hydroxide technical note, Osnabruck, Germany: GEH Wasserchemie Gmbh & Co. KG, 1998.

11. J Simms, J Upton, J Barnes. Arsenic removal studies and the design of a 20,000 m^3 per day plant in the UK. American Water Works Association Inorganic Contaminants Workshop, Albuquerque, NM, 2000.

12. E Rosenblum and DA Clifford. The equilibrium arsenic capacity of activated alumina. U.S. Environmental Protection Agency Report, Cincinnati, OH, 1984.

13. P Frank, DA Clifford. Arsenic III oxidation and removal from drinking water. U.S. Environmental Protection Agency Report, Cincinnati, OH, 1986.

14. DA Clifford, CC Lin. Arsenic III and arsenic V removal from drinking water in San Ysidro, New Mexico. U.S. Environmental Protection Agency Report, Cincinnati, OH, 1991.

15. DA Clifford, CC Lin. Ion-exchange, activated alumina, and membrane processes for arsenic removal from groundwater. Proceedings of 45th Annual Environmental Engineering Conference, University of Kansas, 1995.

16. F Rubel Jr, SW Hathaway. Pilot study for the removal of arsenic from drinking water at the Fallon, Nevada, Naval Air Station. U.S. Environmental Protection Agency Report, Cincinnati, OH, 1985.

17. J Simms, F Azizian. Pilot-plant trials on removal of arsenic from potable water using activated alumina. American Water Works Association Water Quality Technology Conference, Denver, CO, 1997.

18. DA Clifford, JV Matson, R Kennedy. Activated alumina: Rediscovered "adsorbent" for fluoride, humic acids, and silica. Ind Water Eng pp 6–12, 1978.

19. DA Clifford. Ion exchange and inorganic adsorption. In: RD Letterman, ed. Water Quality and Treatment, 5th ed. New York: McGraw-Hill, 1999, pp 9.1–9.91.

20. DA Clifford, GL Ghurye, AR Tripp. Arsenic ion-exchange process with reuse of spent brine. Conference Proceedings of American Water Works Association National Meeting, Dallas, TX, 1998.

21. GL Ghurye, DA Clifford, AR Tripp. Combined nitrate and arsenic removal by ion exchange. J Am Water Works Assoc 91:85–96, 1999.

22. GL Ghurye, DA Clifford. Laboratory study on the oxidation of As(III) to As(V). U.S. Environmental Protection Agency, Office of Research and Development EPA/600/R01/021, 2001.

23. LL Horng. Reaction mechanisms and chromatographic behavior of polyprotic acid anions in multicomponent ion exchange. PhD dissertation, University of Houston, Houston, Texas, 1983.

24. I Najm, R Trussell. NDMA formation in water and wastewater. J Am Water Works Assoc 93:92–99, 2001.

25. TJ Sorg, GS Logsdon. Treatment technology to meet the interim primary drinking water regulation for inorganics: Part 2. J Am Water Works Assoc 70:379–392, 1978.

26. S Chang, H Ruiz, WD Bellamy, CW Spangenberg, D Clark. Removal of arsenic by enhanced coagulation and membrane technology. Critical Issues in Water and Waste Water Treatment, American Society of Civil Engineers, Reston, VA, 1994.

27. M Edwards. Chemistry of arsenic removal during coagulation and Fe-Mn oxidation. J Am Water Works Assoc 86:64–78, 1994.

28. J Tong. Development of an iron (III)-coagulation-microfiltration process for arsenic

removal from ground water. MS thesis, University of Houston, Houston, TX, 1996.

29. JG Hering, M Elimelech. Arsenic Removal by Enhanced Coagulation and Membrane Processes. Report 90706. Denver, CO: American Water Works Association Research Foundation, 1996.

30. P Brandhuber, G Amy. Alternative methods for membrane filtration of arsenic from drinking water. Desalination 117:1–10, 1998.

31. MA Anderson, DT Malotky. The adsorption of protolyzable anions on hydrous oxides at the isoelectric pH. J Coll Interface Sci 72:413–427, 1979.

32. JB Harrison, VE Berkheiser. Anion interaction with freshly prepared hydrous iron oxides. Clays Clay Miner 30:97–102, 1982.

33. DG Lumsdon, AR Fraser, JD Russell, NT Livesey. New infrared band assignments for the arsenate ion adsorbed on synthetic goethite (α-FeOOH). J Soil Sci 35:381–386, 1984.

34. T-H Hsia, S-L Lo, C-F Lin, D-Y Lee. Characterization of arsenate adsorption on hydrous iron oxide using chemical and physical methods. Coll Surf 85:1–7, 1994.

35. RB Robinson, GD Reed, B Frazier. Iron and manganese sequestration facilities using sodium silicate. J Am Water Works Assoc 84:77–82, 1992.

36. RK Iler. The Chemistry of Silica. New York: Wiley, 1979.

37. J Gregory, J Duan. The effect of dissolved silica on the action of hydrolyzing metal coagulants. Water Sci Technol 38:113–120, 1998.

38. X Meng, S Bang, GP Korfiatis. Effects of silicate, sulfate, and carbonate on arsenic removal by ferric chloride. Water Res 34:1255–1261, 2000.

39. PJ Swedlund, JG Webster. Adsorption and polymerization of silicic acid on ferrihydrite and its effect on arsenic adsorption. Water Res 33:3414–3422, 1999.

40. Methods for Evaluating Solid Wastes. SW-846, U.S. Environmental Protection Agency, Cincinnati, OH, 1997.

10

Arsenic Metabolism: Resistance, Reduction, and Oxidation

Simon Silver and Le T. Phung
University of Illinois at Chicago, Chicago, Illinois

Barry P. Rosen
Wayne State University, Detroit, Michigan

I. INTRODUCTION

In this comprehensive monograph on environmental and biological aspects of arsenic chemistry, we have limited our chapter to what we know best because the research has come substantially from our laboratories. Arsenic is ubiquitous, from both natural and human-related activities, and living organisms have evolved metabolic systems for coping with the toxicities of arsenic compounds. For bacteria, highly specific resistance systems that confer resistances to both arsenite [As(III)] and arsenate [As(V)] are found widely, encoded both on chromosomes and on mobile small plasmids. Paradoxically, bacteria—perhaps different bacteria—utilize both arsenite oxidase and arsenate reductase as components of arsenic resistance systems. This is surprising, since for most biological systems arsenite is perhaps 50 times more toxic to bacterial cells than is arsenate. However, the toxicity depends on endogenous factors (other gene products, especially membrane oxyanion uptake and efflux pumps) and exogenous factors (environmental redox potential, precipitating available cations, which can make either arsenite or arsenate more or less mobile in the environment) that are described elsewhere in this volume. This chapter describes the genetically determined mechanisms of arsenic resistance and transformations, plus the two best-characterized enzymatic transformations of inorganic arsenic oxyanions. The appearance of arsenic compounds and the selection pressures that maintain these bacterial systems occur in familiar environments (such as well water in Illinois and Michigan where we live), less familiar unperturbed environments, and in human-perturbed environments as diverse as those in mining geochemistry and in clinical medicine.

Acidic mining environments (1) are of particular concern with regard to arsenic pollution, since it is bacterial energy-yielding and growth activities in oxidizing reduced sulfur to H_2SO_4 that results in very low pH (sometimes approaching pH 1) and in parallel reduced arsenic is oxidized to H_3AsO_4. Both bacterial and Archaean prokaryotes are found in these environments and involved in arsenic transformations (2,3).

Arsenical compounds have been used often in medicine for millennia, in ancient folk medicine, and within the last 150 years especially for protozoal diseases, but also in agriculture and as a poison in crime (4). Paul Ehrlich won the Nobel Prize in medicine in 1909 for use of dyes and arsenicals as chemotherapeutic agents, culminating with the famous arsenical Apräparat 606 compound, "Salvarsan," which was the best of early modern antimicrobial agents (5). The organoarsenical melarsoprol is the trypanosomicidal treatment of choice in Africa since being introduced in 1949 (it is synthesized in Germany by Hoechst and formulated in France by Rhone Poulenc, but is only available in United States and Canada as "an investigational drug"). Arsenic trioxide, As_2O_3 (called "trisenox"), injected intravenously is a new and widely encouraged treatment for some forms of leukemia and myelomas (6). In North America and elsewhere in the developed world, organoarsenicals are added at levels of 23–45 g per tonne to feed for boiler chickens. Considering the number of chickens consumed annually in North America and Europe, the amounts added and then released to the environment as chicken effluent are substantial. It is thought that these organoarsenicals enhance chicken growth, perhaps by limiting diseases, e.g., coccidiosis. Although we exist in a world with many anthropogenic organoarsenicals, there is, however, only a single report concerning bacterial metabolism of organoarsenicals (7).

Inorganic arsenic oxyanions, frequently present as environmental pollutants, are very toxic for most micro-organisms. Many microbial strains possess genetic determinants that confer resistance. In bacteria, these determinants are often found on plasmids, which has facilitated their study to the molecular level. Bacterial plasmids conferring arsenic resistance encode specific efflux pumps able to extrude arsenic from the cell cytoplasm, thus lowering the intracellular concentration of the toxic ions. Recently, apparently similar arsenic membrane transport proteins have been found with yeast, plants, and animals (8,9) (see Sec. V, below).

Arsenic resistance is not the only toxic heavy metal ion resistance system found in bacteria. Bacteria have known plasmid and chromosomal genes for resistances to Ag^+, AsO_2^-, AsO_4^{3-}, Cd^{2+}, Co^{2+}, CrO_4^{2-}, Cu^{2+}, Hg^{2+}, Ni^{2+}, Pb^{2+}, Sb^{3+},

TeO_3^{2-}, Tl^+, and Zn^{2+}. If an ion is toxic, micro-organisms have selected variants possessing highly specific, genetically governed resistance determinants that provide tolerance to higher levels of the toxic compounds. In bacteria, the first heavy metal resistance genes to be studied were usually located on plasmids or transposons. This was because the movement of the plasmids from cell to cell and their small sizes facilitated studies. There were early exceptions, however, such as the chromosomal location for mercury and arsenic resistance in *Bacillus*. In recent years, whenever a new total genome (total genetic complement) of a microbe becomes available, it generally includes systems for toxic inorganic ion resistance, especially that to arsenic compounds. Several bacterial resistance mechanisms to metals have been studied with molecular detail (10,11).

With a few plasmids of gram-negative bacteria, the efflux pump consists of a two-component adenosine triphosphatase (ATPase) complex. ArsA is the ATPase subunit and is associated with an integral membrane subunit, ArsB (see Sec. II, below). Arsenate is enzymatically reduced to arsenite (the substrate of ArsB and the activator of ArsA) by the small cytoplasmic arsenate reductase enzyme, the product of the *arsC* gene (Fig. 1A). On the chromosomes of most gram-negative and gram-positive bacteria and on many plasmids, *arsB* and *arsC* genes (and proteins) are found, but *arsA* is not.

In addition to the widespread (almost universal) chromosomal and plasmid arsenic resistance determinants, a few bacteria confer resistance to arsenite alone with a separate determinant for enzymatic oxidation of more-toxic arsenite to

A. Soluble reductase   B. membrane reductase   C. membrane oxidase

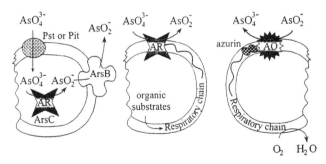

Figure 1 Bacterial reduction of arsenate and oxidation of arsenite. (A). Cytoplasmic arsenate reductase (ArsC) as encoded by bacterial *ars* operons along with chromosomally encoded Pit and Pst phosphate transport systems with arsenate as an alternative substrate. After reduction from arsenate to arsenite, arsenite is removed from the cell by the ArsB membrane protein. (B). Anaerobic periplasmic arsenate reductase. (C). Aerobic periplasmic arsenite oxidase, linked via azurin to the respiratory chain.

less-toxic arsenate (Fig. 1C) (see sec. IV, below) with O_2 as the terminal electron acceptor. This may be coupled to reduced uptake of arsenate by phosphate transport systems, as known with cyanobacteria (12). Still additional bacteria have unusual anaerobic respiratory chains with arsenate as the terminal electron acceptor instead of O_2 (Fig. 1B) (see Chaps. 11 and 12). In contrast to the detailed information on the mechanisms of arsenic resistance and enzymatic reduction and oxidation in bacteria, little work has been reported on this subject in Archaea, algae and fungi.

II. MOLECULAR GENETICS OF BACTERIAL ARSENIC RESISTANCE: VARIETIES OF OPERONS

Fundamentally the same genes (and encoded biochemical mechanism) are found on plasmids in gram-negative and gram-positive bacteria (10,11,13). Closely similar chromosomal gene clusters have been found to determine normal background arsenic resistance both in *Bacillus* and in *Escherichia coli* (10,11,14,15), and as yet genetically uncharacterized systems have different bases for arsenic resistance in environmentally important bacteria (16) (see Chaps. 12–15). Therefore, what we know today may be only the first half of a larger picture of microbial arsenic transformations.

Bacterial resistance to arsenic ions governed by plasmids was first discovered by Novick and Roth (17) in a group of *Staphylococcus aureus* β-lactamase plasmids that determine resistances to antibiotics and also to heavy metals. Arsenic resistance plasmids confer tolerance to both arsenate and arsenite as well as to antimony (III) (18). Resistance to all three ions is inducible and cross-induction among them occurs (18). Arsenic resistance determinants are very common in plasmids of both gram-negative and gram-positive bacteria.

Silver et al. (18) found that resistance to arsenate in both *E. coli* and *S. aureus* was due to lowered uptake of arsenate by resistant cells. As expected, high phosphate concentrations protected cells from arsenate toxicity. Studies of resistant *E. coli* and *S. aureus* showed that the reduced arsenate uptake is due to an accelerated efflux of the toxic ions in a energy-dependent process (19,20). It was initially thought that arsenate was the extruded oxyanion, but later studies showed that arsenate had to be reduced to arsenite prior to extrusion (21,22). Rosen and Borbolla (23) showed that it was actually arsenite efflux in *E. coli* that was ATP-dependent.

Of several dozen currently known sequenced *ars* operons, Figure 2 diagrams the arrangements of genes for six representatives that make several useful points. Starting in the middle of Figure 1, the *ars* operon of plasmid R773 was the first sequenced and remains by far the most thoroughly studied *ars* system. There are five genes named *arsR*, *arsD*, *arsA*, *arsB*, and *arsC* co-transcribed

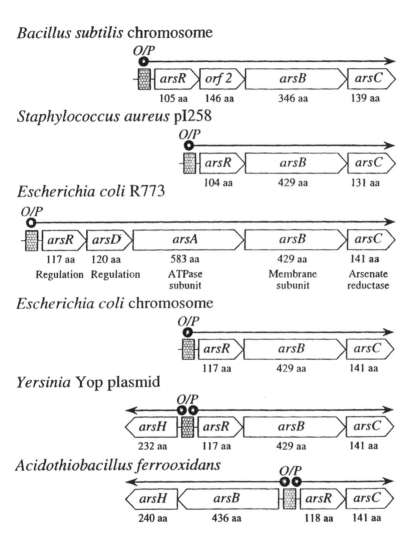

Figure 2 Various genes and gene patterns for As(III), As(V) and Sb(III) *ars* resistance systems. Stippled box: O/P, operator/promoter sites of repressor regulation and initiation of mRNA synthesis (length of mRNA shown by arrows). Open boxes: *ars* genes with lengths of gene products given as amino acids.

from a single operator/promoter (O/P) site for start of messenger ribonucleic acid (mRNA) synthesis; and like all other *ars* operons (as far as known), this system confers resistances to As(III), As(V), and Sb(III) (18). The genes (and their products) are in order *arsR* (which determines a dimeric regulatory repressor protein), *arsD* (which determines a small secondary transcriptional regulatory protein that has been less studied), *arsA* (which determines the large membrane-associated ATPase subunit that contains closely homologous N- and C- halves, with separate ATP binding sites, apparently arising from gene fusion) (Fig. 3), *arsB* (which determines the membrane transport protein that removes arsenite—and antimonite—from the cytoplasm in an energy-dependent pumping process), and *arsC* (which determines the small cytoplasmic arsenate reductase enzyme that all *ars* operons encode). The *ars* operon from a different plasmid R46, with the same five genes, was also sequenced and with protein products apparently differing from those of R773 by only 7–15%. This level of amino acid product "drift" indicates a family of such systems with those of plasmids R773 and R46 separate for perhaps a dozen million years and not two recently spread copies of the same determinant. The more ancient existence of arsenic resistance determinants is demonstrated, however, by the finding that the number of genes can vary and the details of their functions can differ.

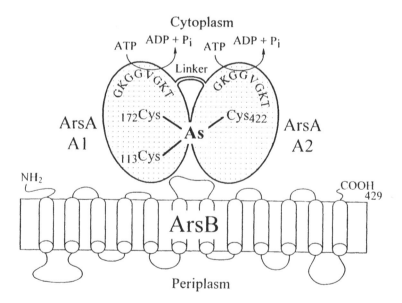

Figure 3 Model of the ArsA ATPase plus ArsB membrane transport protein from plasmid R773. (Adapted from refs. 8 and 11).

After the best-characterized five-gene *arsRDABC* operons found so far only on a few plasmids of gram-negative bacteria (see example R773 in Fig. 2), the next best-studied and certainly most widely found arsenic resistance systems today are the three-gene *arsRBC* operons found on plasmids of gram-positive bacteria (see example pI258 in Fig. 2) and in bacterial chromosomes (see example *Escherichia coli* in Fig. 2) (8,10,11). Although at earlier stages, we thought that only these two patterns existed, more recent efforts have shown additional genes such as *orf2* of the *Bacillus subtilis* chromosome and *arsH* of plasmids of *Yersinia* and the chromosome of *Acidothiobacillus* (previously called *Thiobacillus*) *ferrooxidans* (see Fig. 2). Even the number of operons, that is, transcriptional units, can vary and either *arsH* or *arsH* together with *arsB* can be separately transcribed (see Fig. 2). There is no basis currently for understanding the differences achieved and advantages to the bacterial cell of the extra genes and separate transcriptional control. In addition to a wide range of bacteria, Archaea such as the halophile *Halobacterium* sp. has related *ars* genes, although the gene organization is again different (24,25). The Archaeal *ars* gene products have not been studied.

Given the *ars* operon organization in Figure 1, one can consider the gene products and the regulation of their synthesis. The operons always start with an *arsR* gene and are regulated by the As(III)/Sb(III)-responsive ArsR repressor, which is the dimeric product (two 13-kDa monomers) encoded by the gene (26–28). The operon can be induced by arsenate, arsenite, antimonite, and bismuth in vivo, but arsenate does not bind to the ArsR repressor in vitro, but must be converted to As(III) to function as an inducer. ArsRs from all sources to date are closely related to each other and to other members of a dimeric helix-turn-helix repressor family, which include repressors of other metal resistance systems, including those for Cd(II), Hg(II) (in *Streptomyces* only), Pb(II), and Zn(II) (9) (see below for more details). The next gene, *arsB*, encodes the ArsB membrane protein that confers resistance to arsenite and antimonite by extruding (pumping) the metalloid oxyanions from the cells. Such oxyanion efflux pumps may have evolved more than once (this is called ''independent invention'') (29), but the 45-kDa ArsBs of most examples listed in Figure 2 are similar inner-membrane proteins with N- and C-termini in the cytosol and 12 membrane-spanning segments (see Fig. 3) (30). Even the R773 ArsB from gram-negative *E. coli* and the pI258 ArsB from a gram-positive species are 58% identical in amino acid sequence and the pI258 ArsB can function in *E. coli* (31). A chimera composed of half of the R773 and half of the pI258 ArsB functions in *E. coli* (32). As a transport protein for the thiol-binding arsenite oxyanion, one expected ArsB to contain critical cysteine residues. However, there are no critical cysteine residues in ArsB, avoiding the possibility of arsenite becoming ''jammed'' by binding. The energy for the ArsB efflux pump is in the form of the membrane potential (33) when ArsB is functioning alone, either because of the absence of an *arsA* gene (see Fig. 2) or because the *arsA* gene has been deleted. When ArsA is present, the energy for ArsB-mediated arsenite efflux is no longer the membrane potential but is derived from ATP hydrolysis by the ArsAB complex (see Fig. 3). Arsenate resistance requires an *arsC* gene, which is present in all *ars* operon arsenic resistance systems (see Fig. 1). There are two distinct and unrelated families of ArsC arsenate reductase proteins (Fig. 4), but all appear to be small cytoplasmic proteins that utilize oxidized/reduced cysteine thiol cycling in order reduce arsenate to arsenite, the substrate of the extrusion system. One family is typified by arsenate reductase encoded by the *E. coli* plasmid R773 (22,34). The R773 arsenate reductase uses both glutathione and glutaredoxin as electron source intermediates (Fig. 5) and prefers glutaredoxin 2, the major *E. coli* form (35). The other family of arsenate reductases (see Fig. 4) is typified by that from *S. aureus* plasmid pI258, and uses thioredoxin as electron source (see Fig. 5) (21,36).

In the five-gene *ars* operons there are two additional genes, *arsD* and *arsA*, located between *arsR* and *arsB* (see Fig. 2). The ArsD forms a dimer of two 13-kDa monomers and is a secondary As(III)-responsive transcriptional repressor (37,38). Only six *arsD* genes have been identified to date, all of which are associated with *arsA*, suggesting that the two genes may have been acquired together in a lateral gene transfer event. While ArsR controls the basal level of *ars* operon expression and also regulation from "off" to "on," ArsD functions as a "molecular throttle control" on the upper level of expression, perhaps to prevent overex-

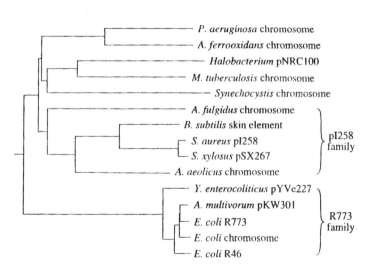

Figure 4 Evolutionary tree of sequence relationships for ArsC arsenate reductase proteins for bacteria, Archaea, and eukaryotic yeast with pI258 and R773 branches marked.

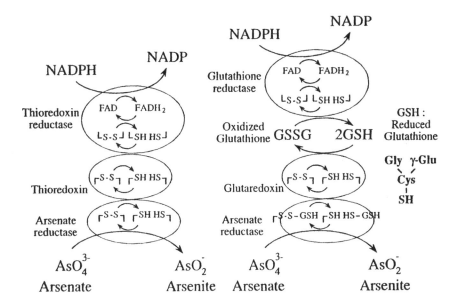

Thioredoxin family   Glutaredoxin family

Figure 5 Comparison of thioredoxin coupled arsenate reductase (pI258 family) and glutaredoxin coupled arsenate reductase (R773 family). GSH: reduced glutathione, the tripeptide consisting of γ-glutamyl-cysteine-glycine. GSSG: S-S bridged oxidized dimer of glutathionine.

pression of ArsB, which is toxic in high amounts (30). ArsA is a 63-kDa ATPase that is allosterically activated by As(III) or Sb(III) (39,40). Detailed stop-flow spectroscopic analysis has led to a model of alternating opened/closed ATP binding sites for the two closely homologous halves of ArsA (40). Binding of either As(III) or Sb(III) at C113 and C172 of the N-half ArsA domain and C422 of the C-half domain (see Fig. 3) fixes the state of the ATPase so that rapid ATP hydrolysis occurs at the C-domain site. The ArsA-ArsB complex is a primary ATP-driven arsenite pump that is thermodynamically more efficient than ArsB alone (33). Hence five-gene *ars* operons confer a higher level of resistance than the three-gene operons.

Each of the five plasmid R773 *ars* gene products has a binding site for arsenic. Yet each is clearly the product of independent evolution, since the binding site in each is different. It is of interest, therefore, to compare these sites. These are described next in the order ArsR, ArsD, ArsA, ArsB, and ArsC.

In ArsR, the binding site is composed of Cys32, Cys34, and Cys37, which form a three-coordinate AsS_3 site (41). R773 ArsR mutants with the three cysteine residues changed to Tyr or Phe all had a noninducible phenotype, and the mutant proteins were still capable of binding to the DNA operator region but responded to inducers very poorly. Thus these cysteines were involved in inducer binding and not in deoxyribonucleic acid (DNA) binding. Binding of As(III) induces a conformational change in ArsR that results in release of the repressor protein from the operator DNA (27). In the absence of direct structural data at this time, one can model the structure of ArsR based on the solved crystal structure of the homologous divalent cation–responding SmtB repressor (42). The SmtB structure was obtained in the absence of metal or DNA, but in a form that could effectively bind DNA. In the ArsR model, the sequence between residues 32 and 37 forms an α-helix-loop-α-helix motif characteristic of DNA binding regions that includes the three critical As(III)-binding cysteines (Fig. 6). From the results of x-ray absorption spectroscopy experiments, each of the three As-S distances is 2.25 Å, similar to the 2.23 Å distance from x-ray diffraction analysis of crystals of As(III) or Sb(III) complexed to small-molecule dithiols (43). The current model of the ArsR dimer has the three sulfur atoms 3.2 Å from each other in a

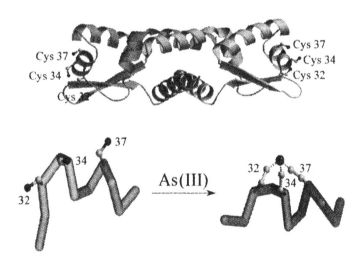

Figure 6 Model for ArsR repressor protein based on crystal structure of the related SmtB repressor of cyanobacterial metallothionein (*smtA*) mRNA synthesis. (Top) Dimer of ArsR with α-helix-loop-α-helix region proposed to bind to operator DNA. (Bottom left) Tri-cysteine region folded as α-helix, with Cys32, Cys34, and Cys37 not suitably spaced to bind As(III). (Bottom right) Tri-cysteine region unfolded and coordinately binding As(III).

trigonal pyramidal structure with As at the peak and the lone electron pair projecting upward (see Fig. 6). In the model of ArsR without arsenite, the minimum distance between the thiolates of Cys32 and Cys37 is greater than 6 Å; and when arsenite is added, the thiols of Cys32, Cys34, and Cys37 rotate and move closer in distance, and positioned to bind As(III) (see Fig. 6). However, residues 32–37 can no longer form an α-helix (see Fig. 6). Thus the ability to bind to DNA would be lost. It is proposed that depression of *ars* operon expression by binding of As(III) is the result of a conformational change in the DNA binding domain of ArsR with loss of the first α-helix.

ArsD is a secondary regulator of *ars* operon transcription (37,38), and its presence or absence has little effect on resistance during laboratory conditions. The existence of ArsD in several different plasmids, however, indicates a role in the environment with selection more subtle than we currently understand. ArsD is the secondary As(III)-responsive repressor of *ars* operon transcription and has a much lower affinity for both DNA and As(III) than does ArsR (38). Plasmid R773 ArsD has eight cysteine residues, more than any of the other proteins encoded by the operon. Six of the cysteines are in vicinal pairs, Cys12-Cys13, Cys112-Cys113, and Cys119-Cys120, and vicinal cysteine pairs are known to form strong As(III) binding sites (44). Recent evidence suggests that each of the Cys12-Cys13 and Cys112-Cys113 vicinal pairs form independent As(III) binding sites, and that both are required for ArsD activity (S. Li, Y. Chen, and B. P. Rosen, unpublished data).

ArsA is a membrane-associated ATPase (see Fig. 3) (45,46) attached to the ArsB inner-membrane protein (30,47) and energizing the arsenite efflux pump by ATP hydrolysis (33,39). Such alternative energy coupling is unique among known bacterial uptake or efflux transport systems. To date, all other systems that have been studied are either membrane potential–driven or ATP-driven transporters. The ArsAB pump is the only one that can be converted from one form of energy coupling to the other by addition or removal of genes. This is a natural phenomenon (8) and also can be reconstructed in laboratory studies (33).

The 583-residue R773 ArsA ATPase is activated by As(III) or Sb(III) (48). It has been purified as a soluble protein in the absence of ArsB. ArsA has two homologous halves, A1 and A2, that are connected by a short linker peptide (49). Both A1 and A2 have consensus nucleotide binding domains (key sequences listed in Fig. 3), and both nucleotide binding domains are required for activity (46,50). There are two cysteine residues in A1, Cys113 and Cys172, and a third in A2, Cys422, that are required for As(III) binding (see Fig. 3) and activation (51). From the results of chemical crosslinking experiments, the thiolates of these three residues were shown to be within 3–6 Å of each other (52). The crystal structure of ArsA at 2.3 Å resolution has recently been determined, and the nature of the As(III)/Sb(III) binding site revealed (53). In the crystal structure, there are three Sb atoms coordinated to a total of six protein amino acid ligands, which include the three cysteine residues, two histidine residues (His148 and His453), and one Ser420. Each Sb atom is coordinated to one amino acid residue in the A1 domain and a second residue in the A2 domain. The third coordination for each Sb(III) in the crystal structure appears to be a nonprotein ligand. Thus, binding of As(III) or Sb(III) brings the A1 and A2 halves of the protein tightly together. This contact facilitates formation of an interface of the two nucleotide binding sites, which results in an acceleration of catalysis (45,54).

Arsenic is a semi-metal or metalloid that has both metallic and nonmetallic properties. In contrast to the other gene products of the operon, ArsB does not use metal-thiol chemistry for catalysis. There is only a single cysteine in ArsB, and this residue is not required for arsenite transport (55).

All *ars* operons encode an ArsC arsenate reductase enzyme. However, as mentioned above, there are two sequence-unrelated families of arsenate reductases whose role is to reduce less toxic As(V) to more toxic As(III) (21,34,36). It is only As(III) and not As(V) that is pumped out from the cells by the ArsB transport protein. It seems counterintuitive from an environmental biology or metabolic chemistry point of view to convert a less toxic compound to a more toxic form. We have speculated that arsenite pumping activity evolved prior to the existence of oxygen in the atmosphere, and ArsC evolved only after an oxidizing atmosphere developed (8). At that point arsenite would spontaneously oxidize to arsenate, presenting a selective pressure for evolution of a reductase. The two types of *arsC* genes are positioned last in all *ars* operons shown in Fig. 2. The *E. coli* chromosomal and plasmid R773 ArsCs are 91% identical in sequence. In the other family of ArsC enzymes, the *S. aureus* plasmid pI258 and *Bacillus* chromosome versions are 60% identical. However, the *S. aureus* pI258 version is less than 15% identical with that from plasmid R773; and the ArsC sequence from *Yersinia* plasmid Yop is only 15% identical to that from *A. ferrooxidans* although the predicted products of the associated ArsB membrane proteins are 80% identical. *Yersinia* and *A. ferrooxidans* are both gram-negative bacteria, and the new ArsB sequence deducted from the released *Pseudomonas aeruginosa* genome is also unrelated to that of plasmid R773. Therefore, *arsC* genes and proteins seem to have evolved independently more than once. The invention of arsenate reductase is thought to have occurred early in the evolution of cellular life (before the division into gram-negative and gram-positive bacteria) (see Fig. 4) in a world rich in geochemically released arsenic, and not to have arisen more recently as a result of human pollution. The most striking difference between the two arsenate reductase enzymes is the source of reducing potential. Both the R773 and pI258 arsenate reductases require small intracellular proteins that function as general disulfide reducing agents. The pI258 enzymes utilized thioredoxin, both in vivo and in vitro (21,36). In contrast, the *E. coli* plasmid R773 enzyme uses glutaredoxin and reduced glutathione (GSH) (see Fig. 5) (34).

There are other differences between the two arsenate reductases that have been studied. *Staphylococcus aureus* arsenate reductase has a high affinity for arsenate with a K_m of 1 μM, whereas with the *E. coli* R773 enzyme has a measured K_m of 8 mM. This difference seems unlikely and might arise from in vitro assay conditions. Phosphate (the intracellular metabolite analog of arsenate) and nitrate (but not sulfate) stimulate pI258 reductase activity, whereas arsenite, antimonite, and tellurite are inhibitors for the *S. aureus* enzyme (36). In contrast, phosphate, sulfate (not nitrate), and arsenite (not antimonite) are competitive inhibitors for the *E. coli* enzyme (34). In the crystal structure of R773 ArsC, sulfate can be seen occupying the anion binding site (P. Martin, B. Edwards, and B. P. Rosen, unpublished data).

III. ARSENATE REDUCTION AND ARSENATE REDUCTASE

ArsC is the only protein of the plasmid *ars* operon that catalyzes a chemical reaction with the substrate oxidized As(V), in the form of arsenate, and the product reduced As(III), arsenite, which is the substrate of another gene product, the ArsB or ArsAB membrane pump. Thus, ArsC is necessary but not sufficient to confer resistance to arsenate. Since arsenite is still more toxic (mol/mol) than arsenate, once arsenite is formed, it must be rapidly extruded from the cell to complete the resistance mechanism. It seems counterintuitive to have a resistance mechanism that converts a less-toxic oxyanion into a more toxic oxyanion. However, with the advantage of hindsight, one can rationalize that bacterial cells have difficulty distinguishing arsenate from the closely similar oxyanion phosphate. Indeed, both phosphate uptake transport systems of *E. coli*, Pit and Pst (see Fig. 1), also accumulate arsenate, and indeed with similar K_ms (18). Thus it might be difficult for bacterial cells to evolve an arsenate efflux system that could discriminate between low (micromolar) levels of the toxic substrate arsenate, and much higher (millimolar) levels of the intracellular nutrient phosphate. Peter Mitchell (before he invented the "chemiosmotic hypothesis" for which he received the Nobel Prize) experimentally determined that once phosphate is accumulated by bacterial cells, under normal conditions it is never released. Given the need to explain the existence of arsenate reductase, it appears that it might be more feasible for evolution to have invented the enzyme arsenate reductase, plus the closely associated ArsB membrane protein, to remove arsenite from the cells.

Although we have tried periodically over the years, we have thus far been unable to demonstrate a physical association of intracellular arsenate reductase with membrane ArsB transport protein (S. Silver and B. P. Rosen, unpublished data). The recently released entire 4.4-million-base-pairs genome of *Mycobacterium tuberculosis* (56), the cause of the major human disease, includes among the 4000 predicted genes one whose long 498 amino acid product appears to be a physical fusion of ArsB (a membrane transport domain) and ArsC (a cytoplasmic enzyme domain). The first two-thirds of the *M. tuberculosis* product is 50% identical with a predicted ArsB product from *Bacillus subtilis*; and after an apparent "linker" of only two amino acids, the remaining sequence of the *M. tuberculosis* protein is 75% identical to ArsC arsenate reductase of *Streptomyces coelicolor* (EMBO accession AL138667). Perhaps surprisingly, this gene cannot be found in the complete *Mycobacterium leprae* genome or in the more distantly related high G + C bacterium *Corynebacterium diphtheriae* (http://www.sanger.ac.uk/Projects/Microbes/; J. Parkhill, personal communication). These alignments are from translation products of recent DNA genome sequences. If direct biochemical analysis shows the long *M. tuberculosis* product is made as appears and has both cytoplasmic reductase and membrane pump activities, then one might expect the association also of the two proteins when (as in all other known sequences) they are not covalently linked.

Arsenate reductases, initially characterized from plasmid R773 of gram-negative bacteria and plasmid pI258 of gram-positive *S. aureus* both reduce arsenate to arsenite and both confer arsenate resistance (21,34,36). However, their in vitro measured properties are very different and their energy coupling is different. As the amino acid sequences are only 15% identical, it appears that arsenate reductase enzymatic activity evolved twice independently among bacterial types (29) (M. Galperin, personal communication) and the initial distinction between R773-class (gram negative) and pI258-class (gram positive) has disappeared, as representatives of both classes of enzymes have been found in both groups of bacteria.

The enzyme arsenate reductase is a small monomeric thiol-chemistry–dependent enzyme that was found initially in both gram-negative (plasmid R773) and gram-positive (plasmid pI258) bacteria (21) and subsequently in essentially all bacteria, in both direct physiological assays and in new completely total genome sequences. Not only are the sequences of the two arsenate reductases very different but enzyme properties as well are different, for example, phosphate (the physiological oxyanion most similar to the substrate arsenate) is a competitive inhibitor of the R773 form but stimulates the pI258 form of the enzyme. Arsenate reductase of the pI258 class derives reducing power from a small protein called thioredoxin (21,36), which is used in various processes of central metabolism of bacteria and higher organisms. In contrast, arsenate reductase of the R773 class uses glutaredoxin (34), which is related to thioredoxin, but a different protein. The small coupling proteins are not exchangeable.

In the R773 arsenate reductase, Cys12 has been shown to be an active site residue (57). This enzyme was crystallized (58) and its structure solved (P. D. Martin, V. S. de Mel, J. Shi, T. B. Gladysheva, B. P. Rosen, and B. F. Edwards, unpublished data). The structure with bound sulfate shows that the initial anion binding site is positively charged with three arginine residues, Arg60, Arg94, and Arg107, binding an oxyanion noncovalently. This has allowed formulation of a detailed reaction pathway for arsenate reduction for the first time (see Fig. 7). After binding facilitated by Arg residues, the next step is formation of a covalent Cys12—S—As(V)O$_3$H$_2$ bond (Fig. 7), which involves movement of Arg60 into the active site, with corresponding movement of Arg94 outward. At this point, reduction occurs in two steps, with one electron from reduced tripeptide glutathione (GSH) and the second from the reduced small protein glutaredoxin (GrxSH) (Fig. 7). A novel Cys12—S—As(O)$_2$H-SG intermediate is predicted, although this species has not been measured. Following full reduction, the product is thought to remain bound in the active site as the very novel Cys12—S—As(III)$^+$—OH, a positively charged trivalent arsenic-enzyme adduct and the oxidized dithiol glutaredoxin-glutathione adduct (GrxS-SG) (Fig. 7) is released. This enzyme-product complex has been identified in the crystal structure following soaking with the product oxyanion arsenite. Finally, the oxyanion product As(OH)$_3$ is released and the reduced enzyme reformed (Fig. 7).

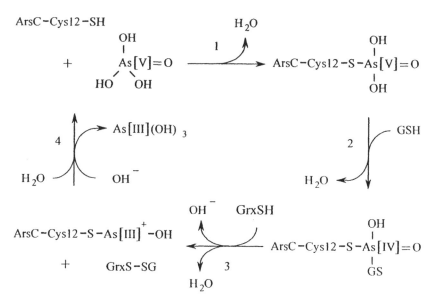

Figure 7 Steps in ArsC arsenate reductase function coupled to glutathione and glutaredoxin. Step 1: binding of arsenate to Cys12 thiol. Step 2: binding of reduced glutathione (GSH) to As(V) with reduction to As(IV) level. Step 3: binding of reduced glutaredoxin (GrxSH) and reduction to As(III) level. Step 4: release of arsenite and reforming of reduced cysteine C12.

The arsenate reductase encoded by plasmid pI258 is not by sequence or reaction mechanism a homologue of the R773 arsenate reductase. Interestingly, however, pI258 arsenate reductase is somewhat homologous to a family of low-molecular-weight proteins called tyrosylphosphate phosphatases. The physical structure and reaction mechanism of arsenate reductase from plasmid pI258 have not been studied in the same depth as those from plasmid R773. The pI258 arsenate reductase utilizes thioredoxin (36) rather than glutathione and glutaredoxin. In addition to the presence of Cys10 in pI258 arsenate reductase equivalent to Cys12 of R773 arsenate reductase, Cys82 and Cys89 of pI258 enzyme have been shown to be essential for enzymatic activity (59). The formation of an intraprotein oxidized disulfide bond between Cys82 and Cys89 has been observed in purified pI258 arsenate reductase (59,60), leading to the hypothesis that this oxidized intraprotein disulfide species is the intermediate in the reaction cycle equivalent to the interpeptide-polypeptide GrxS-SG complex for the R773 arsenate reductase. Thus, nature will have two quite independent "inventions" of remarkably equivalent chemical reaction processes.

IV. ARSENITE OXIDATION: MICROBIAL AND ENZYMATIC

Oxidation of As(III) represents a potential detoxification process that allows micro-organisms to tolerate higher levels of arsenite. Several examples of bacterial oxidation of arsenite to arsenate were being reported as early as 1918 (reviewed in Chaps. 13 and 15). Osborne and Ehrlich (61) and Phillips and Taylor (62) isolated *Alcaligenes* strains able to oxidize arsenite with oxygen as a final electron acceptor. Anderson et al. (63) purified and characterized arsenite oxidase from an *Alcaligenes faecalis* strain (Fig. 8). This enzyme is located on the outer surface of the inner membrane and exhibits arsenite oxidation activity in the presence of azurin and cytochrome *c* as electron acceptor. The purified protein has a molecular weight of 100,000 and occurs as a hetero-dimer containing several metal centers including both [3Fe-4S] HiPIP (high potential iron protein) and Rieske-type [2Fe-2S] centers (63,64). There is a pterinmolybdenum cofactor that can be released on denaturation, like that from xanthine oxidase, which is an arsenite-sensitive molybdenum-protein (see Chap. 15). The arsenite may coordinate to sulfurs bonded to the molybdenum (see Chap. 15).

In addition to plasmid arsenic resistance that is well understood and for which clusters of genes have been isolated and sequenced, there are bacterial arsenic metabolism systems that involve oxidation of arsenite to arsenic. Arsenite oxidation by aerobic pseudomonads was first found with bacteria isolated from cattle dipping solutions where arsenicals were used as agents against ticks around the time of World War I. They were subsequently isolated by Turner and Legge (65) and the reductase activity shown to result from cell-surface reduction (66).

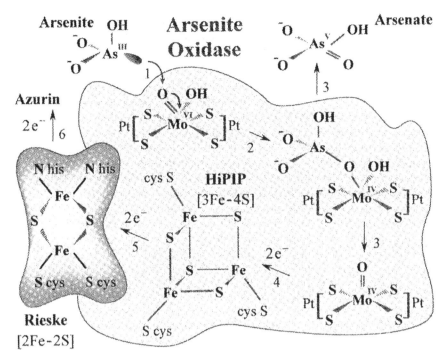

Figure 8 Model and proposed reaction cycle for arsenite oxidase from *A. faecalis*. Reaction steps are (1) binding of arsenite, AsO_2OH^{2-}, to the enzyme, (2) two-electron transfer to Mo, oxidizing As(III) to As(V) and reducing Mo(VI) to Mo(IV), (3) release of arsenate oxyanion, (4) two-electron transfer from Mo(VI) to [3Fe-4S] center, regenerating Mo(IV) reaction center, (5) two-electron transfer from [3Fe-4S] center in large subunit to [2Fe-2S] Rieske center of small subunit, and (6) electron transfer from the [2Fe-2S] center of arsenite oxidase to the associated small copper protein azurin. (Based on Refs. 63 and 64.)

Similar microbes identified as *Alcaligenes* were next isolated by Osborne and Ehrlich (61) and Phillips and Taylor (62). It was from one of the Turner and Legge (65) strains, *Alcaligenes faecalis* strain NCIB 8687 (available in a culture collection), that arsenite oxidase was purified and characterized by Anderson et al. (63) and further studied by McNellis and Anderson (67). The crystal structure of arsenite oxidase from strain NCIB 8687 was recently solved (64) and is described in Chapter 15.

The arsenite oxidase is a molybdenum-containing hydroxylase (63,64) with two [Fe-S] centers (see Fig. 8). It is found in the periplasmic space between the inner and outer membranes of *Alcaligenes*, coupled via the small blue copper protein azurin to cytochrome *c* (63). Arsenite oxidase consists of a large subunit of 825 amino acids residues that contains the single Mo bound between two pterin guanosine dinucleotide cofactors plus an unusual [3Fe-4S] center referred to as HiPIP (for high potential iron protein) that was characterized spectroscopically by electron spin resonance (ESR) by Anderson et al. (63). The smaller subunit of approximately 134 amino acids contains the other iron center, a [2Fe-2S] center, shown by ESR spectroscopy to be of the Rieske type, as in cytochrome bc_1. As shown in Figure 8, the HiPIP center is anchored into the protein through three cysteine thiol bounds, while the Rieske center is anchored by two cysteine thiols to one Fe^{2+} and by two histidine imidazol nitrogens to the other Fe^{2+}.

As diagramed in Figure 8 (and described in more depth in Chapter 15), the arsenite binds to a Mo-associated oxygen bridged between the Mo-pterin complex in a large funnel-shaped cavity in the large subunit. The [3Fe-4S] and the [2Fe-2S] clusters lie approximately 15 Å from the Mo atom, so a multistep electron transport process is needed for reoxidation of the Mo(IV) to Mo(VI) (see Fig. 8). A reaction cycle for arsenite oxidase has been suggested by Ellis et al. (64) based on the structural and spectroscopic analysis. Arsenite, as $AsO_2(OH)^{2-}$, is thought to enter the funnel opening and to bind in the vicinity of the Mo, utilizing several hydrophilic amino acid side chains lining the funnel. By analogy to other Mo-pterin enzymes, it is considered that Mo=O is approached by the arsenite oxyanion. Arsenite is oxidized to arsenate and Mo(VI) reduced to Mo(IV) (see Fig. 8) (see Chap. 15). Arsenate is released, leaving a reduced enzyme that is reoxidized by electron transport from the Mo(IV) to the [3Fe-4S] HiPIP center and from there to the [2Fe-2S] Rieske center on the other subunit (see Fig. 8) (64) (see Chap. 15). From the Rieske [2Fe-2S] center, the electrons are transferred through azurin to the respiratory chain, completing the cycle. A more detailed proposed pathway for electrons is given by Ellis et al. (64).

Our laboratory has recently started cloning and sequencing of the first gene complex for arsenite oxidase, from *Alcaligenes* strain NCIB 8687 provided by G. Anderson and with the preliminary amino acid sequence derived from the crystal structure of Ellis et al. (64). Using as now standard for gene fishing from amino acid sequences, degenerate oligonucleotide primers were used in carefully controlled polymerase chain reactions (PCRs) to obtain a 2-kb DNA fragment that was cloned and sequenced (L. T. Phung, unpublished data). Comparison of the translation results from 656 codons of the PCR product showed 70 differences (11%) from the initial "calls" from the crystal structure, plus another that resulted from an error in the DNA analysis. Half of the differences involved Asp/Asn and Glu/Gln differences, which are know to be difficult to see in electron density maps. Eleven differences were apparent lysines from the DNA sequence but alanines or serines in the electron density predictions. This is consistent with the propensity for lysine side chains to be mobile. These 69 corrections were already incorporated in the structure of Ellis et al. (64). In sum, this is our first experience with isolating a new gene cluster starting with amino acid data.

"Walking" upstream and downstream from the initial 2 kb of DNA data will yield other genes in the proposed operon. The gene for the small Rieske subunit plus the regulatory gene and transcription start site. Other genes are possible in the complex. Given the PCR product from a gene of the large subunit of arsenite oxidase of strain NCIB 8687, Southern blotting DNA/DNA hybridization analysis has identified a closely similar gene in the DNA of the *Alcaligenes* strain of Osborne and Ehrlich (61) (L. T. Phung, unpublished data).

There are two remaining systems for enzymatic changes in arsenicals, the periplasmic reduction of arsenate to arsenite as part of an oxyanion-coupled anaerobic respiration (16,68) (Chapter 13) and the coupled cleavage of carbon-arsenic bonds with oxidation to arsenate (7). These systems appear to be of major environmental concern in arsenic-containing settings, but they have not been approached by molecular genetics as yet.

V. EUKARYOTE ARSENIC RESISTANCE: GENES AND ENZYMES

Transport-efflux based resistances to arsenicals have been described in several lower Eukaryotes (69). The detailed understanding of the functions of the three-gene cluster (*ACR1*, *ACR2*, and *ACR3*) that confers resistance to arsenate and arsenite (70) in *Saccharomyces cerevisiae* has developed (Fig. 9). A surprising new arsenic resistance operon was recently found in yeast (70). This is unusual since homologs of bacterial toxic metal resistance systems have been generally limited to prokaryotic cells and in the past eukaryotic resistance mechanisms have been different and apparently independent in evolutionary origin. The three genes involved form a functional gene cluster, unusual in itself for Eukaryotes, and when cloned as a 4.2-kb fragment from right arm of yeast chromosome XVI to a multicopy plasmid, this gene complex determines resistance to both arsenate and to arsenite. [There is a yeast ArsA homolog encoded as ORF YDL100C on chromosome IV. Whether this gene (product) functions in arsenic resistance is not known.] The first gene, *ACR1*, appears to be transcribed by itself and its product is homologous in sequence to other yeast transcriptional regulators. Disruption of this gene in the chromosome led to hypersensitivity to arsenite and arsenate. Disruption of *ACR3* (homologous to the gram-positive bacterial *arsB*) eliminated both resistances, whereas disruption of *ACR2* (homologous to the bacterial *arsC*) eliminated arsenate resistance alone, as expected (70).

The gene product Acr2p is an arsenate reductase (71,72) (see Fig. 9) that is not sequence homologous to the ArsC arsenate reductases of either the plasmid R773 or the plasmid pI258 families (see above and Fig. 4). However, it is a homolog of the CDC25 family of phosphoprotein phosphatases (73) and may represent now a third independent invention (29) of this enzyme activity. Acr2p

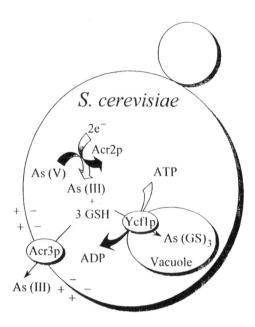

Figure 9 Yeast *S. cerevisiae* cell with small daughter cell "bud" and proteins of arsenate reduction and transport. Acr2p: the yeast cytoplasmic arsenate reductase. Acr3p: the potential-driven membrane arsenite efflux protein, equivalent to bacterial ArsB. Ycf1p: the novel As(III)-3 GSH adduct carrier than transports the adduct complex into the cell vacuole compartment, functioning as an ATPase.

requires both glutathione and glutaredoxin (72), suggesting strongly that thiol chemistry is involved and that the basic reaction mechanism is fundamentally very similar. Acr2p shares the $Cys(X)_5Arg$ motif of the phosphate binding loop with CDC25 homologs. In Acr2p, mutagenesis of either Cys76 or Arg82 results in loss of arsenate resistance and reductase activity, consistent with the hypothesis that these residues are active residues (R. Mukhopadhyay and B. P. Rosen, unpublished data). Thus, phosphatases and arsenate reductases, which may have evolved from a common ancestor, may also utilize common steps in their respective reaction mechanisms.

The *ACR3* gene encodes a membrane protein that is required for arsenite resistance (70,74) by catalyzing arsenite extrusion from the yeast cell (see Fig. 9) (75). When the *S. cerevisiae ACR3* gene was disrupted, the cells were still resistant to low levels of arsenite, suggesting that there might be a second resistance determinant (75). The additional yeast cadmium resistance factor protein, Ycf1p, which is a close homolog of human MRP1 (multidrug-associated resistance protein), encodes a Mg^{2+}-ATP-dependent glutathione S-conjugate transporter (see Fig. 9) responsible for vacuolar sequestration of cadmium and other compounds (76,77). The human MRP also confers arsenite resistance in human cancer cell lines (78). Furthermore, the cadmium sensitivity of a *S. cerevisiae* strain with a disrupted *YCF1* gene was complemented by expression of either the human MRP cDNA (79) or the gene for an MRP homolog from the higher plant *Arabidopsis thaliana* (80). *YCF1*-disrupted yeast is as sensitive to arsenite as a strain with an *ACR3* disruption, and a doubly disrupted yeast strain is hypersensitive to arsenicals (75). Acr3p is localized in the yeast plasma membrane (see Fig. 9) and catalyzes arsenite extrusion from cells. The Ycf1p protein is localized in the vacuolar membrane (see Fig. 9) and catalyzes ATP-coupled sequestration of As(glutathione)$_3$. Thus, Acr3p and Ycf1p are parallel and additive pathways for arsenic resistance. Acr3p is specific for arsenite, while Ycf1p has broad substrate specificity, providing resistance to As(III), Sb(III), Cd(II), and Hg(II). One final point here in this volume on environmental aspects of arsenic biology: findings with bacteria immediately lead to related findings with yeast cells, and then in turn to higher plants and animals, including humans. For arsenic there is no sharp distinction between the biological aspects in lower and higher organisms.

ACKNOWLEDGMENTS

Work on arsenic resistance in our laboratories has been supported by grants from the U.S. Department of Energy and National Institutes of Health. Colleagues have freely exchanged information and ideas over the years, especially Guangyung Ji, Hiranmoy Bhattacharjee, Mallika Ghosh, and Rita Mukhopadhyah.

REFERENCES

1. KJ Edwards, B Hu, RJ Hamers, JF Banfield. A new look at microbial leaching patterns on sulfide minerals. FEMS Microbiol Ecol 34:197–206, 2001.
2. HM Sehlin, EB Lindström. Oxidation and reduction of arsenic by Sulfolobus acidocaldarius strain BC. FEMS Microbiol Lett 93:87–92, 1992.
3. K Suzuki, N Wakao, Y Sakurai, T Kimura, K Sakka, K Ohmiya. Transformation of Escherichia coli with a large plasmid of Acidiphilium multivorum AIU 301 encoding arsenic resistance. Appl Environ Microbiol 63:2089–2091, 1997.
4. J Lenihan. The Crumbs of Creation. Trace Elements in History, Medicine, Industry, Crime and Folklore. Bristol, UK: Adam Hilger, 1988, p 151.
5. F Himmelweit. The Collected Papers of Paul Ehrlich, in four volumes including a complete bibliography. Vol. III. London: Pergamon, 1960.
6. GJ Roboz, S Dias, G Lam, WJ Lane, SL Soignet, RP Warrell, S Rafii. Arsenic trioxide induces dose- and time-dependent apoptosis of endothelium and may ex-

ert an antileukemic effect via inhibition of angiogenesis. Blood 96:1525–1530, 2000.

7. JP Quinn, G McMaullan. Carbon-arsenic bond cleavage by a newly isolated Gram-negative bacterium, strain ASV2. Microbiology 141:721–727, 1995.

8. BP Rosen. Families of arsenic transporters. Trends Microbiol 7:207–212, 1999.

9. C Xu, BP Rosen. Metalloregulation of soft metal resistance pumps. In: B Sarkar, ed. Metals and Genes. New York: Plenum, 1999, p 406.

10. S Silver, LT Phung. Bacterial heavy metal resistance: New surprises. Annu Rev Microbiol 50:753–789, 1996.

11. S Silver. Genes for all metals: A bacterial view of the periodic table. J Indust Microbiol Biotech 20:1–12, 1998.

12. T Thiel. Phosphate transport and arsenate resistance in the cyanobacterium Anabaena variabilis. J Bacteriol 170:1143–1147, 1988.

13. S Silver. The bacterial view of the periodic table: Specific functions for all elements. Rev Mineral 35:345–360, 1997.

14. A Carlin, W Shi, S Dey, BP Rosen. The ars operon of Escherichia coli confers arsenical and antimonial resistance. J Bacteriol 177:981–986, 1995.

15. C Diorio, J Cai, J Marmor, R Shinder, MS DuBow. An Escherichia coli chromosomal ars operon homolog is functional in arsenic detoxification and is conserved in gram-negative bacteria. J Bacteriol 177:2050–2056, 1995.

16. D Ahmann, AL Roberts, LR Krumholz, FM Morel. Microbe grows by reducing arsenic. Nature 371:750, 1994.

17. RP Novick, C Roth. Plasmid-linked resistance to inorganic salts in Staphylococcus aureus. J Bacteriol 95:1335–1342, 1968.

18. S Silver, K Budd, KM Leahy, WV Shaw, D Hammond, RP Novick, GR Willsky, MH Malamy, H Rosenberg. Inducible plasmid-determined resistance to arsenate, arsenite and antimony (III) in Escherichia coli and Staphylococcus aureus. J Bacteriol 146:983–996, 1981.

19. S Silver, D Keach. Energy-dependent arsenate efflux: The mechanism of plasmid-mediated resistance. Proc Natl Acad Sci USA 79:6114–6118, 1982.

20. HL Mobley, BP Rosen. Energetics of plasmid-mediated arsenate resistance in Escherichia coli. Proc Natl Acad Sci USA 79:6119–6122, 1982.

21. G Ji, S Silver. Reduction of arsenate to arsenite by the ArsC protein of the arsenic resistance operon of Staphylococcus aureus plasmid pI258. Proc Natl Acad Sci USA 89:9474–9478, 1992.

22. KL Oden, TB Gladysheva, BP Rosen. Arsenate reduction mediated by the plasmid-encoded ArsC protein is coupled to glutathione. Mol Microbiol 12:301–306, 1994.

23. BP Rosen, MG Borbolla. A plasmid-encoded arsenite pump produces arsenite resistance in Escherichia coli. Biochem Biophys Res Commun 124:760–765, 1984.

24. WV Ng, SA Ciufo, TM Smith, RE Bumgarner, D Baskin, J Faust, B Hall, C Loretz, J Seto, J Slagel, L Hood, S DasSarma. Snapshot of a large dynamic replicon in a halophilic archaeon: Megaplasmid or minichromosome? Genome Res 8:1131–1141, 1998.

25. WV Ng, SP Kennedy, GG Mahairas, B Berquist, M Pan, HD Shukla, SR Lasky, NS Baliga, V Thorsson, J Sbrogna, S Swartzell, D Weir, J Hall, TA Dahl, R Welti,

YA Goo, B Leithauser, K Keller, R Cruz, MJ Danson, DW Hough, DG Maddocks, PE Jablonski, MP Krebs, CM Angevine, H Dale, TA Isenbarger, RF Peck, M Pohlschroder, JL Spudich, KW Jung, M Alam, T Freitas, S Hou, CJ Daniels, PP Dennis, AD Omer, H Ebhardt, TM Lowe, P Liang, M Riley, L Hood, S DasSarma. Genome sequence of Halobacterium species NRC-1. Proc Natl Acad Sci USA 97: 12176–12181, 2000.

26. W Shi, J Dong, RA Scott, MY Ksenzenko, BP Rosen. The role of arsenic-thiol interactions in metalloregulation of the ars operon. J Biol Chem 271:9291–9297, 1996.

27. J Wu, BP Rosen. Metalloregulated expression of the ars operon. J Biol Chem 268: 52–58, 1993.

28. C Xu, BP Rosen. Dimerization is essential for DNA binding and repression by the ArsR metalloregulatory protein of Escherichia coli. J Biol Chem 272:15734–15738, 1997.

29. MY Galperin, DR Walker, EV Koonin. Analogous enzymes: Independent inventions in enzyme evolution. Genome Res 8:779–790, 1998.

30. J Wu, LS Tisa, BP Rosen. Membrane topology of the ArsB protein, the membrane subunit of an anion-translocating ATPase. J Biol Chem 267:12570–12576, 1992.

31. G Ji, S Silver. Regulation and expression of the arsenic resistance operon from Staphylococcus aureus plasmid pI258. J Bacteriol 174:3684–3694, 1992.

32. D Dou, S Dey, BP Rosen. A functional chimeric membrane subunit of an ion translocating ATPase. Antonie van Leeuwenhoek 65:359–368, 1994.

33. S Dey, BP Rosen. Dual mode of energy coupling by the oxyanion-translocating ArsB protein. J Bacteriol 177:385–389, 1995.

34. TB Gladysheva, KL Oden, BP Rosen. Properties of the arsenate reductase of plasmid R773. Biochemistry 33:7288–7293, 1994.

35. J Shi, A Vlamis-Gardikas, F Åslund, A Holmgren, BP Rosen. Reactivity of glutaredoxins 1, 2, and 3 from Escherichia coli shows that glutaredoxin 2 is the primary hydrogen donor to ArsC-catalyzed arsenate reduction. J Biol Chem 274:36039–36042, 1999.

36. G Ji, EA Garber, LG Armes, C-M Chen, JA Fuchs, S Silver. Arsenate reductase of Staphylococcus aureus plasmid pI258. Biochemistry 33:7294–7299, 1994.

37. J Wu, BP Rosen. The arsD gene encodes a second trans-acting regulatory protein of the plasmid-encoded arsenical resistance operon. Mol Microbiol 8:615–623, 1993.

38. Y Chen, BP Rosen. Metalloregulatory properties of the ArsD repressor. J Biol Chem 272:14257–14262, 1997.

39. BP Rosen, H Bhattacharjee, T Zhou, AR Walmsley. Mechanism of the ArsA ATPase. Biochim Biophys Acta 1461:207–215, 1999.

40. AR Walmsley, T Zhou, MI Borges-Walmsley, BP Rosen. A kinetic model for the action of a resistance efflux pump. J Biol Chem 276:6378–6391, 2001.

41. W Shi, J Wu, BP Rosen. Identification of a putative metal binding site in a new family of metalloregulatory proteins. J Biol Chem 269:19826–19829, 1994.

42. WJ Cook, SR Kar, KB Taylor, LM Hall. Crystal structure of the cyanobacterial metallothionein repressor SmtB: A model for metalloregulatory proteins. J Mol Biol 275:337–346, 1998.

43. DB Sowerby. In S Patai, ed. The Chemistry of Organic Arsenic, Antimony and Bismuth Compounds. New York: Wiley, 1994, pp 25–88.

44. RD Hoffman, MD Lane. Iodophenylarsine oxide and arsenical affinity chromatography: New probes for dithiol proteins. Application to tubulins and to components of the insulin receptor-glucose transporter signal transduction pathway. J Biol Chem 267:14005–14011, 1992.

45. P Kaur. The anion-stimulated ATPase ArsA shows unisite and multisite catalytic activity. J Biol Chem 274:25849–25854, 1999.

46. P Kaur, BP Rosen. Mutagenesis of the C-terminal nucleotide-binding site of an anion-translocating ATPase. J Biol Chem 267:19272–19277, 1992.

47. LS Tisa, BP Rosen. Molecular characterization of an anion pump. The ArsB protein is the membrane anchor for the ArsA protein. J Biol Chem 265:190–194, 1990.

48. CM Hsu, BP Rosen. Characterization of the catalytic subunit of an anion pump. J Biol Chem 264:17349–17354, 1989.

49. J Li, BP Rosen. The linker peptide of the ArsA ATPase. Mol Microbiol 35:361–367, 2000.

50. CE Karkaria, CM Chen, BP Rosen. Mutagenesis of a nucleotide-binding site of an anion-translocating ATPase. J Biol Chem 265:7832–7836, 1990.

51. H Bhattacharjee, J Li, MY Ksenzenko, BP Rosen. Role of cysteinyl residues in metalloactivation of the oxyanion-translocating ArsA ATPase. J Biol Chem 270:11245–11250, 1995.

52. H Bhattacharjee, BP Rosen. Spatial proximity of Cys113, Cys172, and Cys422 in the metalloactivation domain of the ArsA ATPase. J Biol Chem 271:24465–24470, 1996.

53. T Zhou, S Radaev, BP Rosen, DL Gatti. Structure of the ArsA ATPase: The catalytic subunit of a heavy metal resistance pump. EMBO J 19:4838–4845, 2000.

54. AR Walmsley, T Zhou, MI Borges-Walmsley, BP Rosen. The ATPase mechanism of ArsA, the catalytic subunit of the arsenite pump. J Biol Chem 274:16153–16161, 1999.

55. Y Chen, S Dey, BP Rosen. Soft metal thiol chemistry is not involved in the transport of arsenite by the Ars pump. J Bacteriol 178:911–913, 1996.

56. ST Cole, R Brosch, J Parkhill, T Garnier, C Churcher, D Harris, SV Gordon, K Eiglmeier, S Gas, CE Barry III, F Tekaia, K Badcock, D Basham, D Brown, T Chillingworth, R Connor, R Davies, K Devlin, T Feltwell, S Gentles, N Hamlin, S Holroyd, T Hornsby, K Jagels, A Krogh, J McLean, S Moule, L Murphy, K Oliver, J Osborne, MA Quail, M-A Rajandream, J Rogers, S Rutter, K Seeger, J Skelton, R Squares, S Squares, JE Sulston, K Taylor, S Whitehead, BG Barrell. Deciphering the biology of Mycobacterium tuberculosis from the complete genome sequence. Nature 393:537–544, 1998.

57. J Liu, TB Gladysheva, L Lee, BP Rosen. Identification of an essential cysteinyl residue in the ArsC arsenate reductase of plasmid R773. Biochemistry 34:13472–13476, 1995.

58. VS de Mel, MA Doyle, TB Gladysheva, KL Oden, PD Martin, BP Rosen, BF Edwards. Crystallization and preliminary X-ray diffraction analysis of the ArsC protein from the Escherichia coli arsenical resistance plasmid, R773. J Mol Biol 242:701–702, 1994.

59. J Messens, G Hayburn, A Desmyter, G Laus, L Wyns. The essential catalytic redox couple in arsenate reductase from Staphylococcus aureus. Biochemistry 38:16857–16865, 1999.

60. J Messens, G Hayburn, E Brosens, G Laus, L Wyns. Development of a downstream process for the isolation of Staphylococcus aureus arsenate reductase overproduced in Escherichia coli. J Chromatogr B Biomed Sci Appl 737:167–178, 2000.

61. FH Osborne, HL Ehrlich. Oxidation of arsenite by a soil isolate of Alcaligenes. J Appl Bacteriol 41:295–305, 1976.

62. SE Phillips, ML Taylor. Oxidation of arsenite to arsenate by Alcaligenes faecalis. Appl Environ Microbiol 32:392–399, 1976.

63. GL Anderson, J Williams, R Hille. The purification and characterization of arsenite oxidase from Alcaligenes faecalis, a molybdenum-containing hydroxylase. J Biol Chem 267:23674–23682, 1992.

64. PJ Ellis, T Conrads, R Hille, P Kuhn. Crystal structure of the 100 kDa arsenite oxidase from Alcaligenes faecalis in two crystal forms at 1.64 Å and 2.03 Å. Structure 9:125–132, 2001.

65. AW Turner, JW Legge. Bacterial oxidation of arsenite. II. Description of bacteria isolated from cattle-dipping fluids. Aust J Biol Sci 7:452–478, 1954.

66. JW Legge, AW Turner. Bacterial oxidation of arsenite. III. Cell-free arsenite dehydrogenase. Aust J Biol Sci 7:496–503, 1954.

67. L McNellis, GL Anderson. Redox-state dependent chemical inactivation of arsenite oxidase. J Inorg Biochem 69:253–257, 1998.

68. JM Macy, K Nunan, KD. Hagen, DR Dixon, PJ Harbour, M Cahill, LI Sly. Chrysiogenes arsenatis gen. nov., sp. nov., a new arsenate respiring bacterium isolated from gold mine wastewater. J System Bacteriol 46:1153–1157, 1996.

69. H Bhattacharjee, M Ghosh, R Mukhopadhyay, BP Rosen. Arsenic transporters from E. coli to humans. In: JK Broome-Smith, S Baumberg, CJ Sterling, FB Ward, ed. Transport of Molecules across Microbial Membranes. Leeds, UK: Society for General Microbiology, 1999, vol 58, pp 58–79.

70. P Bobrowicz, R Wysocki, G Owsianik, A Goffeau, S Ulaszewski. Isolation of three contiguous genes, ACR1, ACR2 and ACR3, involved in resistance to arsenic compounds in the yeast Saccharomyces cerevisiae. Yeast 13:819–828, 1997.

71. R Mukhopadhyay, BP Rosen. Saccharomyces cerevisiae ACR2 gene encodes an arsenate reductase. FEMS Microbiol Lett 168:127–136, 1998.

72. R Mukhopadhyay, J Shi, BP Rosen. Purification and characterization of ACR2p, the Saccharomyces cerevisiae arsenate reductase. J Biol Chem 275:21149–21157, 2000.

73. EB Fauman, JP Cogswell, B Lovejoy, WJ Rocque, W Holmes, VG Montana, H Piwnica-Worms, MJ Rink, MA Saper. Crystal structure of the catalytic domain of the human cell cycle control phosphatase, Cdc25A. Cell 93:617–625, 1998.

74. R Wysocki, P Bobrowicz, S Ulaszewski. The Saccharomyces cerevisiae ACR3 gene encodes a putative membrane protein involved in arsenite transport. J Biol Chem 272:30061–30066, 1997.

75. M Ghosh, J Shen, BP Rosen. Pathways of As(III) detoxification in Saccharomyces cerevisiae. Proc Natl Acad Sci USA 96:5001–5006, 1999.

76. ZS Li, M Szczypka, YP Lu, DJ Thiele, PA Rea. The yeast cadmium factor protein

(YCFl) is a vacuolar glutathione S-conjugate pump. J Biol Chem 271:6509–6517, 1996.

77. ZS Li, YP Lu, RG Zhen, M Szczypka, DJ Thiele, PA Rea. A new pathway for vacuolar cadmium sequestration in Saccharomyces cerevisiae: YCFl-catalyzed transport of bis(glutathionato)cadmium. Proc Natl Acad Sci USA 94:42–47, 1997.

78. SP Cole, KE Sparks, K Fraser, DW Loe, CE Grant, GM Wilson, RG Deeley. Pharmacological characterization of multidrug resistant MRP-transfected human tumor cells. Cancer Res 54:5902–5910, 1994.

79. R Tommasini, R Evers, E Vogt, C Mornet, GJ Zaman, AH Schinkel, P Borst, E Martinoia. The human multidrug resistance-associated protein functionally complements the yeast cadmium resistance factor 1. Proc Natl Acad Sci USA 93:6743–6748, 1996.

80. R Tommasini, E Vogt, M Fromenteau, S Hortensteiner, P Matile, N Amrhein, E Martinoia. An ABC-transporter of Arabidopsis thaliana has both glutathione-conjugate and chlorophyll catabolite transport activity. Plant J 13:773–780, 1998.

11

Bacterial Respiration of Arsenate and Its Significance in the Environment

Ronald S. Oremland
U.S. Geological Survey, Menlo Park, California

Dianne K. Newman
California Institute of Technology, Pasadena, California

Brian W. Kail and John F. Stolz
Duquesne University, Pittsburgh, Pennsylvania

I. INTRODUCTION

Although arsenic is a trace element in terms of its natural abundance, it nonetheless has a common presence within the earth's crust. Because it is classified as a group VB element in the periodic table, it shares many chemical and biochemical properties in common with its neighbors phosphorus and nitrogen. Indeed, in the case of this element's most oxidized ($+5$) oxidation state, arsenate [$HAsO_4^{2-}$ or As (V)], its toxicity is based on its action as an analog of phosphate. Hence, arsenate ions uncouple the oxidative phosphorylation normally associated with the enzyme glyceraldehyde 3-phosphate dehydrogenase, thereby preventing the formation of phosphoglyceroyl phosphate, a key high-energy intermediate in glycolysis. To guard against this, a number of bacteria possess a detoxifying arsenate reductase pathway (the *arsC* system) whereby cytoplasmic enzymes remove internal pools of arsenate by achieving its reduction to arsenite [$H_2AsO_3^-$ or As (III)]. However, because the arsenite product binds with internal sulfhydryl groups that render it even more toxic than the original arsenate, efficient arsenite efflux from the cell is also required and is achieved by an active ion ''pumping'' system (1). The details of this bacterial arsenic detoxification phenomenon have been well established in the literature, and Chapter 10 in this volume provided a thorough review. Here, we discuss bacterial respiration of arsenate and its significance in the environment. As a biological phenomenon, respiratory growth on arsenate is quite remarkable, given the toxicity of the element. Moreover, the consequences of microbial arsenate respiration may, at times, have a significant impact on environmental chemistry.

Much less is understood about the mechanisms of microbial arsenate respiration than about the mechanisms of arsenate detoxification, although these processes face similar challenges, namely: the transport of arsenate into the cell, its reduction to arsenite, the protection of intracellular proteins from arsenite, and the export of arsenite out of the cell. In the case of arsenate-respiring microorganisms, the nature of arsenic transport across the cell membrane is as yet unclear, and we must look to arsenic-resistant organisms as our initial models. Presumably, for gram-negative bacteria, arsenate enters the cells through nonspecific outer-membrane porins, or, in the case of phosphate starvation, through inducible outer-membrane proteins that are designed for phosphate transport but cannot discriminate between arsenate and phosphate (such as PhoE) (2). Upon reaching the periplasm, arsenate may be reduced to arsenite by respiratory reductases (see below) or further transported into the cytoplasm, as is known to occur in *Escherichia coli* and presumably in other As-resistant organisms. Under conditions of phosphate abundance (>1 mM), arsenate can enter the cytoplasm through a low-affinity phosphate transport system (Pit). In *E. coli*, *pitA* is constitutively expressed and couples uptake of inorganic phosphate to the proton motive force (3). In strains dependent on Pit for inorganic phosphate uptake, exposure to arsenate leads to the depletion of intracellular adenosine triphosphate (ATP) stores and the intracellular accumulation of arsenate, demonstrating the direct interference of arsenate in phosphate metabolism (4).

Despite the similarity between arsenate and phosphate, organisms have evolved ways to discriminate between the two compounds. For example, mutations in *pitA* lead to the induction of the high-affinity phosphate transport system (Pst), which differentiates between arsenate and phosphate approximately 100-fold more accurately than PitA. The key components of the Pst system include PstS, PstA, PstB, and PstC. PstS is a periplasmic protein that binds inorganic phosphate with high selectivity and carries it to the high-affinity phosphate transporter (located in the cytoplasmic membrane) comprised of PstA, PstC, and PstB. Transport of phosphate through this complex is coupled to ATP hydrolysis at PstB, making transport of phosphate through the Pst system more costly for the cell than through Pit (2). The switch from the low- to high-affinity phosphate transport systems results in moderate arsenate tolerance (5). Whether arsenate-respiring organisms also rely on phosphate transport systems to uptake arsenic or have evolved arsenate-specific transporters remains an open question.

Once inside the cell, arsenate ions can be reduced to arsenite via membrane-bound or cytoplasmic enzymes. The former are linked to cellular energy conservation and are described in detail later in this chapter and in Chapter 12; the latter are characteristic of As-resistant microbes, do not conserve energy, and have been described in Chapter 10. Before detailing the bioinorganic chemistry of arsenic by micro-organisms, however, we will briefly discuss the incorporation of arsenic into organic compounds. For more details on this subject, we refer the reader to comprehensive reviews by Phillips (6) and Reimer (7).

Depending upon the organism, inorganic arsenic ultimately may be converted into arsenosugars as well as a variety of methylated species. The occurrence of this phenomenon in plants and algae is highly variable. Internal arsenic pools occurring in mosses, for example, are dominated by inorganic species as opposed to organic compounds (8). Once inside the cell, however, arsenate may undergo further reduction to a lower redox state (arsine) where it is incorporated into organic matter in a fashion analogous to that of quaternary nitrogen compounds. For example, arsenobetaine is an analog of the internal osmolyte glycine betaine commonly found in marine organisms (Fig. 1). Arsenobetaine commonly occurs in a variety of marine animals (9), as well as in bacteria found in the arsenic-rich waters of hypersaline Mono Lake, California (10). In the latter case, arsenobetaine may serve a dual purpose of functioning not only as a compatible internal solute in an environment of high osmotic stress, but also as a mechanism for converting potentially toxic internal pools of arsenate and arsenite into an innocuous organic compound. Animals presumably obtain their organoarsenic compounds from the food chain, by eating As-containing plants and algae. The occurrence of organoarsenicals in the tissues of deep-sea hydrothermal vent ani-

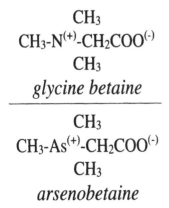

$$CH_3$$
$$CH_3\text{-}N^{(+)}\text{-}CH_2COO^{(-)}$$
$$CH_3$$
glycine betaine

$$CH_3$$
$$CH_3\text{-}As^{(+)}\text{-}CH_2COO^{(-)}$$
$$CH_3$$
arsenobetaine

Figure 1 Molecular structure of the osmolyte glycine betaine and its analog arsenobetaine.

mals suggests that it is also possible for animals to obtain arsenic from autotrophic and/or symbiotic bacterial sources rather than phytoplankton (11). Neff (12) has reviewed the toxicology of various arsenic species in marine ecosystems.

A number of micro-organisms isolated from sediments, macroalgae, and the intestinal tract of chitons have been shown to degrade arsenobetaine to tri-methylarsine oxide [$(CH_3)_3AsO$], dimethylarsinic acid [$(CH_3)_2AsO(OH)$], meth-ylarsonic acid [$CH_3AsO(OH)_2$], arsenite, and arsenate. In this pathway, arsenic is oxidized back to the +5 state (9). In anoxic environments, glycine betaine is cleaved to form acetate and trimethylamine, both of which may enter methano-genic degradation pathways (13). Little is known about the occurrence of trimeth-ylarsine in nature, although methylated arsenic compounds with arsenic in the +5 oxidation state have been detected in the aerobic regions of a number of water bodies (14), presumably from aerobic degradation of arsenobetaine or from partial oxidation of methyl arsines. Methanogenic attack of trimethylarsine would result in the formation of highly toxic arsine gas (AsH_3) rather than ammonia. Relatively little is known about the biogeochemical cycling of organoarsenic compounds in the aerobic or anaerobic environments, making this an area ripe for future investigation.

Although the *arsC* system of bacterial arsenate resistance and the decompo-sition of organoarsenic compounds represent potential mechanisms whereby re-duced inorganic arsenic species (arsenite or arsine) can accumulate in an external aqueous milieu, neither process conserves energy or offers any special evolution-ary advantage to the cells other than survival in a toxic aquatic matrix. It was the report by Ahmann et al. (15) in 1994 that first recognized arsenate as an anaerobic terminal electron acceptor capable of supporting the respiratory growth of new species of Eubacteria. The discovery of this phenomenon has implications across several disciplines of environmental importance, including microbiology, biochemistry, toxicology, and geochemistry, and is the subject of three recent succinct reviews (16–18). Because this is a fast-emerging field and the phenome-non may ultimately impact the health of large human populations in such regions as the Ganges delta of Banagladesh and India (19), we have written this chapter to further update and summarize recent findings.

II.   MICROBIOLOGY AND BIOCHEMISTRY OF ARSENATE RESPIRATION

Currently, in pure culture, there are seven novel species of Eubacteria, most of them isolated from arsenic-rich environments (20–27), that are capable of respir-ing arsenate (Table 1). Although this is only a small number of representative species, it is already clear that the phenomenon is polyphyletic. It occurs in both gram-positive (low G + C) Eubacteria and in at least three subdivisions (delta,

Table 1 Novel Bacterial and Archaeal Isolates That Can Grow by Respiratory Arsenate Reduction

Microbe	Classification	Refs.
Sulfurospirillum arsenophilum	ε-Proteobacteria	15, 20
Sulfurospirillum barnesii	ε-Proteobacteria	20, 21, 22
Chrysiogenes arsenatis	Deep-branch Proteobacteria	23
Desulfotomaculum auripigmentum	Low G + C gram positive	24, 25
Bacillus arsenicoselenatis	Low G + C gram positive	26
Bacillus selenitireducens	Low G + C gram positive	26
Desulfovibrio Ben-RB	δ-Proteobacteria	27
Pyrobaculum arsenaticum	Crenoarchaea	28

epsilon, and gamma) of the gram-negative Proteobacteria, along with another deeply branching representative (*Chrysiogenes*). Preliminary evidence suggests that a marine strain of *Shewanella* sp. also respires arsenate (D. K. Newman, personal communication). Until quite recently there were no reports of Archaea that respire arsenate, although the elevated concentrations of arsenic commonly occurring in hot springs and in some hypersaline lakes suggests a niche in which arsenate-respiring Crenoarchaea and Haloarchaea patiently await discovery. Recently, Huber et al. (28) reported on the ability of a newly isolated hyperthermophilic Crenoarchaea, *Pyrobaculum arsenaticum*, to respire arsenate. Examination of three other members of this genus revealed that one of them, *P. aerophilum*, was also capable of arsenate respiration. Both organisms are facultative autotrophs, being able to grow by respiring arsenate with H_2 as the electron donor and CO_2 as the source of cell carbon. Two organisms, *Bacillus arsenicoselenatis* and *B. selenitireducens* are extremophiles adapted to the high pH and salinity of Mono Lake, California. As yet, there are no reports of extreme acidophilic Archaea or Bacteria that respire arsenic. However, the high concentrations of arsenic in the waters of Iron Mountain, California (29) and the occurrence of mats of iron-oxidizing Archaea in this system (30) would suggest the presence of arsenic-oxidizing and -reducing Archaea as well.

All of the Eubacterial arsenate-respirers reduce arsenate quantitatively to arsenite. No significant gaseous products (methylated arsines or AsH_3) or elemental arsenic are produced. None of the arsenate respirers currently in culture are obligate arsenate-reducers, and all exhibit various degrees of flexibility with regard to their ability to use electron acceptors other than arsenate. For example, most can use nitrate, three can use selenium oxyanions, two are microaerophiles, and two are sulfate-reducers. Dissimilatory nitrate reduction by *Sulfurospirillum barnesii* results in the formation of ammonium rather than N_2 (21). *Desulfotoma-*

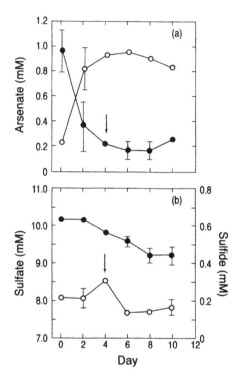

Figure 2 Sequential reduction of (a) arsenate and (b) sulfate during growth of *Desulfoto-maculum auripigmentum*. (From Ref. 24.)

culum auripigmentum respires arsenate first, followed by sulfate (Fig. 2), re-sulting in the precipitation of yellow arsenic trisulfides and the mineral orpiment (24,25). The precedence of As(V) before sulfate can be explained by the higher energy yield ($\Delta G'$) associated with the oxidation of a given electron donor (H_2) with arsenate (-23.03 kJ/mol e$^-$) as opposed to sulfate (-0.42 kJ/mol e$^-$) (16). However, *Desulfomicrobium* strain Ben-RB exhibits concurrent reduction of ar-senate and sulfate, suggesting that there are considerable physiological differ-ences that factor into this phenomenon as well (27).

Growth of *B. arsenicoselenatis* on lactate results in its oxidation to acetate plus CO_2 with the reduction of arsenate to arsenite (Fig. 3). Growth conforms to the equation:

$$\text{lactate}^- + 2\ HAsO_4^- + H^+$$
$$\rightarrow \text{acetate}^- + 2H_2AsO_3^- + HCO_3^-\ \Delta G_f^o = -23.4\ \text{kJ/mole}^- \quad (1)$$

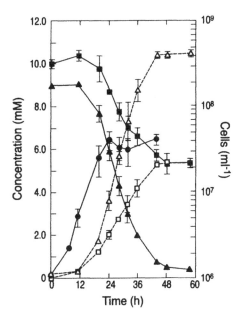

Figure 3 Growth of *Bacillus arsenicoselenatis* with lactate as its electron donor and arsenate as its electron acceptor. Symbols: ▲, arsenate; △, arsenite; ■, lactate; □, acetate; ●, cells. (From Ref. 26.)

Therefore, the reduction of arsenate is highly exergonic, although it is not nearly as potent an oxidant as are selenate, nitrate, or manganic ions (16).

To date, only the respiratory arsenate reductase (DAsR) from *C. arsenatis* has been purified and characterized (31). A periplasmic enzyme, the DAsR is composed of an 87-kDa polypeptide and a 29-kDa polypeptide. It is a heterodimer with a native molecular mass of 123 kDa. The active site is located in the large subunit and contains molybdenum. N-terminal amino acid sequence analysis also revealed a putative [Fe-S] cluster binding site. The small subunit contains at least one [Fe-S] cluster and is believed to be involved in electron transfer and anchoring. A small-molecular-mass *c*-type cytochrome has also been linked to the enzyme complex. The enzyme exhibits a high degree of specificity for arsenate and has a K_m of 300 µM. How arsenate reduction in the periplasm is coupled to the generation of the proton motive force is unclear, however, as arsenate reduction should consume protons and electrons ($HAsO_4^{2-} + 4H^+ + 2e^- \rightarrow HAsO_2 + 2H_2O$), that are initially generated from the oxidation of a substrate like hydrogen ($H_2 \rightarrow 2 H^+ + 2e^-$). More studies are needed to understand the details of the electron transfer pathway to arsenate in this organism, and we refer the reader to Chapter 12 for a more comprehensive discussion of this system.

Work continues on the purification and characterization of the DAsR from *S. barnesii*. This effort has been complicated for two reasons. The first is that the enzyme complex is membrane bound and not readily solubilized. The second is the presence of a membrane-bound bidirectional hydrogenase (17). Initial attempts at purifying the complex were made using native preparative gel electrophoresis. The gels were developed using methyl viologen as the artificial electron donor and arsenate as the electron acceptor. Two activities were observed, a complex with low mobility that barely migrated into the gel, and a complex with high mobility that migrated just above the dye front. The high-mobility complex contained polypeptides of 65, 30, and 22 kDa (Fig. 4). N-terminal amino acid sequence indicated that the 65-kDa polypeptide was the large subunit of a Ni-Fe hydrogenase. This has since been confirmed by metals and Western blot analyses and activity assays (D. K. Newman et al., unpublished data). A BLAST search failed to find identities for either the 30- or 22-kDa polypeptides. Thus they are not the two additional subunits found in membrane-bound Ni-Fe hydrogenases

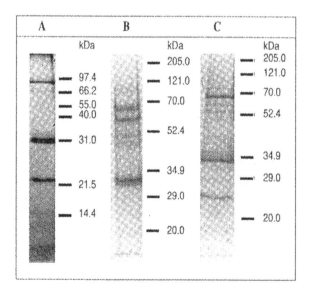

Figure 4 Protein complexes exhibiting arsenate reductase activity from *Sulfurospirillum barnesii* as achieved by using preparative gel electrophoresis (A) and hydroxyapatite column chromatography (B and C). (A) High-mobility complex containing polypeptides of 65, 30, and 22 kDa; (B) fraction that exhibited both hydrogenase and arsenate reductase activity; (C) fraction exhibiting only arsenate reductase activity. The 65-kDa polypeptide in all three complexes is the large subunit of the Fe-Ni hydrogenase. The molecular weight standards for each frame are given to the right of the gels.

(i.e., [Fe-S] cluster protein and cytochrome b). Furthermore, the N-terminal amino acid sequence of a 55-kDa polypeptide that was highly enriched in soluble fractions was identical to the 22-kDa N-terminal, suggesting that the native form of the 22-kDa polypeptide is a dimer.

Recent purification efforts employing column chromatography have yielded two fractions that exhibit arsenate reductase activity. Both fractions contain the large subunit of the Ni-Fe hydrogenase (Fig. 4). The presence of hydrogenase in the DAsR complex raises the question of its function: Does it donate electrons to a separate arsenate reductase, or does it directly reduce arsenate itself? Ni-Fe hydrogenases are known to donate electrons to a number of terminal reductases, including the fumarate reductase from *Wolinella succinogenes* (32). There is also a report of technicium reduction by a Ni-Fe hydrogenase from *E. coli* (33). In addition to *Wolinella succinogenes*, three other bacterial species closely related to *S. barnesii* also contain a similar Ni-Fe hydrogenase: *Helicobacter pylori, Campylobacter jejuni,* and *Desulfovibrio desulfuricans*. Not one of these organisms, however, respires arsenate. Both *C. jejuni* and *D. desulfuricans* do, however, exhibit arsenic resistance (B. Kail and A. Dawson, personal communication). It is tempting to speculate that the hydrogenase in these organisms may serve as the electron donor to the arsenate-reducing enzyme (be it for respiration or resistance), although this remains to be proven. What is clear, however, is that the detoxifying reductases are quite different from the respiratory ones, and that the respiratory reductases themselves are different from one another. This suggests that they arose from independent evolutionary lines.

III. ARSENATE REDUCTION IN ANOXIC ECOSYSTEMS

Discerning the contribution that micro-organisms make toward carrying out the reduction of arsenate in nature is greatly complicated by several chemical factors, which will be just briefly mentioned here. First, in addition to biological reduction, arsenate can also be reduced to arsenite by strong, naturally occurring reductants, such as sulfide, although this is generally favored at low pH rather than under neutral or alkaline conditions (where the reaction rate is slow) (24,34). Conversely, a number of naturally occurring oxidants, such as Fe(III) and Mn(IV), can reoxidize As(III) back to As(V) (35). Secondly, arsenite formed in the presence of free sulfides at low to neutral pH will precipitate as arsenic trisulfide (As_2S_3), provided the concentration of sulfide is within the appropriate range (25). Finally, there is the question of sorption of both arsenic species to mineral phases. Arsenate adsorbs strongly to a number of common minerals, such as ferrihydrite, goethite, chlorite, and alumina, which constrains its mobility in aquifers, soils, and sediments. This is a complicated, pH-dependent phenomenon (36). Phosphate has chemical properties similar to arsenate and is a common anion present in nutrient-rich systems, such as aquifers underlying P-fertilized agricultural soils. Phosphate can therefore compete with arsenate for these adsorptive sites, thereby making arsenate more mobile under conditions of phosphate abundance. Arsenite is more toxic and generally more mobile in nature than is arsenate. Nonetheless, arsenite still has significant sorptive interactions with certain minerals, such as goethite, which constrains its mobility in the environment (37).

Bearing the above constraints in mind, respiratory arsenate reduction can be discerned experimentally in arsenic-contaminated anoxic sediments (38). We offer the results of Dowdle et al. (39) as an instructive example of the complex biological and chemical dynamics of this process in natural systems. In these experiments, the reduction of arsenate required anoxia and was eliminated by autoclaving and by respiratory inhibitors like dinitrophenol and cyanide, thereby proving its biological rather than chemical mode of action. Reduction was speeded by certain electron donors, such as H_2, lactate, or glucose, and mineralization of [^{14}C]acetate could be linked to the presence of arsenate. These results pointed to direct reduction via an anaerobic respiratory process, rather than an indirect reduction via formation of a potent reductant (e.g., sulfide) by bacterial activity. Recovery of arsenite was enhanced by the presence of added electron donors, suggesting that binding of arsenite to ferrihydrite could be eliminated by the latter's reduction to Fe(II) (Fig. 5). Potentially competing electron acceptors (sulfate) or periodic table neighbors (phosphate) did not significantly affect arsenate removal rates (Fig. 6). However, the poor recoveries of arsenite in these experiments point to the precipitation or binding of this reduced product to other sorptive sites. Thus, while sulfate did not inhibit removal of arsenate ions from solution (Fig. 6b), the recovery of arsenite was less than in the controls, presumably due to the formation of insoluble arsenic trisulfides associated with sulfate reduction (25,26,40). Likewise, Fe(III) sorbed arsenate (Fig. 6d), while Mn(IV) probably sorbed arsenite (Fig. 6c). Nitrate appeared to slow the removal of arsenate and totally prevented the accumulation of arsenite (Fig. 6d). This was probably due to its action as a competitive sink for electrons with arsenate coupled with the ability of nitrate to achieve the reoxidation of arsenite to arsenate. Experiments with chloramphenicol in nitrate-respiring sediments revealed that a de novo synthesis of protein was required in order to reduce arsenate, which suggested that the arsenate was not being reduced by the nitrate reductase, but rather by a specific arsenate reductase that required induction. Molybdate did not inhibit any arsenate reduction while tungstate did (Fig. 7). This suggested that sulfate reducers were not responsible (they would have been inhibited by both molybdate and tungstate), but rather that the reduction was mediated by a molybdenum-containing enzyme, like that found for the respiratory arsenate reductase of C. arsenatis (31).

The above example reveals that the activity of respiratory arsenate reducers can be elicited when arsenate anions are present at high concentrations ($\sim$10

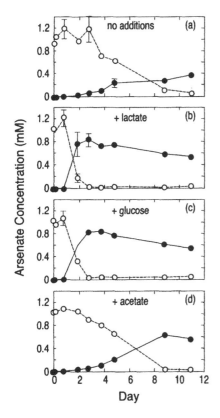

Figure 5 Effect of electron donors on the reduction of arsenate (○) to arsenite (●) in anoxic estuarine sediment slurries incubated (a) without additions, (b) with lactate, (c) with glucose, and (d) with acetate. (From Ref. 39.)

mM) in the aqueous phase of experimental systems. It is important to recognize, however, that the behavior of this consortium (e.g., with respect to inhibition by alternative electron acceptors and other compounds) reflects merely the microbial population at this particular site and cannot be generalized. Indeed, other studies with field sediment/soil slurries have suggested that a significant fraction of the organisms responsible for microbial arsenate reduction in different locales include sulfate reducers (38) (J. Sharp and D. K. Newman, unpublished data).

Much less is known about the ability of microbes to reduce arsenate at lower ambient concentrations, especially when it is associated with the solid rather than liquid phase, either as a component of an insoluble mineral or as an adsorbed anion. The mechanisms that have been investigated thus far include

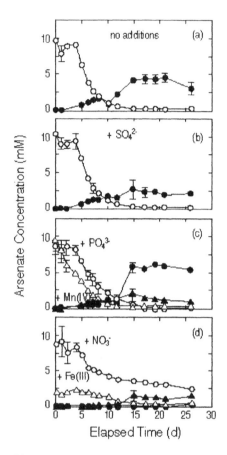

Figure 6 Effect of electron acceptors on the reduction of arsenate (○) to arsenite (●) in estuarine sediment slurries incubated (a) without additions, (b) with sulfate ions, (c) with phosphate or manganic ions, and (d) with nitrate or ferric ions. (From Ref. 39.)

bacterial destruction of the primary sorptive matrix (e.g., ferrihydrite), direct reduction of adsorbed arsenate, and reduction mediated by low-molecular-weight electron shuttles produced by bacteria (Fig. 8).

Cummings et al. (41) worked with the Fe(III)-respiring *Shewanella alga* and followed scorodite (FeAsO$_4$2H$_2$O) reductive dissolution in culture as well as by adding this bacterium to sterilized, metal-contaminated sediments. In both cases, this microbe was able to release arsenate into solution by breaking down the crystalline structure of scorodite via reducing Fe(III) to Fe(II). Presumably, this solubilized arsenate would then undergo bacterial reduction to arsenite by any arsenate respirers present in the natural flora. However, Newman et al. (24)

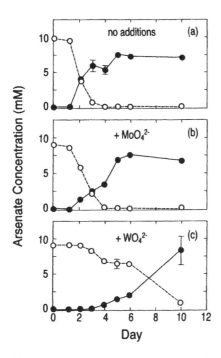

Figure 7 Effect of group VIA oxyanion inhibitors on the reduction of arsenate (○) to arsenite (●) in estuarine sediments slurries incubated (a) without additions, (b) with molybdate, and (c) with tungstate. (From Ref. 39.)

found that *Desulfotomaculum auripigmentum* was able to directly reduce the arsenate contained in scorodite, thereby releasing As(III) into solution and achieving growth as well. Ahmann et al. (42) working with ferric arsenate–amended sediments from the As-contaminated Aberjona watershed were able to demonstrate the microbially catalyzed release of both Fe(II) and As(III) by the natural microbiota, as well as by adding *S. arsenophilum* (strain MIT-13) or an As(V)-respiring enrichment to sterilized sediments. By comparison, addition of the Fe(III)-respiring *Geobacter metallireducens* resulted in little detectable As(III) production.

Most of the above examples followed the dissolution of arsenic in complex sediment systems amended with scorodite and cultures of iron- or arsenate-respiring bacteria. More work was needed with simpler, defined systems to unravel the biological mechanisms from the adsorptive chemical phenomena. Zobrist et al. (43) studied the reductive dissolution of ferrihydrite that was coprecipitated with arsenate. *Sulfurospirillum barnesii* was the organism of choice because it respires both Fe(III) and As(V). Washed cell suspensions of *S. barnesii* simultaneously reduced Fe(III) as well as As(V) (Fig. 9). However, As(III) still had a

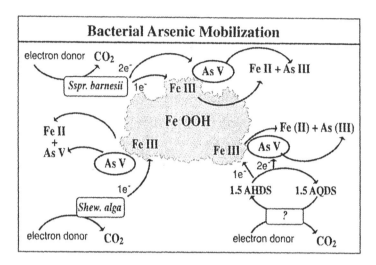

Figure 8 Schematic representation of three possible microbial mechanisms for mobilization of arsenic oxyanions adsorbed to ferrihydrite surfaces by respiratory reduction. Bottom left: *Shewanella alga* reduces Fe(III) to Fe(II), thereby releasing As(V) into solution (41). Lower right: bacterially reduced electron "shuttle" molecules pass electrons to solid-phase As(V) and Fe(III) (48). Top: *Sulfurospirillum barnesii* directly reduces both As(V) and Fe(III) (43).

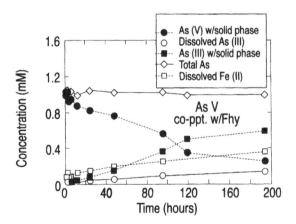

Figure 9 Reduction of As(V) and Fe(III) in aggregates of arsenate coprecipitated with ferrihydrite by cell suspensions of *Sulfurospirillum barnesii*. (From Ref. 43.)

strong capacity for adsorption to the remaining unreacted ferrihydrite, and although there was significant mobilization of arsenite into the aqueous phase, most of the As(III) that was formed remained associated with the solid phase. This interpretation was verified in abiotic experiments by the observation that the coprecipitated ferrihydrite still had a sorptive capacity for As(III) but did not desorb As(V) from its matrix. Another experiment was conducted using As(V) coprecipitated with $Al(OH)_3$ (Fig. 10). In this case, the respiratory reduction of sorbed As(V) was balanced by the increase in levels of As(III), but because abiotic experiments showed that As(III) does not adsorb to this mineral, all the As(III) was found in solution with none occurring in association with the solid phase. However, because *S. barnesii* does not reduce Al(III), the $Al(OH)_3$ matrix was not broken down and the interior coprecipitated As(V) was unavailable for continued respiratory reduction. This resulted in less overall As(V) reduction than was the case for ferrihydrite (see Fig. 9).

These experiments pointed out that respiratory reduction of As(V) sorbed to solid phases can indeed occur in nature, but its extent and the degree of mobilization of the As(III) product is constrained by the type of minerals present in a given system. What remains unclear is whether micro-organisms can actually reduce As(V) while it is attached to the mineral surface, or if they attack a monolayer of aqueous As(V) that is in equilibrium with the As(V) adsorbed onto the surface layer. If, as is the case for dissimilatory metal-reducing bacteria such as *Geobacter sulfurreducens* and *Shewanella oneidensis* (44,45), components of the electron transport chain are localized to the outer-membrane of some arsenate-respiring bacteria, direct reductive dissolution of insoluble arsenate minerals may be possible by attached bacteria. Too little is known at present about the topology

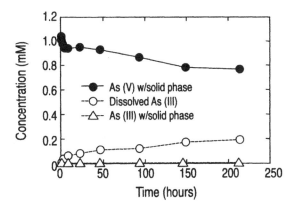

Figure 10 Reduction of As(V) in aggregates of arsenate coprecipitated with aluminum hydroxide by cell suspensions of *Sulfurospirillum barnesii*. (From Ref. 43.)

of the arsenate reductases, however, to resolve this question, and further research is merited. It is noteworth, however, that Langer and Inskeep (46) did not observe reductive dissolution of As(V) adsorbed onto ferrihydrite in the presence of an unidentified glucose-fermenting organism. This organism reduced As(V) by the detoxification rather than by the respiratory mechanism, and proved capable of only reducing As(V) that was dissolved in the aqueous phase. Although the cellular location of the *arsC*-type reductase is not known for this isolate, in well-studied bacterial systems they are associated with the cytoplasm rather than the membrane (1). The hypothesis that reduction of insoluble arsenate minerals requires surface-associated proteins can be tested by identifying the cellular location of arsenate reductases in organisms that grow on such minerals.

Another possible mechanism for reductive dissolution of arsenate adsorbed onto mineral surfaces would involve low-molecular-weight organic substances that act as electron shuttles. These molecular "shuttles" have been found to act as electron acceptors for anaerobic growth of diverse metal-respiring bacteria (47), and can transfer these electrons to insoluble, oxidized metals thereby achieving their dissolution (48,49). Bacteria are known to utilize exogenous humic substances in such a capacity (47), and it has also been proposed that some may even excrete their own shuttling compounds (50), which would enable electron transfer to mineral matrices that would otherwise be inaccessible. A reduced humic analog, anthrahydroquinone-2,6-disulfonate (AHDS), is commonly used as a model compound for such investigations. It was unable to reduce soluble As(V) directly, although it could also serve as an electron donor for *Wollinella succinogenes* and *S. barnesii* to achieve the reduction of arsenate directly (51). However, no work has been done on the ability of AHDS to reduce As(V) adsorbed onto or coprecipitated with minerals such as ferrihydrite, making this an area still ripe for future investigations.

The presence of high concentrations of arsenic in the environment results from anthropogenic sources (e.g., mine tailings, pesticide production, animal hide tanning) as well as from natural sources (weathering of arsenic minerals, hydrothermal waters). Some regions of the United States, such as the Carson Desert of western Nevada, have a particularly high natural abundance of arsenic, with some of its shallow groundwaters exhibiting concentrations as high as ~ 13 μM (52,53). Serious human health problems are common in the population that ingests water from these aquifers as their primary drinking supply (54). In order to determine if the arsenic contained in these groundwaters is a consequence of in situ reduction of As(V) to As(III) occurring in aquifer materials, an assay of arsenate reductase activity is needed. Fortunately, [73]As(V) is a commercially available, gamma-emitting radiotracer ($t_{1/2} = 80.3$ days). In theory, its use would allow investigations of arsenic biogeochemistry to be conducted at ambient concentrations of this element. However, preliminary attempts to employ this radiotracer in aquifer sediments have been unsuccessful. This failure has been due,

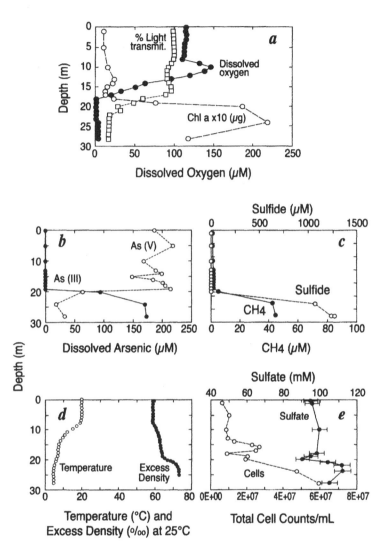

Figure 11 Vertical depth profiles taken in July, 1999 of various physical, chemical, and biological properties of Mono Lake, California, when the lake was in a meromictic condition. (a) Dissolved oxygen, light transmissivity, and chlorophyll a content; (b) arsenate and arsenite; (c) methane and sulfide; (d) temperature and density; (e) direct counts of bacterial cells and sulfate. (From Ref. 57.)

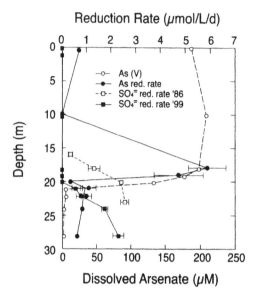

Figure 12 Vertical depth profiles of arsenate with rates of respiratory arsenate and sulfate reduction in the water column of meromictic Mono Lake, California, made during October, 1999. Sulfate-reduction profiles from the last period of meromixis (1986) when the lake was 4 m shallower are shown for comparison. (From Ref. 57.)

in part, to the differential chemical sorptive behavior of both arsenate and arsenite with the various types of solid matrices assayed, which results in qualitative rather than quantitative recoveries of the two species (P. Dowdle and R. Oremland, unpublished data). The nonquantitative recovery of these two species makes employment of this radiotracer as yet unsuitable for the determination of rate constants in sediments, soils, and aquifer materials.

Anoxic water samples, because they contain little in the way of particles, are far easier than aquifer materials to develop radioassays for the measurement of arsenate reduction. Arsenic speciation quantitatively changes from arsenate to arsenite with vertical transition from the surface oxic waters to the anoxic bottom depths of stratified lakes and fjords (55,56). This also occurs in Mono Lake, California (57), a transiently meromictic, alkaline (pH = 9.8), and hypersaline (salinity = 70–90 g/L) soda lake located in eastern California (Fig. 11). The combined effects of hydrothermal sources coupled with evaporative concentration have resulted in exceptionally high (~200 μM) dissolved arsenate concentrations in its surface waters. Haloalkaliphilic arsenate-respiring bacteria have been isolated from the lake sediments (26), and sulfate reduction, achieved with

[^{35}S]sulfate radiotracer, had been measured in the lake's anoxic waters during a past episode of meromixis in 1986. In situ measurements made with ^{73}As(V) during 1999 revealed that arsenate reduction was a significant microbial activity in the water column of the lake (57), with the highest activity occurring at a depth of 18 m (Fig. 12). Much lower activity occurred at depths below 20 m, mainly because the arsenate pools were so low (<10 μM) even though rate constants were high (~0.3 day^{-1}). Although sulfate reduction rates increased with depth below 20 m, the abundance of arsenate respirers was 10-fold higher than that of sulfate reducers as determined by MPN (most probable number) methods. An important finding was that microbial arsenate respiration in the water column could account for as much as 14% mineralization of annual primary productivity, with sulfate reduction accounting for as much as 41%. This finding was the first report that arsenate respiration can make a significant contribution to the carbon cycle of an ecosystem.

IV. CONCLUSIONS

The exciting discovery of the process of respiratory reduction of arsenate by prokaryotes has thus far resulted in identification of this unique means of respiration in seven new and highly diverse species of Eubacteria, and in one new and one previously isolated species of Crenoarchaea. Relatively little is known as yet about the respiratory arsenate reductases of these organisms, since they have only been preliminarily characterized in C. arsenatis and in S. barnesii. Clearly, more detailed biochemical, biophysical, and genetic characterization of these DAsR enzymes from these organisms as well as from some of the extremophiles capable of growing on arsenate is needed. Such work may eventually determine if these enzymes were derived from a common ancestral gene and propagated through microbial communities by such means as lateral gene transfer, whether they evolved from detoxifying arsC reductases (or vice versa), or if they arose independently in several distantly related taxonomic groups of prokaryotes by convergent evolution. Regardless of their origin or their biochemical structure, microbial enzymes that reduce arsenate appear to play an important role in the mobility of this element in aqueous matrices and in the formation and destruction of various minerals. In addition to more isolations and enzymological studies, there is a need for future research that is oriented to the adaptation of molecular tools based on polymerase chain reaction technology to identify natural assemblages of active arsenate-respiring micro-organisms in the environment. When such techniques are coupled with sensitive bioassays for DAsR using ^{73}As, they will broaden our understanding of the significance of arsenate reduction in a number of anoxic ecosystems, such as sediments, hot springs, freshwater and marine systems, as well as in geological materials.

ACKNOWLEDGMENTS

We thank Prof. H. L. Ehrlich and Prof. L. Young for helpful comments on an earlier draft of this manuscript.

REFERENCES

1.   C Cervantes, G Ji, JL Ramírez, S Silver. Resistance to arsenic compounds in microorganisms. FEMS Microbiol Rev 15:355–367, 1994.
2.   S Silver, M Walderhaug. Gene regulation of plasmid- and chromosome-determined inorganic ion transport in bacteria. Microbiol Rev 56:195–228, 1992.
3.   NN Rao, A Torriani. Molecular aspects of phosphate transport in Escherichia coli. Mol Microbiol 4:1083–1090, 1990.
4.   GR Willsky, MH Malamy. Effect of arsenate on inorganic phosphate transport in *Escherichia coli.* J Bacteriol 144:366–374, 1980.
5.   DH Nies, S Silver. Ion efflux systems involved in bacterial metal resistances. J Indust Microbiol 14:186–199, 1995.
6.   DJH Phillips. The chemical forms of arsenic in aquatic organisms and their interrelationships. In: JO Nriagu, ed. Arsenic in the Environment. Part I: Cycling and Characterization. New York: John Wiley & Sons, 1994, pp 263–288.
7.   KJ Reimer. The methylation of arsenic in marine sediments. Appl Organomet Chem 3:475–490, 1989.
8.   I Koch, L Wang, CA Ollson, WR Cullen, KJ Reimer. The predominance of inorganic arsenic species in plants from Yellowknife, Northwest Territories, Canada. Environ Sci Technol 34:22–26, 2000.
9.   K Hanaoka, H Yamamoto, K Kawashima, S Tagawa, T Kaise. Ubiquity of arsenobetaine in marine animals and degradation of arsenobetaine by sedimentary microorganisms. Appl Organometal Chem 2:371–376, 1988.
10. R Ciulla, MR Diaz, BF Taylor, MF Roberts. Organic osmolytes in aerobic bacteria from Mono Lake, an alkaline, moderately hypersaline environment. Appl Environ Microbiol 63:220–226, 1997.
11. EH Larsen, CR Quétel, R Munoz, A Fiala-Medioni, OFX Donard. Arsenic speciation in shrimp and mussel from the Mid-Atlantic hydrothermal vents. Marine Chem 57:341–346, 1997.
12. JM Neff. Ecotoxicology of arsenic in the marine environment. Environ Toxicol Chem 16:917–927, 1997.
13. GM King. Utilization of hydrogen, acetate, and "non-competitive" substrates by methanogenic bacteria in marine sediments. Geomicrobiol J 3:275–306, 1984.
14. LCD Anderson, KW Bruland. Biogeochemistry of arsenic in natural waters: The importance of methylated species. Environ Sci Technol 25:420–427, 1991.
15. D Ahmann, AL Roberts, LR Krumholz, FMM Morel. Microbe grows by reducing arsenic. Nature 371:750, 1994.
16. DK Newman, D Ahmann, FMM Morel. A brief review of microbial arsenate respiration. Geomicrobiol J 15:255–268, 1998.

17. JF Stolz, RS Oremland. Bacterial respiration of arsenic and selenium. FEMS Microbiol Rev 23:615–627, 1999.
18. RS Oremland, JF Stolz. Dissimilatory reduction of selenate and arsenate in nature. In DR Lovley, ed. Environmental Microbe-Metal Interactions, Washington DC: ASM Press, 2000, pp 199–224.
19. R Nickson, J MacArthur, W Burgess, KM Ahmed, P Ravenscroft, M Rahman. Arsenic poisoning of Bangaladesh groundwater. Nature 395:338, 1998.
20. JF Stolz, DJ Ellis, J Switzer Blum, D Ahmann, DR Lovley, RS Oremland. *Sulfurospirillum barnesii* sp. nov., *Sulfurospirillum arsenophilum* sp. nov., and the *Sulfurospirillum* clade in the Epsilon Proteobacteria. Int J Syst Bacteriol 49:1177–1180, 1999.
21. RS Oremland, J Switzer Blum, CW Culbertson, PT Visscher, LG Miller, P Dowdle, FE Strohmaier. Isolation, growth, and metabolism of an obligately anaerobic, selenate-respiring bacterium, strain SES-3. Appl Environ Microbiol 60:3011–3019, 1994.
22. AM Laverman, J Switzer Blum, JK Schaefer, EJP Philips, DR Lovley, RS Oremland. Growth of strain SES-3 with arsenate and other diverse electron acceptors. Appl Environ Microbiol 61:3556–3561, 1995.
23. JM Macy, K Nunan, KD Hagen, DR Dixon, PJ Harbour, M Cahill, LI Sly. *Chrysiogenes arsenatis* gen. nov., sp. nov., a new arsenic-respiring bacterium isolated from gold mine wastewater. Int J Syst Bacteriol 46:1153–1157, 1996.
24. DK Newman, EK Kennedy, JD Coates, D Ahmann, DJ Ellis, DR Lovely, FMM Morel. Dissimilatory arsenate and sulfate reduction in *Desulfotomaculum auripigmentum* sp. nov. Arch Microbiol 168:380–388, 1997.
25. DK Newman, TJ Beveridge, FMM Morel. Precipitation of arsenic trisulfide by *Desulfotomaculum auripigmentum*. Appl Environ Microbiol 63:2022–2028, 1997.
26. J Switzer Blum, A Burns Bindi, J Buzelli, JF Stolz, RS Oremland. 1998. *Bacillus arsenicoselenatis*, sp. nov., and *Bacillus selenitireducens* sp. nov.: Two haloalkaliphiles from Mono Lake, California that respire oxyanions of selenium and arsenic. Arch Microbiol 171:19–30, 1998.
27. JM Macy, JM Santini, BV Pauling, AH O'Neill, LI Sly. Two new arsenate/sulfate-reducing bacteria: Mechanisms of arsenate reduction. Arch Microbiol 173:49–57, 2000.
28. R Huber, M Sacher, A Vollmann, H Huber, D Rose. Respiration of arsenate and selenate by hyperthermophilic Archaea. Syst Appl Microbiol 23:305–314, 2000.
29. DK Nordstrom, C.N. Alpers. Negative pH, efflorescent mineralogy, and consequences for environmental restoration at the Iron Mountain Superfund site, California. Proc Natl Acad Sci USA 96:3455–3462.
30. KJ Edwards, PL Bond, TM Gihring, JF Banfield. A new iron-oxidizing, extremely acidophilic archaea is implicated in an extreme acid mine drainage generation. Science 287:1796–1799, 2000.
31. T Krafft, JM Macy. Purification and characterization of the respiratory arsenate reductase of *Chrysiogenes arsenatis*. Eur J Biochem 255:647–653, 1998.
32. R Gross, J Simon, F Theis, A Kroeger. Two membrane anchors of *Wolinella succinogenes* hydrogenase and their function in fumarate and polysulfide respiration. Arch Microbiol 170:50–58, 1997.

33. JR Lloyd, JA cole, LE Macaskie. Reduction and removal of heptavalent technicium from solution by *Escherichia coli*. J Bacteriol 179:2014–2021, 1997.

34. JA Cherry, AU Shaikh, DE Tallman, RV Nicholson. Arsenic species as an indicator of redox conditions in groundwater. J Hydrol 43:373–392, 1979.

35. DW Oscarson, PM Hunag, C Defosse, A Herbillon. Oxidative power of Mn(IV) and Fe(III) oxides with respect to As(III) in terrestrial and aquatic environments. Nature 91:50–51, 1981.

36. CC Fuller, JA Davis, GA Waychunas. Surface chemistry of ferrihydrite. Part 2: Kinetics of arsenate adsorption and coprecipitation. Geochim Cosmochim Acta 57: 2271–2282, 1993.

37. BA Manning, SE Fendorf, S. Goldberg. Surface structures and stability of arsenic (III) on goethite: Spectroscopic evidence for inner-sphere complexes. Environ Sci Technol 32:2383–2388, 1998.

38. JM Harrington, SE Fendorf, RF Rosenzweig. Biotic generation of arsenic (III) in metal(loid)-contaminated freshwater lake sediment. Environ Sci Technol 32:2425–2430, 1998.

39. PR Dowdle, AM Laverman, RS Oremland. Bacterial reduction of arsenic (V) to arsenic (III) in anoxic sediments. Appl Environ Microbiol 62:1664–1669, 1996.

40. KA Rittle, JI Drever, PJS Colberg. Precipitation of arsenic during bacterial sulfate reduction. Geomicrobiol J 13:1–11, 1995.

41. DE Cummings, F Caccovo Jr., S Fendorf, RF Rosenzweig. Arsenic mobilization by the dissimilatory Fe(III)-reducing bacterium Shewanella alga BrY. Environ Sci Technol 33:723–729, 1999.

42. D Ahmann, LR Krumholz, H Hemond, DR Lovley, FMM Morel. Microbial mobilization of arsenic from sediments of the Aberjona Watershed. Environ Sci Technol 31:2923–2930, 1997.

43. J Zobrist, PR Dowdle, JA Davis, RS Oremland. Mobilization of arsenite by dissimilatory reduction of adsorbed arsenate. Env Sci Technol 34:4747–4753, 2000.

44. S Gaspard, F Vazquez, and C Holliger. Localization and solubilization of the iron (III) reductase of Geobacter sulfurreducens. J Bacteriol 64:3188–3194.

45. AS Beliaev, D Saffarini. *Shewanella putrefaciens* mtrB encodes an outer membrane protein required for Fe(III) and Mn(IV) reduction. J Bacteriol 180:6292–6297, 1998.

46. HW Langer, WP Inskeep. Microbial reduction of arsenate in the presence of ferrihydrite. Environ Sci Technol 34:3131–3136, 2000.

47. DR Lovley, JD Coates, EL Blunt-Harris, EJP Philips, JC Woodward. Humic substances as electron acceptors for microbial respiration. Nature 382:445–448, 1996.

48. DR Lovley, JL Fraga, EL Blunt-Harris, LA Hayes, EJP Hilips, JD Coates. Humic substances as a mediator for microbially catalyzed metal reduction. Acta Hydrochim Hydrobiol 26:152–157, 1998.

49. KP Nevin, DR Lovley. Potential for nonenzymantic reduction of Fe(III) via electron shuttling in subsurface sediments. Environ Sci Technol 34:2472–2478, 2000.

50. DK Newman, R Kolter. A role for excreted quinones in extracellular electron transfer. Nature 405:94–97, 2000.

51. DR Lovley, JL Fraga, JD Coates, EL Blunt-Harris. Humics as an electron donor for anaerobic respiration. Environ Microbiol 1:89–98, 1999.

52. AH Welch, MS Lico. Factors controlling As and U in shallow ground water, southern Carson Desert, Nevada. Appl Geochem 13:521–539, 1998.

53. AH Welch, DB Westjohn, DR Helsel, RB Wanty. Arsenic in ground water of the United States: Occurrence and geochemistry. Groundwater 38:589–604, 2000.

54. PA Glancy. Geohydrology of the basalt and unconsolidated sedimentary aquifers in the Fallon area, Churchill County, Nevada. U.S. Geological Survey Water-Supply Paper 2263, 1986.

55. ML Peterson, R Carpenter. Biogeochemical processes affecting total arsenic and arsenic species distributions in an intermittently anoxic fjord. Marine Chem 12:295–321, 1983.

56. P Seyler, JM Martin. Biogeochemical processes affecting arsenic speciation in a permanently stratfieid lake. Environ Sci Technol 23:1258–1263, 1989.

57. RS Oremland, PR Dowdle, S Hoeft, JO Sharp, JK Schaefer, LG Miller, J Switzer Blum, RL Smith, NS Bloom, D Wallschlaeger. Bacterial dissimilatory reduction of arsenate and sulfate in meromictic Mono Lake, California. Geochim Cosmochim Acta 64:3073–3084, 2000.

12

Unique Modes of Arsenate Respiration by *Chrysiogenes arsenatis* and *Desulfomicrobium* sp. str. Ben-RB

Joan M. Macy[†] **and Joanne M. Santini**
La Trobe University, Melbourne, Victoria, Australia

I. INTRODUCTION

Arsenic-contaminated water presents a serious worldwide health problem. In Australia, New Zealand, and the United States, many sources of surface water and groundwater contain extremely high levels of arsenic. This water is generally associated with arsenopyrite (FeAsS)-containing ores that have been exposed to air and water because of extensive mining activity. In Victoria, Australia, arsenic (a mixture of arsenate and arsenite) levels as high as 12,000 µg/L have been reported in groundwater of old mine shafts and 300,000 µg/L in surface water from abandoned mines. Arsenic-containing groundwater or surface water, although not generally used as a source of drinking water in these countries, must be removed to permit further mining. However, before it is expelled into the environment, the arsenic levels must be reduced to <500 µg/L. It is therefore critical that a simple, effective, and inexpensive method be found for arsenic removal from such water. The method must also be environmentally sound and not result in large amounts of arsenic-containing sludge. The use of arsenic-

---

† Deceased.

Table 1 Comparisons of *Chrysiogenes arsenatis* and *Desulfomicrobium* sp. str. Ben-RB with Other Arsenate Respirers

	Chrysiogenes arsenatis	*Desulfomicrobium* sp. str. Ben-RB	*Sulfurospirillum barnesii* str. SES-3	*Sulfurospirillum arsenophilum* str. MIT-13	*Desulfotomaculum auripigmentum* str. OREX-4	*Bacillus selenitireducens* str. MLS10	*Bacillus arsenicoselenatis* str. E1H
Subgroup of the Bacteria	Chrysiogenetes	δ	ε	ε	Low G + C gram positive	Low G + C gram positive	Low G + C gram positive
Morphology	Curved rod	Rod	Vibrio	Vibrio	Slightly curved rod	Rod	Rod
Length (µm)	1.0–2.0	1.5–2.2	1.5	1.0	2.5	2.0–6.0	3.0–10.0
Diameter (µm)	0.5–0.75	0.8	0.3	0.1–0.3	0.4	0.5	0.5–1.0
Motility	+	+	+	+	–	–	–
Gram reaction	Negative	Negative	Negative	Negative	Positive	Positive	Positive
Temperature optimum (°C)	25–30	25–30	33	20	25–30	ND[a]	ND
Relationship to oxygen	Strictly anaerobic	Strictly anaerobic	Microaerophilic	Microaerophilic	Strictly anaerobic	Microaerophilic	Strictly anaerobic
Doubling time on arsenate (hr)	4	9	5	~14	~20	ND	ND
Electron acceptors							
Arsenate	+	+	+	+	+	+	+
Nitrate	+	–	+	+	–	+	+
Nitrite	+	–	–	–	+	+	–
Sulfate	–	+	–	–	+	+	–
Thiosulfate	–	ND	+	+	–	–	–
Fe(III)	–	ND	+	ND	–	–	+
Selenate	–	–	+	–	–	–	+
Electron donors							
Hydrogen	–	ND	+[b]	–	+	ND	ND
Formate	–	ND	+[b]	+	–	–	–
Acetate	+	–	–	–	–	–	–
Pyruvate	+	+	+	+	+	+	+
Lactate	+	ND	+	+	+	+	ND
Fumarate	+	ND	+	+	–	ND	+
Succinate	+	ND	–	ND	+	–	–
Malate	+	ND	+	ND	+	–	+
Citrate	–	ND	+	ND	–	–	+
Benzoate	–	ND	ND	–	–	ND	ND
Methanol	–	ND	ND	ND	–	–	ND
Glucose	–	ND	ND	ND	–	+	–

[a] ND, not determined.
[b] Growth only occurs in the presence of acetate.

metabolizing bacteria may present a novel approach for the removal of arsenic in the environment. This chapter, however, concentrates on arsenate reduction.

The reduction of arsenate [As(V)] to arsenite [As(III)] is known to occur in anoxic environments (1,2). Until recently, however, the organisms responsible for this reduction were not known. A number of different bacteria have been isolated that are able to respire with arsenate, reducing it to arsenite. With one exception, these organisms use the nonrespiratory substrate lactate as the electron donor (3–6) and are listed in Table 1. Two of them, *Desulfotomaculum auripigmentum* str. OREX-4 (7,8) and *Desulfomicrobium* sp. str. Ben-RB (9), also respire with sulfate as the terminal electron acceptor. None are able to use the respiratory substrate acetate as the electron donor for arsenate respiration. The only organism known able to do so is *Chrysiogenes arsenatis* (10).

Phylogenetically, *C. arsenatis* differs from the other arsenate-respiring bacteria and from other phyla of the Bacteria and so is the first representative of a new phylum (11) (see Table 1). The other arsenate-respiring bacteria fall within three different divisions of the Bacteria (see Table 1). The two *Bacillus* species are, however, unrelated to *D. auripigmentum* even though they are all members of the low G + C gram-positive bacteria.

Preliminary biochemical studies of the enzyme that catalyzes arsenate reduction in *Sulfurospirillum* barnesii have been conducted and the information only cited in two review articles (12,13). This enzyme is an integral membrane protein with a calculated mass of 100 kDa consisting of three different subunits; 65, 31, and 22 kDa. The equivalent enzyme from *Desulfomicrobium* sp. str. Ben-RB (9) is discussed in Section III.D. The only respiratory arsenate reductase that has been studied in detail is that of *C. arsenatis* (14), discussed in Section II.D.

This chapter concentrates on arsenate respiration by *Chrysiogenes arsenatis* and *Desulfomicrobium* sp. str. Ben-RB. The evidence indicates that they have specific respiratory arsenate reductases involved in energy generation. The isolation, phylogeny, physiology, and biochemistry of arsenate reduction are described separately for each organism.

II. CHRYSIOGENES ARSENATIS

A. Isolation

Mud samples were taken from a reed bed near the Ballarat goldfields in Ballarat, Victoria, Australia, for the purpose of isolating an arsenate-respiring bacterium (10). Anaerobic enrichments were made in a minimal salts medium containing arsenate as the terminal electron acceptor and the respiratory substrate acetate as the electron donor. After a number of days, the arsenate was reduced to arsenite, the acetate oxidized to carbon dioxide, and the pH of the cultures increased [Eq. (1)].

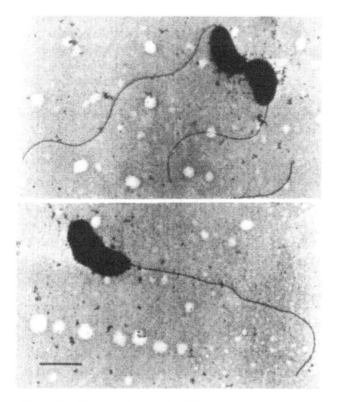

Figure 1   Electron micrograph of *Chrysiogenes arsenatis*. Bar = 1 μm. (From Ref. 10.)

$$2HAsO_4^{2-} + 2H_2AsO_4^- + 5H^+ + CH_3COO^- \rightarrow 4H_3AsO_3 + 2HCO_3^- \quad (1)$$
$$(\text{at pH 7}, \Delta G^{0\prime} = -252.6 \text{ kJ/mol of acetate})$$

After three subcultures, a small (1 μm), gram-negative, curved bacterium designated BAL-1 was isolated from agar (Fig. 1) (10). It is strictly anaerobic and
very motile, powered by a single polar flagellum.

B.   Phylogenetic Characterization

The complete 16S rRNA gene sequence of strain BAL-1 was determined (10).
Phylogenetic analysis indicated that BAL-1 branched deeply with *Synechococcus*
sp. str. 6301 and with "*Flextistipes sinusarabici*" [an anaerobic, gram-negative,
flexible, rod-shaped thermophilic bacterium isolated from brine water samples

taken from the Red Sea at a depth of 2000 m (15)] (Fig. 2). BAL-1 showed only 74.8–81.8% sequence similarity with the sequences of all other major phyla of the Bacteria. The highest similarities were with *Lactobacillus casei* (81.8%), "*F. sinusarabici*" (81.7%), *Bacillus subtilis* (81.7%), and *Synechococcus* sp. str. 6301 (81%).

These data therefore support the conclusion that BAL-1 is phylogenetically unique and at present is the first representative of a new deeply branched lineage of the Bacteria and was thus named *Chrysiogenes arsenatis* (*Chrysiogenes*, sprung from a gold mine; *arsenatis*, of arsenate) (10). In fact, *C. arsenatis* represents the first representative of a new phylum of the Bacteria, Chrysiogenetes (11). The nearest relative of *C. arsenatis*, "*F. sinusarabici*," is physiologically distinct.

C. Physiological Characteristics

When grown in a minimal medium containing arsenate as the terminal electron acceptor and acetate as the electron donor, *C. arsenatis* has a generation time of 4 hr (10). The arsenate is reduced to arsenite, the acetate is oxidized to carbon dioxide, and the pH of the medium increases from 7 to 9.4. Growth does not occur in the absence of either acetate or arsenate. This organism can also grow on arsenate bound to iron oxide. It can tolerate levels of up to 30 mM arsenate but is inhibited once the concentration of arsenite reaches 10 mM (11).

Apart from acetate, *C. arsenatis* will grow on pyruvate, L/D-lactate, fumarate, succinate, and malate as electron donors (Table 1). However, in these cases, the electron acceptor arsenate is essential (10). When grown with acetate, nitrate and nitrite can serve as terminal electron acceptors. Nitrate is reduced to nitrite and then to ammonia. Yeast extract (0.1%) stimulates the rate and extent of growth on acetate plus in the presence of arsenate and/or nitrate (11). Moreover, the presence of hydrogen stimulates the rate and extent of growth with arsenate only (10).

C. arsenatis is a physiologically unique bacterium differing from the other known arsenate-reducing bacteria (see Sec. I and Table 1) in that it grows with the respiratory substrate acetate as the electron donor. The ability of *C. arsenatis* to use acetate as the electron donor and carbon source when reducing arsenate to arsenite suggests that arsenate respiration supports the growth of this bacterium [see Eq. (1)].

D. Biochemistry of Arsenate Reduction

Arsenate reductase (Arr), the enzyme that catalyzes the reduction of arsenate to arsenite in *C. arsenatis*, has been purified and characterized (14). It is soluble, located in the periplasm of *C. arsenatis*. The enzyme consists of two different

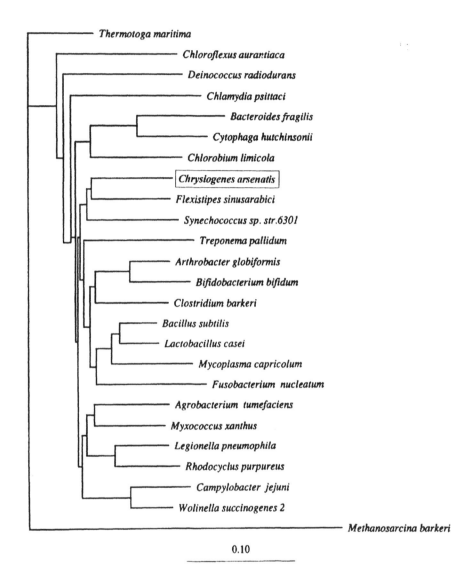

Figure 2 Phylogenetic neighbor-joining dendrogram showing the relationships of *Chrysiogenes arsenatis* with representatives of the phyla of the Bacteria. Bar = 0.1% sequence difference. (From Ref. 10.)

subunits, ArrA (87 kDa) and ArrB (29 kDa). The native molecular weight was estimated to be 123 kDa based on gel filtration chromatography and so it appears to be a heterodimer ($\alpha_1\beta_1$). Using reduced benzyl viologen as the artificial electron donor, the K_m for arsenate was 0.3 mM. The V_{max} for arsenate reduction was found to be 7013 μmol arsenate reduced min^{-1} mg protein^{-1}, which corresponds to an enzyme turnover of 14377 sec^{-1} (based on the molecular weight of 123 kDa). The enzyme is specific for arsenate, as nitrate, sulfate, selenate, and fumarate do not serve as alternative electron acceptors. Synthesis of the enzyme appears to be regulated as no enzyme activity can be detected when *C. arsenatis* is grown with nitrate solely as the terminal electron acceptor, whereas essentially 100% activity was detected when the organism is grown with arsenate or arsenate/nitrate.

Arr contains molybdenum, iron, and acid-labile sulfur (14). It appears to be a member of the family of mononuclear molybdenum enzymes (16). Approximately 14 $\pm$ 0.4 mol equivalents of iron and 16.4 $\pm$ 0.8 mol equivalents of acid-labile sulfur are present, suggesting that Arr contains several iron-sulfur clusters as prosthetic groups.

The N-terminal sequences of both subunits, ArrA and ArrB have been determined (Figs. 3 and 4) (14). The N-terminal sequence of the ArrA subunit is similar to molybdenum-containing subunits of a number of enzymes (see Fig. 3), including PsrA of the polysulfide reductase from *Wolinella succinogenes* and NapA of the periplasmic nitrate reductase of both *Escherichia coli* and *Pseudomonas* sp. G-179. Each of these is the catalytic subunit of the respective enzymes and all contain a cysteine cluster at the N-terminus (see Fig. 3) and molybdopterin guanosine dinucleotide as the organic component of their molybdenum cofactor (17).

The class of mononuclear molybdoenzymes can be divided into three groups based on the structure of their molybdenum centers: (1) the xanthine oxidase family, which is the largest and most diverse family (the molybdenum hydroxylases) and catalyzes the hydroxylation of a broad range of aldehydes and

```
C.a    ArrA    1-  QTGTGASAMGEAEGKWIPSTCQGCTTWCP
W.s    PsrA    1-  GALEKQEIKGSA--KFVPSICEMCTSSCT
E.c    NapA    1-  PGVARAVVGQQEAIKWDKAPCRFCGTGCG
P.sp   NapA    1-  QSVPGGVAAL--EIKWSKAPCRFCGTGCG
```

Figure 3 Sequence alignment of the N-termini of ArrA and molybdenum-containing proteins. The sequences belong to the arsenate reductase of *Chrysiogenes arsenatis* (*C.a* ArrA) (14), the polysulfide reductase of *Wolinella succinogenes* (*W.s.* PsrA), and the periplasmic nitrate reductases of *Escherichia coli* (*E.c.* NapA) and *Pseudomonas* sp. G-179 (*P.s.* sp. NapA). Boxed amino acids show identity to *C. arsenatis* ArrA.

```
C.a ArrB    1-  ----AKYGMAIDLHKCAGCGACGLACKTQNNTDD
W.s PsrB    1-  --MAKKYGMIHDENLCIGCQACNIACRSENKIPD
B.s NarH    1-  MKIKAQIGMVMNLDKCIGCHACSVTCKNTWTNRS
E.c DmsB    1-  --MTTQYGFFIDSSRCTGCKTCELACKDYKDLTP
N.h NorB    1-  MDIRAQVSMVFHLDKCIGCHTCSIACKNIWTDRK
S.t PhsB    2-  NHLTNQYVMLHDEKRCIGCQACTVACKVLNDVPE
E.c FdnH   24-  RDYKAEVAKLIDVSTCIGCKACQVACSEWNDIRD
```

Figure 4 Sequence alignment of the N-termini of ArrB and iron-sulfur-containing proteins. The sequences belong to the arsenate reductase of *Chrysiogenes arsenatis* (*C.a.* ArrB) (14), the polysulfide reductase of *Wolinella succinogenes* (*W.s.* PsrB), the nitrate reductase of *Bacillus subtilis* (*B.s.* NarH), the dimethyl sulfoxide reductase of *Escherichia coli* (*E.c.* DmsB), the nitrite oxidoreductase of *Nitrobacter hamurggensis* (*N.h.* NorB), the thiosulfate reductase of *Salmonella typhimurium* (*S.t.* PhsB), and the nitrate-inducible formate dehydrogenase of *E. coli* (*E.c.* FdnH). Boxed amino acids show sequence identity to *C. arsenatis* ArrB.

aromatic heterocyclics; (2) the sulfite oxidase family (the eukaryotic oxo trans-ferases) and the assimilatory nitrate reductases, which catalyze oxygen atom transfer to or from a substrate; and (3) the DMSO reductase family, which is exclusively found in eubacteria and catalyzes either oxygen atom transfer or other oxidation-reduction reactions.

The N-terminal sequence of ArrA suggests that Arr is a member of the DMSO reductase family. These enzymes feature molybdenum coordinated by two bidentate dithiolene ligands contributed by two equivalents of the molybdopterin cofactor (16,18).

The N-terminal sequence of the ArrB subunit is similar to iron-sulfur proteins that contain the same type of cysteine cluster within their N-termini (Fig. 4). Most of these iron-sulfur proteins are subunits of molybdenum-containing enzymes, for example, PsrB of the polysulfide reductase from *W. succinogenes* and DmsB of the dimethyl sulfoxide reductase from *E. coli* (see Fig. 4). The proteins listed in Figure 4 are all components of different electron transport systems and are therefore involved in electron transfer. The DmsB contains four cysteine clusters (each with four cysteine residues) which ligate four [4Fe-4S] centers (19). The degree of similarity of ArrB to the iron-sulfur proteins suggests that this subunit is also an iron-sulfur protein and may be involved in the transfer of electrons to the molybdenum cofactor of the ArrA subunit.

E. Conclusion

Growth of *C. arsenatis* is dependent upon energy conserved during the reduction of arsenate to arsenite, the electrons for this reduction coming from the oxidation of acetate to carbon dioxide (10). The ability of *C. arsenatis* to utilize acetate as the electron donor excludes the possibility that adenosine triphosphate (ATP) is formed via any substrate-level phosphorylation process as might be the case when lactate is used as the electron donor. This therefore suggests that this organism gains energy from arsenate respiration alone. The arsenate reductase presumably functions as a terminal reductase linked in a manner as yet unknown to an electron transport chain in the membrane. This is supported by the finding that some arsenate reductase activity was always found associated with the membranes (14). This association may occur via a third membrane-bound subunit or via a membrane-bound electron-carrying protein (e.g., cytochrome) loosely bound to the arsenate reductase.

III. DESULFOMICROBIUM SP. STR. BEN-RB

A. Isolation

Black mud samples were taken from an arsenic-contaminated reed bed in Bendigo, Victoria, Australia, for the purpose of isolating an arsenate/sulfate-reducing bacterium (9). Bendigo is found approximately 100 km east of Ballarat (see Sec. II.A). Anaerobic enrichments were made in a minimal medium containing sulfate as the terminal electron acceptor and lactate as the electron donor. Arsenate was not included in the medium, as whenever this was done *C. arsenatis* was found to be the predominant organism. The enrichment was subcultured twice into the same medium and then a third time into a medium also containing arsenate. After two additional subcultures, a gram-negative sulfate-reducing bacterium was isolated from agar (Fig. 5) (9). It was designated Ben-RB and is a motile rod powered by means of a single polar flagellum.

B. Phylogenetic Characterization

The phylogenetic analysis of the 16S rRNA gene sequence of strain Ben-RB showed that it belonged to a strongly supported branch of the δ-Proteobacteria containing all the described species of the genus *Desulfomicrobium* as well as two unspeciated isolates (Fig. 6) (9). Members of this group were closely related to each other, with sequence similarities in the range of 97.3–99.8%. The closest phylogenetic relative of strain Ben-RB was the unidentified sulfate-reducing bacterium strain Äspö 1 (98.7% sequence similarity). This organism was isolated from a deep underground bore hole at the Äspö hard rock laboratory in Sweden and was shown to be similar to *Desulfomicrobium baculatum* (20). At this level of sequence similarity it is not known to which species of *Desulfomicrobium* that Ben-RB (designated *Desulfomicrobium* sp. str. Ben-RB) belongs.

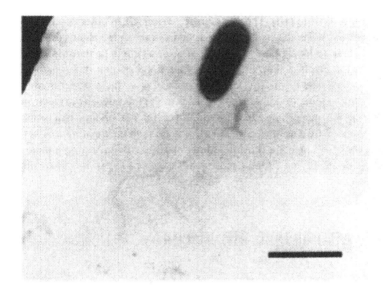

Figure 5   Electron micrograph of *Desulfomicrobium* sp. str. Ben-RB. Bar = 2 μm.

C.   Physiological Characteristics

Desulfomicrobium sp. str. Ben-RB can use sulfate and/or arsenate as terminal electron acceptors, using lactate as the electron donor instead of acetate as does *C. arsenatis* (see Sec. II.C). Lactate alone does not support growth; however, with sulfate as the terminal electron acceptor a doubling time of 6 hr was observed (9). Sulfate was reduced to sulfide and lactate oxidized to acetate. When grown with both arsenate and sulfate as terminal electron acceptors, growth was slower (doubling time of 8 hr). Both sulfate and arsenate were reduced concomitantly to sulfide and arsenite, respectively. *Desulfomicrobium* sp. str. Ben-RB could also grow in a medium containing only arsenate as the terminal electron acceptor, the doubling time of which was 9 hr. Only a small amount of arsenate was reduced (decreasing from 4 to 2.8 mM) as once the concentration of arsenite reached 1.5 mM the culture lysed. This organism was also found to be highly resistant to arsenate (30 mM) when grown in a medium that contained both sulfate and arsenate as terminal electron acceptors. However, the presence of arsenite resulted in slower growth.

D.   Biochemistry of Arsenate Reduction

The enzyme of *Desulfomicrobium* sp. str. Ben-RB that reduces arsenate to arsenite is located in the membrane (9). The membrane fraction also contained most

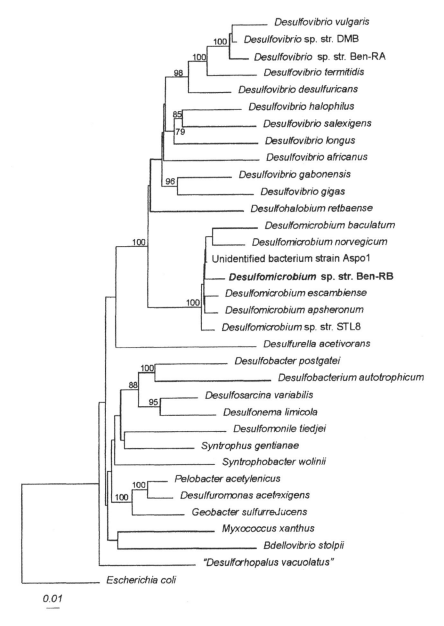

Figure 6 Neighbor-joining tree showing the phylogenetic relationship of *Desulfomicrobium* sp. str. Ben-RB with species of the genus *Desulfomicrobium* and other members of the δ-Proteobacteria. The sequence of *Escherichia coli* was used as the outgroup. Significant bootstrap values from 100 analyses are shown at the branch points of the tree. Bar = 0.01% sequence difference. (From Ref. 9.)

of the cytochromes present in the cell, the predominant being a cytochrome c_{551}. The soluble fraction did not contain significant amounts of cytochromes. When, under anoxic conditions, the membranes were reduced with dithionite, the addition of arsenate resulted in the reoxidation of the cytochromes (Fig. 7). These results indicate that arsenate is reduced via electrons transported from cytochromes (presumably c_{551}) of a membrane-bound electron transport system.

Like the Arr of *C. arsenatis*, the synthesis of the *Desulfomicrobium* sp. str. Ben-RB arsenate reductase appears to be regulated (9). The highest enzyme activity (100%) was detected when the organism was grown with sulfate and arsenate (0.192 U/mg protein). When grown with sulfate alone as the terminal electron acceptor only 11% of the total activity was detected (0.021 U/mg protein).

The enzyme of *Desulfomicrobium* sp. str. Ben-RB is either a *c*-type cytochrome or is associated in the membrane with such a cytochrome. This enzyme is therefore different from that of *C. arsenatis* (see Sec. II.D). If it is a *c*-type cytochrome, it may be similar to a number of multiheme *c*-type metal reductase cytochromes found in other sulfate-, sulfur-, and iron (III)-reducing bacteria. These metal reductase activities are specific to the polyheme cytochrome *c* from class III as defined by Ambler (21). Such activity was first demonstrated for the soluble cytochrome c_3 of *Desulfovibrio vulgaris*, shown to reduce both chromium

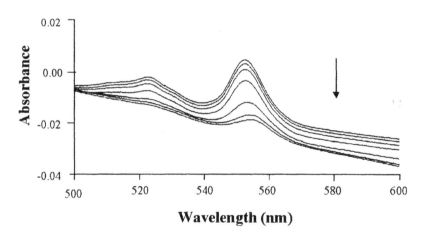

Figure 7 Reduced-oxidized difference spectrum of the membrane fraction from *Desulfomicrobium* sp. str. Ben-RB in 50 mM MES (pH 6.5; 0.54 mg protein/ml). The reference cuvette contained oxidized membrane fraction. The experimental cuvette contained the membrane fraction in an anoxic cuvette under an atmosphere of 100% nitrogen. Cytochromes were reduced with sodium dithionite (0.1 mg/ml). Thereafter, 10 mM arsenate was added to the experimental anaerobic cuvette to oxidize the cytochromes; sequential scans were done (each scan: 1.1 min). Arrow indicates the direction in which the absorbance is changing. (From Ref. 9.)

(VI) and uranium (VI) (22,23). More recently, it has been found that various multiheme *c*-type cytochromes of sulfate-reducing bacteria all function as iron (III) reductases (24). These cytochromes are usually found in the periplasm and include the cytochrome c_7 of *Desulfuromonas acetoxidans* (25,26), cytochrome c_3 of *Desulfovibrio vulgaris* Hildenborough, octahemic cytochrome c_3 (Mr 26,000) and tetrahemic cytochrome c_3 (Mr 13,000) of *Desulfovibrio desulfuricans* Norway, and cytochrome c_3 of *Desulfovibrio gigas*. In addition, the cytochrome c_7 of *D. acetoxidans* reduces manganese (IV), vanadium (V), chromium (VI), iron (III)-oxide, polysulfide, and elemental sulfur (25–27). Finally, the cytochrome c_{552} of *Geobacter sulfurreducens* also has metal reductase activities. This protein acts as an iron (III) reductase for electron transfer to insoluble iron hydroxides or to sulfur and manganese dioxide (28,29).

As opposed to the multiheme cytochrome *c*-type metal reductases described above, which are soluble, usually periplasmic proteins, the *c*-type cytochrome involved in arsenate reduction by *Desulfomicrobium* sp. str. Ben-RB is in the membrane. Whether the enzyme that catalyzes the reduction of arsenate to arsenite is in fact a *c*-type cytochrome or associated with one remains to be determined.

E. Conclusion

The only other organism known to respire with both arsenate and sulfate as terminal electron acceptors is *D. auripigmentum* (7,8). However, unlike *Desulfomicrobium* sp. str. Ben-RB, which can reduce both electron acceptors concomitantly, the presence of arsenate inhibited sulfate reduction by *D. auripigmentum*. Sulfate reduction only occurred when all the arsenate present in the medium had first been reduced to arsenite (7,8). The doubling time of *D. auripigmentum* on either arsenate or sulfate appeared to be much greater than for *Desulfomicrobium* sp. str. Ben-RB, in the order of 1 or 2 days (7,8). These differences suggest that for *Desulfomicrobium* sp. str. Ben-RB, arsenate and sulfate reduction are separate processes. It would appear that sulfate and arsenate reduction in *D. auripigmentum* are not separate processes and that in fact the ATP sulfurylase may also be involved in arsenate reduction. This is based on the finding that arsenate reduction is inhibited by molybdate, which is a known inhibitor of sulfate reduction (8). The mechanism by which *D. auripigmentum* reduces arsenate remains unknown.

IV. OVERALL CONCLUSIONS

Two new organisms that respire anaerobically using arsenate as the terminal electron acceptor have been isolated from arsenic-contaminated areas in Victoria, Australia. One of these organisms, *C. arsenatis*, is the first representative of a new phylum of the Bacteria and uses the respiratory substrate acetate as the electron donor. It is the only organism of its kind ever isolated. *Desulfomicrobium* sp. str. Ben-RB, a sulfate-reducing bacterium, uses lactate as the electron donor while reducing arsenate. This organism, when grown with both sulfate and arsenate as terminal electron acceptors reduces both concomitantly. Other substrates used by these bacteria and comparisons to the other known arsenate-reducing bacteria can be seen in Table 1.

These two organisms are not only unrelated phylogenetically and physiologically, but the mechanism by which they reduce arsenate is also different. *C. arsenatis* reduces arsenate via a soluble periplasmic arsenate reductase (123 kDa) consisting of two different subunits. The enzyme, a heterodimer, contains molybdenum, iron, and acid-labile sulfur as cofactors and also, based on the N-terminal sequence of both subunits, is probably a molybdenum-cofactor-containing enzyme. To date, it is the only respiratory arsenate reductase ever purified and characterized. Preliminary sequence analysis of the gene encoding the large subunit of the Arr (ArrA) supports the conclusion that the Arr is similar to molybdenum-containing enzymes, such as the polysulfide reductase of *W. succinogenes* (J. M. Santini and J. M. Macy, personal communication). On the other hand, the arsenate reductase of *Desulfomicrobium* sp. str. Ben-RB appears to be a membrane-bound enzyme, either a cytochrome c_{551} or associated with such a cytochrome. Further studies of this enzyme are needed to allow comparisons to made with the Arr of *C. arsenatis* and the *c*-type metal reductases. Both enzymes appear to be different from that of *S. barnesii*, which is an integral membrane protein (100 kDa) with three subunits of 65, 31, and 22 kDa (12,13). The synthesis of both arsenate reductases appears to be regulated, as both organisms must be grown in the presence of arsenate for complete induction of enzyme synthesis. The level at which this regulation occurs remains to be determined.

The two arsenate-respiring bacteria described in this chapter together with the arsenite-oxidizing bacteria (see Chap. 14) may prove suitable for the removal of arsenic from arsenic-contaminated water. The first step of the bioremediation system is the oxidation of arsenite to arsenate (using the arsenite-oxidizing bacteria); the arsenic now is in the form of arsenate. Using both *C. arsenatis* and *Desulfomicrobium* sp. str. Ben-RB in an anaerobic reactor containing both arsenate and sulfate, the arsenate will be reduced to arsenite and the sulfate reduced to sulfide. The arsenite and sulfide then react chemically forming an insoluble compound, orpiment (As_2S_3), which can be removed from the water by filtration. The water, essentially free of arsenic, can be released back into the environment.

ACKNOWLEDGMENTS

This work was supported by an Australian Research Council Grant (A09925054) and two Central Large La Trobe University grants to JMM. We would like to thank S. C. Dawbarn and A. G. Wedd for proofreading of the chapter and R. D.

Schnagl for carrying out the electron microscopy of *Desulfomicrobium* sp. str. Ben-RB.

REFERENCES

1. PR Dowdle, AM Laverman, RS Oremland. Bacterial dissimilatory reduction of arsenic (V) to arsenic (III) in anoxic sediments. Appl Environ Microbiol 62:1664–1669, 1996.
2. SL McGeehan, DV Naylor. Sorption and redox transformation of arsenite and arsenate in two flooded soils. Soil Sci Soc Am J 58:337–342, 1994.
3. D Ahmann, AL Roberts, LR Krumholz, FMM Morel. Microbe grows by reducing arsenic. Nature 371:750, 1994.
4. JF Stolz, DJ Ellis, JW Blum, D Ahmann, DR Lovley, RS Oremland. *Sulfurospirillum barnesii* sp. nov. and *Sulfurospirillum arsenophilum* sp. nov., new members of the *Sulfurospirillum* clade of the ε Proteobacteria. Int J Syst Bacteriol 49:1177–1180, 1999.
5. RS Oremland, JS Blum, CW Culbertson, PT Visscher, LG Miller, P Dowdle, FE Strohmaier. Isolation, growth, and metabolism of an obligately anaerobic, selenate-respiring bacterium, strain SES-3. Appl Environ Microbiol 60:3011–3019, 1994.
6. JS Blum, AB Bindi, J Buzzelli, JF Stolz, RS Oremland. *Bacillus arsenicoselenatis,* sp. nov., and *Bacillus selenitireducens,* sp. nov: Two haloalkliphiles from Mono Lake, California that respire oxyanions of selenium and arsenic. Arch Microbiol 171:19–30, 1998.
7. DK Newman, TJ Beveridge, FMM Morel. Precipitation of arsenic trisulfide by *Desulfotomaculum auripigmentum.* Appl Environ Microbiol 63:2022–2028, 1997a.
8. DK Newman, EK Kennedy, JD Coates, D Ahmann, DJ Elis, DR Lovley, FMM Morel. Dissimilatory arsenate and sulfate reduction by *Desulfotomaculum auripigmentum* sp. nov. Arch Microbiol 168:380–388, 1997b.
9. JM Macy, JM Santini, BV Pauling, AH O'Neill, LI Sly. Two new arsenate/sulfate-reducing bacteria: Mechanisms of arsenate reduction. Arch Microbiol 173:49–57, 2000.
10. JM Macy, K Nunan, KD Hagen, DR Dixon, PJ Harbour, M Cahill, LI Sly. *Chrysiogenes arsenatis* gen. nov., sp. nov., a new arsenate-respiring bacterium isolated from gold mine wastewater. Int J Syst Bacteriol 46:1153–1157, 1996.
11. JM Macy, T Krafft, LI Sly. Genus I. *Chrysiogenes* Macy, Nunan, Hagen, Dixon, Harbour, Cahill, and Sly 1996, 1156[VP]. In: DR Boone, RW Castenholz, GM Garrity, eds. Bergey's Manual of Systematic Bacteriology. 2d ed. Vol. 1. New York: Springer-Verlag, 2001, pp 412–415.
12. DK Newman, D Ahmann, FMM Morel. A brief review of microbial arsenate respiration. Geomicrobiology 15:255–268, 1998.
13. JF Stolz, RS Oremland. Bacterial respiration of arsenic and selenium. FEMS Microbiol Rev 23:615–627, 1999.
14. T Krafft, JM Macy. Purification and characterization of the respiratory arsenate reductase of *Chrysiogenes arsenatis.* Eur J Biochem 255:647–653, 1998.
15. F Fiala, CR Woese, TA Langworthy, KO Stetter. *Flexistipes sinusarabici,* a novel

genus and species of eubacteria occurring in the Atlantis II Deep brines of the Red Sea. Arch Microbiol 154:120–126, 1990.

16. C Kisker, H Schindelin, D Baas, J Rétey, RU Meckenstock, PMH Kroneck. A structural comparison of molybdenum cofactor-containing enzymes. FEMS Microbiol Rev 22:503–521, 1999.

17. C Kisker, H Schindelin, DC Rees. Molybdenum-cofactor-containing enzymes: Structure and mechanism. Annu Rev Biochem 66:233–267, 1997.

18. R Hille, J Rétey, U Bartlewski-Hof, W Reichenbecher, B Schink. Mechanistic aspects of molybdenum-containing enzymes. FEMS Microbiol Rev 22:489–501, 1999.

19. CA Trieber, RA Rothery, JH Weiner. Engineering a novel iron-sulfur cluster in the catalytic subunit of *Escherichia coli* dimethyl-sulfoxide reductase. J Biol Chem 271: 4620–4626, 1996.

20. K Pedersen, J Arlinger, S Edendahl, L Hallbeck. 16S rRNA gene diversity of attached and unattached bacteria in boreholes along the access tunnel to the Äspö hard rock 1 laboratory Sweden. FEMS Microbiol Ecol 19:249–262, 1996.

21. RP Ambler. Sequence variability in bacterial cytochrome c. Biochim Biophys Acta 1058(1):42–47, 1991.

22. DR Lovley, PK Widman, JS Woodward, EJP Phillips. Reduction of uranium by cytochrome c_3 of *Desulfovibrio vulgaris*. Appl Environ Microbiol 59:3572–3576, 1993.

23. DR Lovley, EJP Phillips. Reduction of chromate by *Desulfovibrio vulgaris* and its c_3 cytochrome. Appl Environ Microbiol 60:726–728, 1994.

24. E Lojou, P Bianco, M Bruschi. Kinetic studies on the electron transfer between various c-type cytochromes and iron (III) using a voltametric approach. Electrochimica Acta 43:2005–2013, 1998a.

25. IAC Pereira, I Pacheco, M-Y Liu, J LeGall, AV Zavier, M Teixeira. Multiheme cytochromes from the sulfur-reducing bacterium *Desulfuromonas acetoxidans*. Eur J Biochem 248:323–328, 1997.

26. C Aubert, E Lojou, P Bianco, M Rousset, M-C Durand, M Bruschi, A Dolla. The *Desulfuromonas acetoxidans* triheme cytochrome c_7 produced in *Desulfovibrio desulfuricans* retains its metal reductase activity. Appl Environ Microbiol 64:1308–1312, 1998.

27. E Lojou, P Bianco, M Bruschi. Kinetic studies on the electron transfer between bacterial c-type cytochromes and metal oxides. J Electroanal Chem 452:167–177, 1998b.

28. S Gaspard, F Vazquez, C Hollinger. Localization and solubilization of the iron (III) reductase of *Geobacter sulfurreducens*. Appl Environ Microbiol 64:3188–3194, 1998.

29. S Seeliger, R Cord-Ruwisch, B Schink. A periplasmic and extracellular c-type cytochrome of *Geobacter sulfurreducens* acts as a ferric iron reductase and as an electron carrier to other acceptors or to partner bacteria. J Bacteriol 180:3686–3691, 1998.

13

Bacterial Oxidation of As(III) Compounds

Henry L. Ehrlich
Rensselaer Polytechnic Institute, Troy, New York

I. INTRODUCTION

Arsenite [As(III)] enters the environment naturally from arsenic-containing minerals as well as from industrial wastes, as herbicide, or as insecticide. A number of bacteria have the ability to oxidize arsenite. They include heterotrophs and at least three autotrophs. The autotrophs have the ability to use arsenite as their sole source of energy. At least some of the heterotrophs seem able to use arsenite as an auxiliary source of energy, but others seem to oxidize arsenite to arsenate [As(V)] merely as a means of detoxification, arsenite being more toxic than arsenate (1). More study is needed to distinguish clearly between bacterial arsenite oxidation as an auxiliary source of metabolic energy and bacterial arsenite oxidation merely as a means of detoxification. Toxicity of arsenate is based chiefly on its ability to uncouple oxidative phosphorylation, whereas toxicity of arsenite is due to its inhibition of dehydrogenases and some other enzymes because of its ability to react with their functional —SH groups.

Whether dissolved arsenite occurs in sufficient quantities in the environment to serve as an energy-yielding substrate for bacteria depends on where, how, and in what quantities it originates. Arsenic-containing minerals are the major natural source. Solid or liquid arsenic-containing industrial wastes are a major anthropogenic source. Oxidation of some arsenic-containing minerals by *Thiobacillus ferrooxidans* yields arsenite and arsenate among the products. Two examples are the oxidation of orpiment (As_2S_3) and of arsenopyrite (FeAsS). On the other hand, oxidation of enargite (Cu_3AsS_4) by *T. ferrooxidans* yields only arsenate among the products. The extent to which some other arsenic-containing minerals, such as realgar (As_2S_2), loellingite ($FeAs_2$), seligmannite (2PbS · CuS · As_2S_3), and a number others (2), may be oxidized by bacteria and what the products of oxidation are, has yet to be determined.

This chapter summarizes the present state of knowledge concerning the role of bacteria in the oxidation of dissolved arsenite to arsenate, and in generating dissolved arsenite and arsenate from arsenic-containing minerals.

II. OXIDATION OF ARSENITE IN SOLUTION

A. Freshwater and Sewage

The ability of some bacteria to oxidize dissolved arsenite to arsenate was first reported from South Africa in 1918 by Green (3). He isolated an arsenite-oxidizing bacterium from a cattle-dipping solution in which the function of the arsenite was to protect against insect bites. He named his isolate *Bacillus arsenoxidans*. The oxidation of the arsenite resulted in loss of potency of the cattle-dipping solution. Green's findings were largely ignored until 1949, when Turner (4) reported the isolation of 15 strains of heterotrophic bacteria from cattle-dipping solution in Queensland, Australia. All these strains were capable of oxidizing 0.1 M arsenite to arsenate. In a more detailed taxonomic study, Turner (5) assigned three of the strains to the genus *Pseudomonas*, one to the genus *Xanthomonas*, and the remaining to the genus *Achromobacter*. One of the isolates closely resembled Green's *B. arsenoxidans* (4). In a further study of one of the isolates, *Pseudomonas arsenoxidans-quinque*, its physiology of arsenite oxidation was examined in detail. Using washed cell suspensions, Turner and Legge (6) demonstrated that arsenite oxidation was maximal with 3- to 4-day-old cells and that arsenite was oxidized only by cells previously grown in the presence of arsenite. Oxidation was optimal in air but was also observed anaerobically in the presence of 2,6-dichlorophenol indophenol, phenol blue, *m*-carboxyphenol-indo-2,6-dibromophenol, and ferricyanide as terminal electron acceptors. Cyanide, azide, fluoride, and pyrophosphate inhibited the oxidation, whereas iodoacetate, diethyldithiocarbamate, α,α-dipyridyl, and urethane did not. The optimal pH for the oxidation was 6.4 and the optimal temperature was 40°C. Using cell-free preparations, Legge and Turner (7) located "arsenite dehydrogenase" activity mainly in a soluble fraction obtained by differential centrifugation of cell extract obtained by grinding intact cells in an Utter-Werkman mill with ground glass. Enzyme activity in the soluble fraction was measured colorimetrically under anaerobic conditions in Thunberg tubes with 2,6-dichlorophenol indophenol as terminal electron acceptor. It was also measured manometrically by following CO_2 evolution from bicarbonate buffer in the presence of ferricyanide as terminal electron acceptor. The CO_2 was evolved in a reaction between the acid that resulted from arsenite oxidation,

$$AsO_2^- + 0.5O_2 + H_2O \rightarrow HAsO_4^{2-} + H^+$$
$$\Delta G^{0\prime} = -38.15 \text{ kcal or } -160 \text{ kJ} \quad (1)$$

and the bicarbonate of the buffer,

$$H^+ + HCO_3^- \rightarrow CO_2(g) + H_2O \quad (2)$$

Phenyl mercuric nitrate, iodoacetate, thiourea, and ammonium sulfate stimulated arsenite-oxidizing activity, whereas cupric ion, and p-chloromercuribenzoate inhibited it significantly. Legge (8) showed that cytochromes associated with the solid fraction from broken cells of *P. arsenoxydans-quinque* played a role in aerobic oxidation of arsenite.

Phillips and Taylor (9) isolated several strains of *Alcaligenes faecalis* from raw sewage enriched with arsenite, which oxidized the added arsenite to arsenate. One of these strains, labeled YE56, which oxidized arsenite rapidly, was studied in detail. It grew in yeast extract medium and in a mineral salts medium containing vitamins and glutamic or aspartic acid. Washed, resting cells of strain YE56 suspended in 0.05 M tris-hydrochloride buffer at pH 7.35 oxidized 0.02 M arsenite stoichiometrically to arsenate at a rate of 3.1 μmol arsenite oxidized hr^{-1} mg^{-1} protein, measured in terms of oxygen consumption. A 2–4 hr lag in stoichiometric arsenite-oxidizing activity occurred when mid–log phase cells growing in the absence of arsenite were switched to arsenite-containing medium. Depending on medium composition, arsenite-oxidizing activity first appeared in growing, unadapted cells either at the end of the exponential phase or in the beginning of the stationary phase. An electron transport system appeared to be involved in conveying electrons from the arsenite to oxygen during arsenite oxidation. The strain oxidized arsenite aerobically in the presence of nitrate. Neither nitrate, tellurite, nor sulfite could replace oxygen as terminal electron acceptor in arsenite oxidation, and arsenite could not replace nitrate as electron acceptor anaerobically. Anaerobic nitrate respiration of induced cells was inhibited by the presence of arsenite. Since addition of arsenite did not stimulate growth or increase cell yield, the authors concluded that strain *A. faecalis* YE56 did not conserve energy from arsenite oxidation.

Weeger et al. (9a) isolated a strain labeled ULPAs1 from an arsenic-contaminated aquatic environment. The strain is capable of efficient oxidation of arsenite to arsenate. It is a gram-negative motile rod that showed phylogentic affinity to *Zoogloea* in preliminary tests. Although the authors concluded from their growth experiments that strain ULPAs1 oxidizes arsenite during its exponential growth phase, the data in Figure 3 of their paper show that significant oxidation did not occur until late into the exponential growth phase, similar to the previously mentioned observations of Turner and Legge (6) with *P. arsenoxy-dans-quinque* and Phillips and Taylor (9) with strain YE56 of *A. faecalis*. Strain ULPAs1 exhibited significant Cd and Pb resistance.

B. Soil

In 1953, Quastel and Scholefield (10) observed bacterial oxidation of arsenite to arsenate in soil perfusion experiments, using Audus's modification (11) of the air-lift column designed by Lees and Quastel (12). They noted that sodium arsenite in aqueous solution at 2.5×10^{-3} M concentration when perfused through soil from Cardiff (Wales) became oxidized to arsenate. In the initial perfusion, a lag was observed before arsenite oxidation occurred. No lag was observed on reperfusion of the same soil column. Addition of sulfanilamide increased the initial lag in arsenite oxidation but not the oxidation on reperfusion. The oxidation was inhibited when they added 0.1% sodium azide to the arsenite solution. The initial lag and the effect of the azide indicated to the investigators that the oxidation was biological, but they made no attempt to isolate the arsenite-oxidizing organisms from the soil. They did show that ammonia was not oxidized in these columns.

Isolation from soil of a strain labeled ANA, which resembled *Alcaligenes faecalis*, was reported by Osborne and Ehrlich (13). The strain oxidized arsenite stoichiometrically to arsenate. Washed cell suspensions consumed 0.5 mol of oxygen for every mol of arsenate produced from arsenite. The pH optimum for the reaction was 7.0. Like Turner and Legge's *Pseudomonas arsenoxidans-quin-que* (6), the cells of this strain oxidized arsenite only if they had been pregrown in the presence of arsenite, i.e., arsenite induced the process. Arsenite oxidation in whole cell suspensions was inhibited by 1 mM $HgCl_2$, 1 mM *p*-chloromercuri-benzoate, 10 mM dinitrophenol, 10 mM KCN, and 10 mM NaN_3. Arsenite oxidation in crude cell extracts was inhibited by 5 mM atabrine, 0.83 µM dicoumarol, and 0.15 mM thenoyltrifluoroacetone. Electron transport from arsenite to oxygen in crude extracts involved *c*-type cytochrome and cytochrome oxidase. The involvement of the electron transport chain suggests that the cells may gain some energy from arsenite oxidation. Fractionation of crude extract by differential centrifugation yielded a soluble fraction that required 2,6-dichlorophenol indophenol (DCIP) as electron carrier in arsenite oxidation. The particulate fraction was nearly inactive under the same conditions. The combination with the two fractions eliminated the need for DCIP in the oxidation of arsenite. Further study by Welch (MS thesis, 1977, Rensselaer Polytechnic Institute) showed that arsenite-oxidizing activity of cells in the exponential growth phase was only about one-tenth that of cells in the late exponential/early stationary phase (14). This is qualitatively similar to what had been reported by Turner and Legge (6) and Phillips and Taylor (9). Welch showed that arsenite-grown cells survived longer when stored at pH 7.0 in arsenite-supplemented tris buffer than in tris buffer without arsenite, suggesting that the cells are able to derive energy of maintenance from arsenite oxidation (14).

C. Marine Environment

Few investigations on arsenite oxidation in the marine environment have been published (15,16). Andreae (15) found lower concentrations of arsenite in surface water samples than in deep water samples from the Pacific Ocean. He found arsenite concentrations in surface waters in the range of 0.15–0.01 parts per billion (ppb), whereas below 400 m he found the average concentration to be 7.9 parts per thousand (ppt). He attributed this difference in concentration between surface and deep water to biological uptake and transport. Andreae's measurement of an observed ratio of As(V)/As(III) in deep water of 2.5×10^2 indicated to him a thermodynamic disequilibrium, because at equilibrium the expected ratio was 10^{15} based on a pE of 8.0 and standard activity coefficients of the two arsenic species in seawater. The observed ratio would be expected at a pE of 2.0. Andreae (16) suggested that the disequilibrium is a result of competing bio-oxidation of As(III) to As(V) (e.g., by bacteria) and bioreduction of As(V) to As(III) (e.g., by marine phytoplankton and bacteria). Andreae (16) found that interstitial water in marine sediments contained higher concentrations of arsenate than arsenite, in contradiction to thermodynamic predictions. Various biological and chemical processes can account for this disequilibrium. Andreae (16) also found evidence of influx of arsenate from seawater into sediments and efflux of arsenite from sediments into seawater.

Scudlark and Johnson (17) examined biological oxidation of arsenite at a concentration of 1.3 μM in seawater from Narragansett Bay. They found that bacterial oxidation proceeded at an initial rate of 1100 nmol L^{-1} day^{-1}. This oxidation was inhibited by $HgCl_2$ (200 ppm), NaN_3 (400 ppm), and NaCN (0.001 M). No oxidation occurred in water filtered through a 0.1-μm filter or in autoclaved water. The organisms were planktonic, but the authors were unable to isolate them in pure culture on agar-gelled seawater. They were able to show, however, that after three successive subculturings in artificial seawater amended with 3μM $NH_4H_2PO_4$ and 0.1% dextrose, the bacteria were still able to oxidize AsO_2^-. The bacteria oxidized initial arsenite concentrations as high as 91.8 μM. The kinetics of arsenite oxidation followed an exponential pattern because the bacteria may have multiplied during the extended incubations, lasting for hours or days at 22°C.

III. CHEMOLITHOTROPHIC (AUTOTROPHIC) ARSENITE OXIDIZERS

The bacteria in the studies cited in the previous sections were either shown or assumed to be heterotrophs, which obtain their energy from the oxidation of organic carbon, using oxygen in air as terminal electron acceptor. They use a portion of the total organic carbon they consume for assimilation. It is possible that some of those organisms may have been mixotrophic, obtaining a portion of the energy they need for growth from the oxidation of arsenite to arsenate, and the remainder from the oxidation of organic carbon. However, this has so far not been demonstrated with any of these cultures. Although Welch found that the *Alcaligenes faecalis* strain ANA (13) oxidized arsenite strongly only at the end of the exponential phase and the beginning of the stationary phase (14), he obtained evidence that suggested that the organism used some of the energy available from arsenite oxidation for cell maintenance in the resting phase (14).

Chemolithotrophic (autotrophic) strains that can derive all their energy needs from the oxidation of arsenite and their carbon from CO_2 are known. Arsenite oxidation is exergonic, yielding enough energy around neutrality to satisfy growth requirements [see Eq. (1)].

The first chemolithotrophic culture was isolated in the former Soviet Union by Ilyaletdinov and Abdrashitova (18). The organism, named *Pseudomonas arsenitoxidans* by them, was associated with a gold-arsenic deposit (sulfidic gold ore). The culture consisted of aerobic, gram-negative, motile, straight rods (0.4 × 1.2–1.5 μm) that possessed a capsule but did not form spores. It grew in a medium containing (in g L^{-1}): $NaAsO_2$, 2.0; $(NH_4)_2SO_4$, 1.0; KH_2PO_4, 0.5; KCl, 0.05; $Ca(NO_3)_2$, 0.1; and $NaHCO_3$, 0.5. The initial pH of the medium was 7.5–8.0. The optimal growth temperature was 28–35°C. The culture was slow growing in the test medium.

Much more recently, isolation of two new chemolithotrophic strains, N-25 and N-26, of aerobic, arsenite-oxidizing bacteria was reported by Santini et al. (19). Like *P. arsenitoxidans*, these organisms were isolated from arsenopyrite from a gold mine, this one located in the Northern Territory of Australia. Both strains consisted of small, gram-negative rods (0.5 × 1.0 μm). They were motile, possessing two flagella inserted subterminally. Strain N-26 was studied in more detail. Phylogenetically both strains belonged to the *Agrobacterium-Rhizobium* branch of the α-Proteobacteria. Strain N-26 grew in a medium consisting of (in g L^{-1}): $Na_2SO_4 \cdot 10H_2O$, 0.07; KH_2PO_4, 0.17; KCl, 0.05; $MgCl_2 \cdot 6H_2O$, 0.04; $CaCl_2 \cdot 2H_2O$, 0.05; KNO_3, 0.15; $(NH_4)_2SO_4$, 0.1; and $NaHCO_3$, 0.5 plus trace elements and vitamins. Sodium arsenite was added to a final concentration of 5 mM. The initial pH of the medium was around 8. The doubling time (g) of the culture in this medium was 7.6 ± 0.2 hr. Growth slowed below pH 7.3–7.4. The presence of small amounts of organic carbon did not inhibit growth. Indeed, addition of yeast extract to a final concentration of 0.004% accelerated growth, the value of (g) being lowered to 2.8 ± 0.4 hr. Strain N-26 could also grow heterotrophically with acetate, succinate, fumarate, pyruvate, malate, mannitol, sucrose, glucose, arabinose, fructose, trehalose, raffinose, maltose, xylose, galactose, lactate, salicin, glycerol, lactose, and inositol. It could not grow on citrate, sorbitol, or rhamnose. Since the authors found that arsenite was oxidized by strain N-26 in the presence of yeast extract and that growth was stimulated by it, they inferred that the strain can grow mixotrophically as well as chemolithotrophically. Arsenite oxidase (Aro) activity in strain N-26 was demonstrated in cell-free extracts. Aro activity was highest if the cells were grown in the absence of yeast extract supplement to the minimal medium. Most of the Aro activity (88%) was located in the periplasm of strain N-26, the remainder (12%) was associated with spheroplasts.

IV. ARSENITE OXIDATION BY ARCHAEA

All arsenite-oxidizing bacteria discussed in this chapter up to this point belong to the domain of Bacteria. Not yet considered in this discussion are organisms belonging to the domain of Archaea. At least one archaeon, *Sulfolobus acidocaldarius* strain BC, does exhibit arsenite-oxidizing activity (20,21). Aro in this organism is inducible, i.e., the level of activity in a culture that had been growing in the presence of arsenite was eightfold greater than in a culture that had been pregrown in the absence of arsenite. Maximal Aro activity was found to occur at the end of exponential growth if arsenite was added in the middle of the exponential phase. Since some activity also occurs during early phases of growth, Aro seems not to be completely absent in uninduced cells, but they do not oxidize arsenite measurably when in stationary phase. Arsenite does not appear to be an energy source for *S. acidocaldarius* BC; tetrathionate served that function in these experiments. Arsenite oxidation was not detected in growing cultures until the energy source tetrathionate was depleted. This is probably related to the observation that arsenate was found to be reduced by *S. acidocaldarius* growing on tetrathionate (21). However, the reduction of arsenate by this culture is not an enzymatic process but is due to formation of a transient, unidentified metabolite(s) from tetrathionate that acts as reductant. The rate of the nonenzymatic reduction of arsenate was found to be about 20 times as fast as the rate of enzymatic oxidation of arsenite.

V. ARSENITE OXIDASE

A. Eubacterial Oxidase

The inducible arsenite oxidase from the Eubacterium *Alcaligenes faecalis* (NCIB 8687) has been purified and characterized (22–24). Anderson et al. (24) isolated the enzyme from a sonicate of washed, lysozyme-treated cells that had been harvested in their late exponential growth phase. The sonicate was fractionated by gel filtration through DEAE-sepharose and active fractions concentrated by ultrafiltration. The purified enzyme was found to be monomeric with a molecular mass of 85 kDa. It consisted of two polypeptide chains in an approximate ratio of 70:30. The enzyme structure included one molybdenum, five or six iron atoms, and sulfide. Purification of the oxidase also led to recovery of azurin, a blue protein, which was rapidly reduced by arsenite in the presence of catalytic amounts of Aro, and a red protein. The red protein was a *c*-type cytochrome, which was reduced by arsenite in the presence of catalytic amounts of Aro and azurin. No reduction of the cytochrome occurred in the absence of Aro, but it did occur in the absence of azurin. Denaturation of Aro led to the release of a pterin cofactor characteristic of molybdenum hydroxylases. In intact cells of *A. faecalis*, the enzyme resides on the outer surface of the inner (plasma) membrane. The cytochrome and azurin may be part of an electron transfer pathway in the periplasm.

B. Arsenite Oxidase in Archaea

A limited study of arsenite oxidase in the acidophile *S. acidocaldarius* strain BC was published by Sehlin and Lindström (21). In this work, cell-free extracts from an uninduced culture were prepared in a French press. After differential centrifugation of the extracts, arsenite-oxidizing activity was found in the P_{50} but not in the S_{50} fraction derived from an S_{20} fraction. Proteinase treatment of the P_{50} fraction destroyed arsenite-oxidizing activity, suggesting that the observed activity was enzymatic. Highest arsenite-oxidizing activities were exhibited at pH 2.0 and 4.0, whereas 40- to 50-fold lower activity was noted at pH 7.0. Lowering the pH of an enzyme-reaction mixture from 7.0 increased its activity.

VI. OXIDATION OF As(III)-CONTAINING MINERALS

A. Orpiment

The oxidation state of arsenic in orpiment (As_2S_3) is +3. The mineral has been shown to be oxidized by *Ferrobacillus* (now *Thiobacillus*) *ferrooxidans* strain TM (25) in the absence of added ferrous iron. The experiments were performed in 125-ml Erlenmeyer flasks containing finely ground (63-μm particle size) orpiment to which 30 ml of iron-free mineral salts solution at pH 3.5 had been added. The mineral salts solution had the composition of the 9K iron medium of Silverman and Lundgren (26), but minus the iron. The orpiment contained 63.8% arsenic and 0.03% iron (sulfur content was not determined) and came from Nevada (United States). The inoculum of *T. ferrooxidans* TM had been adapted to orpiment by growth in a percolation column of orpiment (27) followed by passage in 9K iron medium. The cells used for inoculation had been washed to remove as much of the iron as possible so that the 1-ml inoculum ($\sim 10^7$ cells) contained only 110 μg Fe. The flasks were incubated at 25°C. In 42 days, about 410 μg total dissolved As per ml was found in the inoculated flask and about 120 μg ml^{-1} in the uninoculated flask in one of several experiments. This amounted to 3.4 times as much dissolved arsenic in the inoculated flask than in the uninoculated flask. Qualitative determination showed the dissolved arsenic to consist of arsenite and arsenate in the inoculated flask but only of arsenite in the uninoculated flask. A quantitative determination of the two arsenic species was not performed, however. The total dissolved iron concentration in 42 days was about 10 μg ml^{-1} in the inoculated flask but unmeasurable in the uninoculated flask. The pH dropped from 3.5 to 2.0 in 35 days in the inoculated flask and rose from 3.5 to 5.0 in the uninoculated flask. The pH change in the inoculated flask can be explained on the assumption that bacteria oxidized the orpiment to arsenite and sulfate:

$$As_2S_3 + 4H_2O + 6O_2 \rightarrow 2AsO_2^- + 3SO_4^{2-} + 8H^+$$
$$\Delta G^{0\prime} = -489.6 \text{ kcal or } -2047 \text{ kJ} \quad (3)$$

The rise in pH in the uninoculated flask suggests a difference in the course of the reaction from that in the inoculated flask.

Since *T. ferrooxidans* TM does not oxidize arsenite to arsenate (28) (Osborne, unpublished data), the arsenate probably resulted from autoxidation. Chemical oxidation by Fe(III) from the orpiment cannot be ruled out (29,30), but the iron content of the mineral specimen used was very low and the amount of dissolved iron produced by the bacteria was quite low as a result. Whatever the mechanism, the oxidation of arsenite to arsenate in the inoculated flask [see Eq. (1)] resulted in additional acidification.

In the uninoculated flask, the orpiment can be assumed to have autoxidized, forming arsenious acid ($HAsO_2$) and elemental sulfur (S^0):

$$As_2S_3 + 1.5O_2 + H_2O \rightarrow 2HAsO_2 + 3S^0 \quad (4)$$

Autoxidation of S^0 to sulfuric acid is very slow.

Although *T. ferrooxidans* cannot oxidize As(III) to As(V), *S. acidocaldarius* strain BC [now *S. metallicus* (31)] can do it (21). Whether the latter organism can oxidize orpiment seems not to have been determined as yet. It seems likely, however, that it can.

Interestingly, *T. ferrooxidans* TM did not oxidize realgar (As_2S_2) (25).

B. Arsenopyrite

Although arsenopyrite (FeAsS or $FeS_2 \cdot FeAs_2$) is a mineral in which As has an oxidation state of -1, it is considered here because the arsenic is subject to oxidation by some acidophilic members of the domains Bacteria and Archaea, which may be of significance in As pollution problems. The mineral has attracted a lot of attention in the precious metals industry because it, along with iron pyrite, may contain gold that is encapsulated in them. Such gold is inaccessible to extractants like cyanide and thiourea solutions used in commercial treatment of sulfidic gold ores. Thus, to expose the gold and, at the same time to limit the consumption of cyanide extractant when used, the arsenopyrite and pyrite need to be at least partially degraded by oxidation, which can be achieved pyrometallurgically or biohydrometallurgically. In the latter case, acidophilic iron-oxidizing prokaryotes, such as *T. ferrooxidans* (32), *Leptospirillum ferrooxidans*, or *Sulfobacillus thermosulfidooxidans* (33) in the domain Bacteria, or *S. acidocaldarius* strain BC [now *S. metallicus* (31)] in the domain Archaea may be employed (31).

Ehrlich (34) first showed that *Ferrobacillus* (now *Thiobacillus*) *ferrooxidans* could oxidize arsenopyrite. Experiments were set up like those for orpiment oxidation described above. The arsenopyrite was ground to a particle size no greater than 63 μm. It contained 24.7% Fe, 32.8% arsenic, and 0.2% copper. The sulfur content was not determined. Bacterial action was tested in 125-ml Erlenmeyer flasks containing 30 ml of iron-free 9K medium and 0.50 g of the arsenopyrite. The 1-ml inoculum consisted 6.8×10^7 washed *T. ferrooxidans* cells and contained 129.5 μg of total iron. Inoculated and uninoculated flasks were incubated at 30°C. After 21 days, about 1000 μg total As (arsenite and arsenate) per ml and 100 μg total Fe (ferrous and ferric) per ml appeared in solution in inoculated flasks in contrast to about 550 μg total As (arsenite and arsenate) per ml and 100 μg total iron per ml in uninoculated flasks. Since *T. ferrooxidans* TM does not oxidize arsenite, the arsenate probably resulted from oxidation of arsenite by ferric iron (28–30) (see also discussion below), although autoxidation of arsenite cannot be ruled out. In 21 days of incubation, the pH in inoculated and in uninoculated flasks dropped from 3.5 to 2.5, but by 40 days it had risen to 4.3 in uninoculated flasks while remaining at 2.5 in inoculated flasks. This difference in pH behavior must reflect a difference in chemical changes in the inoculated and uninoculated flasks in the later stages. Indeed, a marked increase in the rate of soluble iron production occurred in inoculated flasks at about day 30, whereas in uninoculated flasks there was a gradual decrease in soluble iron.

The arsenic and iron in solution did not reflect the full extent to which the arsenopyrite had been oxidized. Acidification of the culture medium in each flask with 1 ml of concentrated HCl at the end of the experiment increased the arsenic concentration in solution 1.6-fold and the iron concentration 4.4-fold in uninoculated flasks and 1.6- and 7.2-fold, respectively, in inoculated flasks. The increase in dissolved As and Fe on acidification suggests that a portion of the mobilized iron and arsenic was precipitated as iron arsenate and arsenite in inoculated as well as uninoculated flasks. The weight ratios of Fe/As were always higher over 21 days in uninoculated flasks than in inoculated flasks, and in both types of flasks dropped in the first few days of incubation and then increased again. Precipitation of ferric arsenate (scorodite) as well as potassium jarosite [KFe_3 $(SO_4)_2(OH)_6$] in bacterial arsenical pyrite oxidation was reported by Carlson et al.(35).

Based on experimental observations by Monroy-Fernández et al. (30), the mechanism of microbial attack of arsenopyrite can be explained as follows. Two surface reactions take place initially that are promoted by attached bacteria (80%) and planktonic bacteria (20%) in the exponential growth phase:

$$2FeAsS + 2.5O_2 + 4H^+$$

$$\rightarrow 2Fe^{2+} + 2HAsO_2 + 2S^0_{(surface)} + H_2O \text{ (eq. amended by Ehrlich)} \quad (5)$$

$$2S^0_{(surface)} + 3O_2 + 2H_2O \rightarrow 2SO_4^{2-} + 4H^+ \quad (6)$$

These reactions are followed by a reaction promoted by attached bacteria (45–75%) in the beginning of the stationary phase:

$$4Fe^{2+} + 4H^+ + O_2 \rightarrow 4Fe^{3+} + 2H_2O \quad (7)$$

and three chemical reactions, for which exponentially growing planktonic bacteria (15–55%) recycle ferrous iron:

$$FeAsS + 7Fe^{3+} + 4H_2O \rightarrow H_3AsO_4 + 8Fe^{2+} + S^0_{(surface)} + 5H^+ \quad (8)$$

$$HAsO_2 + 2Fe^{3+} + 2H_2O \rightarrow H_3AsO_4 + 2Fe^{2+} + 2H^+ \quad (9)$$

$$H_3AsO_4 + Fe^{3+} \rightarrow FeAsO_{4(surface)} + 3H^+ \quad (10)$$

Finally with attached (10%) and planktonic (90%) bacteria in stationary phase, Eq. (8) and the following chemical reaction become dominant:

$$H_3AsO_4 + Fe^{3+} \rightarrow FeAsO_{4(precipitate)} + 3H^+ \quad (11)$$

Cassity and Pesic (36) found that arsenate but not arsenite stimulated dissolved Fe^{2+} oxidation by *T. ferrooxidans* through precipitation of Fe^{3+} as ferric arsenate.

Fe^{3+} at elevated concentrations is known to cause product inhibition in *T. ferrooxidans*. In contrast to the findings from other experiments (34), mostly arsenate instead of arsenite was formed from arsenopyrite by the strain of *T. ferrooxidans* used in Cassity and Pesic's experiments. They also found that an initial presence of arsenite resulted in a lag in arsenopyrite oxidation.

Simple chemical oxidation of arsenite by ferric iron at acid pH has been questioned by Barrett et al. (37). They found experimentally that Fe^{3+} could not oxidize AsO_2^- chemically at pH 1.3 at either 70 or 45°C in the presence of a mixed culture capable of growing on Fe^{2+} and pyrite (FeS_2). However, when they added pyrite to the reaction mixture, the bacteria did promote oxidation of arsenite at 45°C. They explained the effect of the pyrite as a heterogeneous catalyst, the role of the bacteria being the regeneration of a clean catalytic surface on the pyrite and the reoxidation of Fe^{2+} generated in the oxidation of arsenite by Fe^{3+}. Mandl and Vyskovsky (38) developed a kinetic model for the catalytic role of pyrite in this form of bacterial arsenite oxidation by Fe^{3+}. They performed the experiments on which they based their model with *T. ferrooxidans* strain CCM 4253.

C. Enargite

The arsenic in enargite (Cu_3AsS_4 or $3Cu_2S \cdot As_2S_5$) is pentavalent. Its susceptibility to attack by acidophilic bacteria is discussed here because it involves oxidation that leads to mobilization of the arsenic, but in this instance as arsenate (AsV). This could be of significance in some As pollution problems. Ehrlich (34) first reported enargite oxidation by *T. ferrooxidans* TM. He ran experiments set up like those with arsenopyrite (see above). The enargite contained 38.2% copper, 7.2% arsenic, and 11.2% iron. Like the arsenopyrite, it was ground to a mesh size no larger than 63 µm. The 1-ml inoculum of washed cells for one of these experiments contained 4.8×10^7 cells with a residual iron concentration of 106.0 µg per ml. Incubation of inoculated and uninoculated flasks was at 30°C.

In this experiment, enargite oxidation in the inoculated flasks resulted in the solubilization of 6 times as much Cu and As than in the uninoculated controls in 21 days. Only small amounts of iron were solubilized with or without inoculum during this time, despite a significant presence of iron in the ore. The pH in uninoculated flasks dropped from 3.5 to 2.5–3.0, and in inoculated flasks from 3.5 to 2.0–2.5. After 40 days of incubation, the pH in uninoculated flasks had risen to 3.5–4.0, whereas in inoculated flasks it had dropped to pH 2.0. The dissolved copper was cupric, the iron was ferric, and arsenic was arsenate. No visible precipitate was noted in either inoculated or uninoculated flasks. In both inoculated and uninoculated flasks, As appeared to be more readily solubilized than Cu, based on a comparison of Cu/As weight ratios in solution with the Cu/As ratio in the original mineral. Acidification with 1 ml of concentrated HCl at the end of a 21-day experiment brought additional Cu and As in solution in the inoculated flasks such that the Cu/As weight ratio in solution was like the Cu/As ratio in the original mineral. However in uninoculated flasks, such treatment increased only the soluble arsenic concentration, and the resultant Cu/As weight ratio in solution deviated from that in the original ore even more than before acidification.

An overall equation describing the oxidation of enargite is:

$$2Cu_3AsS_4 + 19O_2 + 2H_2O$$
$$\rightarrow 6Cu^{3+} + 2HAsO_4^{2-} + 8SO_4^{2-} + 2H^+ \quad (12)$$

This equation does not explain the pH rise in uninoculated flasks noted after 40 days. It may reflect occurrence of secondary reactions not seen in the inoculated flasks.

Bacterial oxidation of enargite has become of interest in commercial bioleaching. One example is in the treatment of refractory enargite-pyrite gold concentrate with *T. ferrooxidans* (39) Another example is the leaching of enargite-containing copper ore with *Sulfolobus* BC [now *S. metallicus* (31)] (40).

VII. CONCLUSION

The foregoing shows that arsenite in aerobic environments can undergo bacterial oxidation to arsenate. Since, as shown in the chapter on arsenate reduction, some anaerobic bacteria have the ability to reduce As(V) to different lower oxidation states, bacterial arsenite oxidation must represent part of a microbial arsenic cycle. Microbial activity can also mobilize arsenic in some minerals as arsenite and/or arsenate. These microbial activities have to be considered in any assessment of environmental arsenic pollution.

REFERENCES

1. M-N Collinet, D Morin. Characterization of arsenopyrite oxidizing *Thiobacillus*. Tolerance to arsenite, arsenate, ferrous and ferric iron. Antonie van Leeuwenhoek 57:237–244, 1990.
2. SC Carapella Jr. Arsenic: Element and geochemistry. In: RW Fairbridge, ed. The Encyclopedia of Geochemistry and Environmental Sciences. Encyclopedia of Earth Sciences Series. Vol. IVA. New York: Van Nostrand Reinhold, 1972, pp 41–42.
3. HH Green. Description of a bacterium which oxidizes arsenite to arsenate, and of one which reduces arsenate to arsenite, isolated from a cattle-dipping tank. S Afr J Sci 14:465–467, 1918.
4. AW Turner. Bacterial oxidation of arsenite. Nature 164:76–77, 1949.

5. AW Turner. Bacterial oxidation of arsenite. I. Description of bacteria isolated from arsenical cattle-dipping fluid. Aust J Biol Sci 7:452–478, 1954.

6. AW Turner, JW Legge. Bacterial oxidation of arsenite. II. The activity of washed suspensions. Aust J Biol Sci 7:479–495, 1954.

7. JW Legge, AW Turner. Bacterial oxidation of arsenite. III. Cell-free arsenite dehydrogenase. Aust J Sci 7:496–503, 1954.

8. JW Legge. Bacterial oxidation of arsenite. IV. Some properties of the bacterial cytochromes. Aust J Sci 7:504–514, 1954.

9. SE Phillips, ML Taylor. Oxidation of arsenite to arsenate by *Alcaligenes faecalis.* Appl Environ Microbiol 32:392–399, 1976.

9a. W Weeger, D Lièvremont, M Perret, F Lagarde, J-C Hubert, M Leroy, M-C Lett. 1999. Oxidation of arsenite to arsenate by a bacterium isolated from an aquatic environment. BioMetals 12:141–149.

10. JH Quastel, PG Scholefield. Arsenite oxidation in soil. Soil Sci 75:279–285, 1953.

11. LJ Audus. A new soil perfusion apparatus. Nature 158:419, 1946.

12. H Lees, JH Quastel. A new technique for the study of soil sterilization. Chem Indust 26:238–239, 1944.

13. FH Osborne, HL Ehrlich. Oxidation of arsenite by a soil isolate of *Alcaligenes.* J Appl Bacteriol 41:295–305, 1976.

14. HL Ehrlich. Inorganic energy sources for chemolithotrophic and mixotrophic bacteria. Geomicrobiol J 1:65–83, 1978.

15. MO Andreae. Distribution and speciation of arsenic in natural waters and some marine algae. Deep-Sea Res 25:391–402, 1978.

16. MO Andreae. Arsenic speciation in seawater and interstitial waters: The influence of biological-chemical interactions on the chemistry of a trace element. Limnol Oceanogr 24:440–452, 1979.

17. JR Scudlark, DL Johnson. Biological oxidation of arsenite in seawater. Estuar Coast Shelf Sci 14:693–706, 1982.

18. AN Ilyaletdinov, SA Abdrashitova. Autotrophic oxidation of arsenic by a culture of *Pseudomonas arsenitoxidans.* Mikrobiologiya 50:197–204 (Engl transl pp 135–140), 1981.

19. JM Santini, LI Sly, RD Schnagl, JM Macy. A new chemolithotrophic arsenite-oxidizing bacterium isolated from a gold mine: Phylogenetic, physiological, and preliminary biochemical studies. Appl Environ Microbiol 66:92–97, 2000.

20. EB Lindström, HM Sehlin. Toxicity of arsenic compounds to the sulphur-dependent archaebacterium *Sulfolobus.* In: J Salley, RGL McCready, PL Wichlacz, eds. Biohydrometallurgy 1989, CANMET, Ottawa, Canada, 1989, pp 59–70.

21. HM Sehlin, EB Lindström. Oxidation and reduction of arsenic by *Sulfolobus acidocaldarius* strain BC. FEMS Microbiol Lett 93:878–92, 1992.

22. SJ Rinderle, J Schrier, JW Williams. Purification and properties of arsenite oxidase (abstr 3742). Fed Proc 43:2060, 1984.

23. JW Williams, SJ Rinderle, JA Schrier, LJ Alvey, K Tseng. Arsenite oxidase: A molybdenum-containing iron-sulfur protein (abstr 1050). Fed Proc 45:1660, 1986.

24. GL Anderson, J Williams, R Hille. The purification and characterization of arsenite oxidase from *Alcaligenes faecalis,* a molybdenum-containing hydroxylase. J Biol Chem 267:23674–23682, 1992.

25. HL Ehrlich. Bacterial action on orpiment. Econ Geol 58:991–994, 1963.
26. M Silverman, DG Lundgren. Studies on the chemoautotrophic iron bacterium *Ferrobacillus ferrooxidans*. I. An improved medium and a harvesting procedure for securing high cell yields. J Bacteriol 77:642–647, 1959.
27. HL Ehrlich. Observations of microbial association with some mineral sulfides. In: ML Jensen, ed. Symposium on the Biogeochemistry of Sulfur Isotopes. New Haven, CT: Yale University, 1962, pp 153–168.
28. N Wakao, H Koyatsu, Y Komai, H Shimokawara, Y Sakurai, H Shiota. Microbial oxidation of arsenite and occurrence of arsenite-oxidizing bacteria in acid mine water from a sulfur-pyrite mine. Geomicrobiol J 6:11–24, 1988.
29. JF Braddock, HV Luong, EJ Brown. Growth kinetics of *Thiobacillus ferrooxidans* isolated from mine drainage. Appl Environ Microbiol 48:48–55, 1984.
30. MG Monroy-Fernández, C Mustin, P de Donato, J Berthelin, P Barion. Bacterial behavior and evolution of surface oxidized phases during arsenopyrite oxidation by *Thiobacillus ferrooxidans*. In: T Vargas, CA Jerez, JV Wiertz, H Toledo, eds. Biohydrometallurgical Processing. Vol. 1. Santiago: University of Chile, 1995, pp 57–66.
31. PR Norris. Thermophiles and bioleaching. In: DE Rawlings, ed. Biomining: Theory, Microbes and Industrial Process. Berlin: Springer-Verlag, 1997, pp 247–252.
32. JA Brierley, RY Wan, DL Hill, TC Logan. Biooxidation-heap pretreatment technology for processing lower grade refractory gold ores. In: T Vargas, CA Jerez, JV Wiertz, Toledo H, eds. Biohydrometallurgical Processing. Vol 1. Santiago: University of Chile, 1995, pp 253–262.
33. JA Brierley, CL Brierley. Present and future applications of biohydrometallurgy. In: R Amíls, A Ballester, eds. Biohydrometallurgy and the Environment Toward the Mining of the 21st Century, Part A. Amsterdam: Elsevier, 1999, pp 81– 89.
34. HL Ehrlich. Bacterial oxidation of arsenpyrite and enargite. Econ Geol 59:1306–1312, 1964.
35. L Carlson, EB Lindström, KB Hallberg, OH Tuovinen. Solid-phase products of bacterial oxidation of arsenical pyrite. Appl Environ Microbiol 58:1046–1049, 1992.
36. WD Cassity, B Pesic. Interactions of *Thiobacillus ferrooxidans* with arsenite, arsenate and arsenopyrite. In: R Amils, A Ballester, eds, Biohydrometallurgy and the Environment Toward the Mining of the 21st Century, Part A. Amsterdam: Elsevier, 1999, pp 521–532.
37. J Barrett, DK Ewart, MN Hughes, RK Poole. Chemical and biological pathways in the bacterial oxidation of arsenopyrite. FEMS Microbiol Rev 11:57–62, 1993.
38. M Mandl, M Vyskovsky. Kinetics of arsenic (III) oxidation by iron (III) catalysed by pyrite in the presence of *Thiobacillus ferrooxidans*. Biotechnol Lett 16:1199–1204, 1994.
39. F Acevedo, C Canales, JC Gentina. Biooxidation of an enargite-pyrite gold concentrate in aerated columns. In: R Amils, A Ballester, eds. Biohydrometallurgy and the Environment Toward the Mining of the 21st Century, Part A. Amsterdam: Elsevier, 1999, pp 301–308.
40. B Escobar, E Huenupi, I Godoy, J Wiertz. Arsenic precipitation in the bioleaching of enargite by *Sulfolobus* BC. Biotechnol Lett 22:205–209, 2000.

14

Characteristics of Newly Discovered Arsenite-Oxidizing Bacteria

Joanne M. Santini, Rachel N. vanden Hoven, and Joan M. Macy†
La Trobe University, Melbourne, Victoria, Australia

I. INTRODUCTION

Arsenic is notorious as a toxic semi-metal and the trivalent form, arsenite [As(III)] is considered to be the most toxic (1) as it inactivates the sulfhydryl groups of cysteine residues in proteins (2,3). Organisms have adapted mechanisms to convert arsenite to the less toxic pentavalent form, arsenate [As(V)]. Some of these organisms can use arsenite as their sole source of energy (4) whereas others oxidize it to arsenate as part of a detoxification mechanism (5).

Bacterial oxidation of arsenite to arsenate was first described in 1918 (6). This organism, *Bacillus arsenoxydans*, was isolated from an arsenical cattle dip in South Africa by including organic matter in the form of dung extract in the medium. Unfortunately, this organism was lost before it could be tested for its ability to grow using oxygen as the terminal electron acceptor, arsenite as the electron donor, and carbon dioxide as the sole carbon source (i.e., chemolithoautotrophically). Subsequently, Turner, also working with cattle dips but in Australia, described the isolation of 15 arsenite-oxidizing bacterial strains (7,8). These organisms were also isolated by including organic matter in the medium and were therefore heterotrophic arsenite oxidizers. One isolate, presumably the most rapid

---

† Deceased.

arsenite oxidizer, was studied in further detail. This organism, *Pseudomonas arsenoxydans-quinque* is considered synonymous with *Alcaligenes faecalis* (2).

A number of other heterotrophic arsenite-oxidizing bacteria have been isolated from raw sewage (9) and soil (5), most of which were also identified as *A. faecalis*. However, none have been demonstrated to oxidize arsenite chemolithoautotrophically and oxidize arsenite to arsenate in late logarithmic or stationary phases of growth. For this reason, arsenite oxidation by these organisms is considered a detoxification mechanism rather than one that supports growth despite the fact that the reaction is exergonic [see Eq. 1].

$$2H_3AsO_3 + O_2 \rightarrow HAsO_4^{2-} + H_2AsO_4^- + 3H^+$$
$$(\Delta G^{0\prime} = -256 \text{ kJ/Rx}) \tag{1}$$

The mechanism by which *A. faecalis* oxidizes arsenite has been studied in considerable detail. The relevant enzyme, arsenite oxidase, is located on the outer surface of the cytoplasmic membrane and has been purified and characterized (10). It is a monomeric (85 kDa) oxomolybdoenzyme containing three redox-active centers, including a molybdopterin cofactor and two different iron-sulfur clusters (10,11). The electron transport chain consists of the arsenite oxidase, azurin, a specific cytochrome *c*, and presumably a cytochrome *c* oxidase (10,11).

In contrast to the arsenite oxidizers described above, only two bacteria have been described as being able to grow using the energy gained from arsenite oxidation. The first of these, *Pseudomonas arsenitoxidans*, was isolated from a gold-arsenic deposit and has since been lost (12). It was found to grow chemolithoautotrophically with oxygen as the terminal electron acceptor, arsenite as the electron donor, and carbon dioxide as the sole carbon source with a doubling time in the order of 48 hr. The second chemolithoautotrophic arsenite oxidizer, designated NT-26, was isolated from the Granites gold mine in the Northern Territory, Australia (4). This organism is the fastest arsenite oxidizer reported to date with a doubling time of 7.6 hr when grown chemolithoautotrophically.

This chapter concentrates on arsenite oxidation by NT-26 as well as a number of other recently isolated arsenite-oxidizing bacteria. It will be apparent that all but one of these organisms is phylogenetically distinct from the arsenite oxidizers previously identified. The isolation, phylogeny, physiology, and biochemistry of arsenite oxidation are described.

II. ISOLATION

For the isolation of arsenite-oxidizing bacteria, samples were taken from two different gold mine sites in Australia (13). Moist samples of arsenopyrite (FeAsS) from the Granites gold mine in the Northern Territory, Australia (4), and arsenic-contaminated water from the Central Deborah gold mine in Bendigo, Victoria, Australia, were used for the enrichments (13). Aerobic enrichments were made in a minimal salts medium containing arsenite as the electron donor and carbon dioxide–bicarbonate as the sole carbon source. After a few days, arsenite oxidation was indicated by a decrease in the pH of the medium [see Eq. (1)]. After several subcultures, a number of arsenite-oxidizing bacteria were isolated from agar.

Nine bacteria were isolated from the Northern Territory samples and were designated NT-2, NT-3, NT-4, NT-5, NT-6, NT-10, NT-14, NT-25, and NT-26. Two organisms were isolated from the Bendigo samples and were designated BEN-4 and BEN-5. All of these bacteria oxidize arsenite to arsenate, however they vary in this ability (see Sec. IV). One bacterium, NT-26, a small (1–2 μm), motile by means of two subterminal flagella, gram-negative rod (Fig. 1) was found to be the fastest chemolithoautotrophic arsenite oxidizer and was therefore studied in greater detail (4).

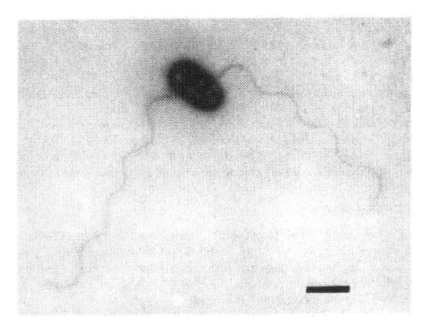

Figure 1 Electron micrograph of NT-26. Bar = 1 μm. (From Ref. 4.)

III. PHYLOGENETIC CHARACTERIZATION OF ARSENITE-OXIDIZING BACTERIA

The phylogenetic analysis of 16S rRNA gene sequences of the above-mentioned arsenite-oxidizing bacteria showed that they were dispersed in either the α-Proteobacteria (Fig. 2) or the β-Proteobacteria (Fig. 3) (13).

A. α-Proteobacteria

Strain BEN-5 was found to be most closely related to *Agrobacterium vitis* (97.7% sequence similarity). The relationship was confirmed by phenotypic tests that showed that BEN-5 exhibited 0.407 similarity with *A. vitis* but, at this level (<0.5), could not be definitely identified as belonging to this species. It is therefore possible that BEN-5 represents a new species of *Agrobacterium*. Strains NT-2, NT-3, and NT-4 exhibited 99.4% sequence similarity and belonged to a well-supported novel branch within the genus *Sinorrhizobium*. These three strains were most closely related to *S. fredii* and *S. xinjiangensis* (99.3% sequence similarity). Due to the high sequence similarities with species in the genus *Sinorrhizobium*, DNA-DNA hybridization will be necessary to determine their separate species identity. Strains NT-25 and NT-26 exhibited 99.8% sequence similarity and their nearest known phylogenetic relatives are *Rhizobium huautlense* (96.2%), *R. galegae* (96.6%), *R. gallicum* (97.4%), and a misidentified "*Acinetobacter*" sp. strain IF-19 (97.4%) isolated from a deep subsurface mine gallery (4). It is likely that strains NT-25 and NT-26 belong to a new species of *Rhizobium*. However, as the sequence similarity to known species of *Rhizobium* is >97%, DNA-DNA hybridization is required to confirm their separate species status.

   The ability of these organisms to oxidize arsenite to arsenate constitutes a novel feature that has not been previously described in the genera *Agrobacterium*, *Rhizobium*, and *Sinorrhizobium*. These strains are therefore the first examples of members of these genera that are able to use arsenite oxidation for growth.

B. β-Proteobacteria

The arsenite oxidizing bacteria belonging to this group have three different phylogenetic affiliations. Strains NT-5, NT-6, and NT-14 belong to a strongly supported branch closely related to species of the genus *Hydrogenophaga*. The three strains have 99.7% sequence similarity and share 97.1–98% similarity with existing members of the genus *Hydrogenophaga*.

   Strain NT-10 belongs to a strongly supported lineage that includes the genera *Alcaligenes*, *Bordetella*, and *Achromobacter* but, due to the low bootstrap support, it is impossible to be more specific about its generic affiliation. The separation of *Bordetella avium* from the main *Bordetella* cluster affects the ability

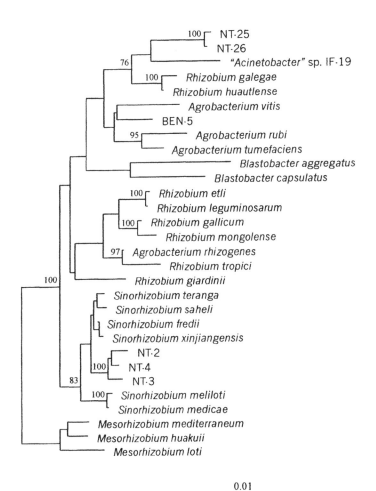

0.01

Figure 2 Phylogenetic neighbor-joining tree showing the phylogenetic relationship of arsenite oxidizing isolates BEN-5, NT-2, NT-3, NT-4, NT-25, and NT-26 with species belonging to the α-Proteobacteria. The sequence of *Mesorhizobium loti* was used as the outgroup. Significant bootstrap values from 100 analyses are shown at the branch points of the trees. Bar = 0.01% sequence difference.

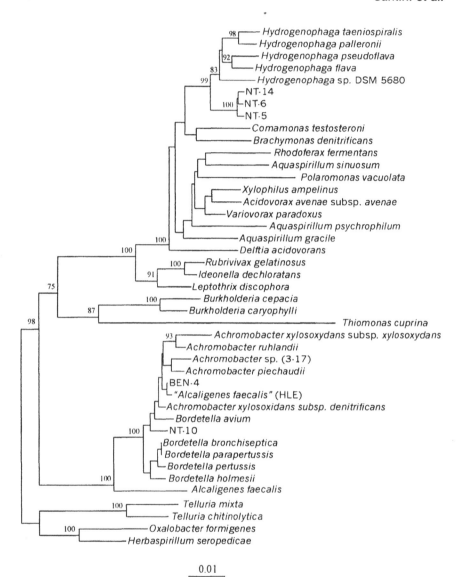

Figure 3 Phylogenetic neighbor-joining tree showing the phylogenetic relationship of arsenite-oxidizing isolates BEN-4, NT-5, NT-6, NT-10, and NT-14 with species belonging to the β-Proteobacteria. The sequence of *Telluria mixta* was used as the outgroup. Significant bootstrap values from 100 analyses are shown at the branch points of the trees. Bar = 0.01% sequence difference.

to assign this organism to either *Achromobacter* or *Bordetella*. Arsenite oxidation has not been previously reported for members of the genus *Bordetella*, although it has been described in strains currently assigned to the genus *Achromobacter*. Strain BEN-4 belongs to the *Achromobacter* lineage. This lineage contains species that have had an uncertain taxonomy and have in the past been assigned to the genera *Alcaligenes* and *Achromobacter*. Strain BEN-4 is phylogenetically most closely related to Ehrlich's arsenite-oxidizing strain previously misidentified as *"Alcaligenes faecalis"* (5) and is designated here *"A. faecalis"* HLE to distinguish it from *A. faecalis* (see Fig. 3), which does not oxidize arsenite. The close relationship indicated by a sequence similarity of 99.8% between the two strains is supported by their common physiological and phenotypic characteristics (13). Consequently, it appears that both BEN-4 and *"A. faecalis"* HLE belong to the genus *Achromobacter*. However, because of the high 16S rRNA sequence similarities between the species of this genus, DNA-DNA hybridization will be required to determine their species identity.

IV. PHYSIOLOGICAL CHARACTERISTICS

Six of the 11 arsenite-oxidizing bacteria (all members of the α-Proteobacteria) were able to grow chemolithoautotrophically (i.e., aerobically in a minimal salts medium with arsenite as the electron donor and carbon dioxide–bicarbonate as the sole carbon source) (Table 1) (13). The time required for these bacteria to oxidize 5 mM arsenite varied from 3 days (for NT-25 and NT-26) to more than 5 days (for BEN-5). The arsenite was oxidized to an equivalent amount of arsenate and the pH of the medium decreased from 8.0 to 6.5. For all of the abovementioned bacteria, no growth occurs in the absence of arsenite (i.e., only in an aerobic minimal salts medium containing carbon dioxide–bicarbonate).

Growth of the arsenite-oxidizing bacteria mentioned above is stimulated by the presence of organic matter (in the form of yeast extract). Not only is the rate of growth stimulated but so is arsenite oxidation. On the other hand, organic matter is essential for arsenite oxidation by NT-5, NT-6, NT-10, NT-14, BEN-4, and *"A. faecalis"* HLE as they are unable to oxidize arsenite chemolithoautotrophically. Interestingly, the chemolithoautotrophs are members of the α-Proteobacteria and the heterotrophic arsenite oxidizers are members of the β-Proteobacteria. The significance of this point is not known. However, a similar relationship also occurs with respect to the arsenite-oxidizing enzymes (see Sec. V).

Two of these organisms have been studied in detail with respect to their arsenite-oxidizing abilities, representing one from each phylogenetic group. They are the chemolithoautotroph NT-26 (α-Proteobacteria) (4) and the heterotrophic

Table 1 Comparisons of Chemolithoautotrophic Arsenite-Oxidizing Ability of Newly Isolated Arsenite-Oxidizing Bacteria with "*A. faecalis*" HLE

Bacterium[a]	Chemolithoautotrophic growth	Time required (days) for chemolithoautotrophic arsenite oxidation[b]
NT-2	+	4–5
NT-3	+	4–5
NT-4	+	4–5
NT-5	−	NA[c]
NT-6	−	NA[c]
NT-14	−	NA[c]
NT-25	+	3
NT-26	+	3
NT-10	−	NA[c]
BEN-4	−	NA[c]
BEN-5	+	>5
"*A. faecalis*" HLE	−	NA[c]

[a] "BEN" = isolated from the Central Deborah Mine in Bendigo; "NT" = isolated from the Granites Gold Mine in the Northern Territory.
[b] Aerobic growth in a minimal salts medium containing 5 mM arsenite as the electron donor and carbon dioxide–bicarbonate as the sole carbon source.
[c] NA = not applicable as these organisms require organic matter for growth and arsenite oxidation.

arsenite oxidizer NT-14 (β-Proteobacteria) (R. N. vanden Hoven, J. M. Santini, and J. M. Macy, personal communication).

When grown chemolithoautotrophically with arsenite (5 mM) as the electron donor, NT-26 has a generation time of 7.6 ± 0.2 hr. Prior to this study, the only chemolithoautotrophic arsenite oxidizer known was *P. arsenitoxidans* (12). As this organism has been lost, direct comparisons with NT-26 and the other chemolithoautotrophic arsenite oxidizers cannot be made.

The presence of yeast extract (0.04% and 0.004%) stimulated both the rate of growth and arsenite oxidation rate of NT-26. The doubling time decreased from 7.6 ± 0.2 to 2.7 ± 0.3 (0.04%) and 3.2 ± 0.2 hr (0.004%). While NT-14 does not grow chemolithoautotrophically with arsenite as the electron donor, it does oxidize arsenite to arsenate and gain some energy from the oxidation in the presence of organic matter (in the form of yeast extract). When grown in a minimal medium with arsenite and yeast extract (0.04 and 0.004%) the generation time is much faster than that of NT-26, with doubling times of 1.9 ± 0 (0.04%)

and 2.07 ± 0.13 hr (0.004%) (R. N. vanden Hoven, J. M. Santini, and J. M. Macy, personal communication). For both NT-26 and NT-14, arsenite is oxidized to equimolar amounts of arsenate throughout growth, which differs from the heterotrophic arsenite oxidizer "*A. faecalis*" HLE (see Sec. I). In addition, growth experiments with NT-26 and NT-14 with varying amounts of yeast extract present in the medium display greater growth when arsenite is present than when it is absent. This clearly indicates that energy is gained by NT-26 and NT-14 from arsenite oxidation as well as from the oxidation of yeast extract.

In addition to chemolithoautotrophic growth on arsenite, NT-26 can grow on a number of other substrates consistent with those used by *Rhizobium* (4). More interestingly, NT-26 can grow aerobically using sulfur-containing compounds as electron donors, such as thiosulfate and hydrogen sulfide (J. M. Santini and J. M. Macy, personal communication). Growth on these compounds has not been previously reported for the Rhizobia. It is not surprising that NT-26 can use these compounds considering the environment, an arsenopyrite (FeAsS)-containing gold mine, in which the organism was isolated. It would be interesting to test the other arsenite-oxidizing bacteria for this ability as this may shed some light on their roles in their native environments.

V. BIOCHEMISTRY OF ARSENITE OXIDATION

Preliminary studies of the enzymes that catalyze the oxidation of arsenite to arsenate in NT-4, NT-14, NT-26, and BEN-5 have been conducted. To date, only the NT-26 arsenite oxidase (Aro) has been purified and characterized (J. M. Santini and J. M. Macy, personal communication). The optimum buffers, pH, and cellular locations of the respective arsenite-oxidizing enzymes are listed in Table 2. Interestingly, the enzymes of the bacteria in the α-Proteobacteria are all peri-

Table 2 Comparisons of Optimum Buffer, pH, and Cellular Location of Arsenite Oxidases from Different Arsenite-Oxidizing Bacteria

Bacterium	Subgroup of the Proteobacteria	Optimum buffer[a] and pH	Cellular location
NT-4	α	NaAc, 5.0	Periplasmic
NT-26	α	MES, 5.5	Periplasmic
BEN-5	α	NaAc, 5.0	Periplasmic
NT-14	β	MES, 5.5	Membrane
"*A. faecalis*" HLE	β	MES, 6.0	Membrane

[a] The concentrations of the buffers, NaAc (sodium acetate) and MES (2-[N-morpholino]ethanesulfonic acid), were 50 mM.

plasmic whereas those of NT-14 and "*A. faecalis*" HLE, members of the β-Proteobacteria, are membrane bound. The significance of this as yet is not known as the NT-14 enzyme has not been purified and therefore cannot be compared with that of "*A. faecalis*" HLE. Preliminary data, however, on the NT-4 and BEN-5 arsenite-oxidizing enzymes suggest that they are similar to that of NT-26.

The NT-26 periplasmic Aro consists of two different subunits, AroA (98 kDa) and AroB (14 kDa). The native molecular weight was estimated by gel filtration chromatography to be 214 kDa and so may have an $\alpha_2\beta_2$ configuration (J. M. Santini and J. M. Macy, personal communication).

Aro contains molybdenum and preliminary data suggest the presence of iron. Further analyses, however, are required to confirm the presence of iron and determine whether acid-labile sulfur is also present in the enzyme. The Aro therefore appears to be a member of mononuclear molybdenum enzymes (15). In addition, the N-terminal sequence of the large subunit (AroA) has sequence similarities with molybdenum-containing subunits of a number of enzymes (Fig. 4), including FdhA of the formate dehydrogenase from *Wollinella succinogenes* and FdnG of the nitrate-inducible formate dehydrogenase from *Escherichia coli*. These proteins are the catalytic subunits of the respective enzymes and all contain a cysteine cluster at the N-terminus and molybdopterin guanosine dinucleotide as the organic component of the molybdenum cofactor (14). Like the arsenate reductase of *C. arsenatis* (see Chapter 12, Sec. II.D), the NT-26 Aro may also belong to the third group of molybdenum-containing enzymes, the DMSO reductase family.

The N-terminal sequence of the small subunit (AroB) showed no significant sequence similarities to any other proteins. The role of this subunit in arsenite oxidation is therefore not clear. The arsenite oxidase genes are currently being cloned and sequenced, and this information may shed some light on the role of this subunit in arsenite oxidation.

Synthesis of both the NT-26 and NT-14 arsenite oxidases appear to be regulated (Table 3). The regulation varies for each enzyme, although no enzyme activity is detected when either organism is grown in the absence of arsenite. It

```
NT-26 AroA   1- AFKRHIDRLPIIPADAKKHNVTCHFCIVGCGYHAYTWP
W.s.  FdhA  38- LRPATKQELIEKYPVSKKVKTICTYCSVGCGYYGLPWP
E.c.  FdnG  27- PKQALAQARNYKLLRAKEIRNTCTYCSVGCGLLMYSLG
```

Figure 4 Sequence alignment of the N-termini of AroA and molybdenum-containing proteins. The sequences belong to the arsenite oxidase of NT-26 (NT-26 AroA), the formate dehydrogenase of *Wolinella succinogenes* (W.s. FdhA), and the nitrate-inducible formate dehydrogenase of *Escherichia coli* (E.c. FdnG). Boxed amino acids show identity to NT-26 AroA.

Table 3 Comparisons of Specific Activities of the Arsenite Oxidases of NT-26 and NT-14 When Grown Under Different Conditions

Growth condition[a]	NT-26 U/mg	NT-14 U/mg
As(III)	0.24	NA[c]
As(III) + YE (0.004%)	0.14	0.017
As(III) + YE (0.4%)	0.06	0.13
YE (0.04%)[b]	<0.0001	<0.0001

[a] Growth medium was a minimal salts medium containing carbon dioxide–bicarbonate as carbon source.
[b] YE = yeast extract.
[c] NA = not applicable as NT-14 cannot grow chemolithoautotrophically.

therefore appears that enzyme synthesis is induced by the presence of arsenite. With respect to NT-26, the highest Aro activity (100%) detected is when the organism was grown in a minimal salts medium with arsenite only (i.e., no yeast extract). When grown with arsenite in the presence of yeast extract (0.004%) 58% of chemolithoautotrophic activity was detected, whereas when grown with arsenite and a 10-fold higher level of yeast extract (0.04%) only 25% activity was detected. These results suggest that (1) when grown in the presence of yeast extract and arsenite, energy is gained from the oxidation of both and (2) synthesis of the Aro is up-regulated. Conversely, the highest NT-14 enzyme activity (100%) was detected when the organism was grown with arsenite and the higher level of yeast extract (0.04%). The enzyme activity decreased to 13% of maximal activity when the organism was grown with arsenite and the lower level of yeast extract (e.g., 0.004%). At this point, an explanation for these results is not evident.

The only other arsenite-oxidizing enzyme purified and characterized is that of "*A. faecalis*"HLE (see Sec. I). In crude cell extracts, the enzyme activity was approximately threefold and sixfold lower than that of NT-26 and NT-14, respectively, grown with arsenite and the higher level of organic matter. Like the arsenite-oxidizing enzymes of NT-26 and NT-14, synthesis of the "*A. faecalis*" HLE enzyme requires induction with arsenite (i.e., the organism must be grown in the presence of arsenite) (10).

VI. CONCLUSIONS

The discovery of six new chemolithoautotrophic arsenite-oxidizing bacteria isolated from gold mines in different regions of Australia demonstrates that energy for growth can be conserved during the oxidation of arsenite to arsenate [see Eq. (1)]. Interestingly, all of these organisms fall within the α-subgroup of the Proteobacteria. Arsenite oxidation has not been previously described for members in this group. The time required for chemolithoautotrophic arsenite oxidation varies. The most rapid arsenite oxidizer presently known, NT-26, has been studied in detail (4).

Five heterotrophic arsenite-oxidizing bacteria were also isolated from gold mine environments in different regions of Australia and fall within the β-subgroup of the Proteobacteria. BEN-4, isolated from Bendigo, was found to be phylogenetically and phenotypically identical to the previously characterized heterotrophic arsenite oxidizer, "A. faecalis" HLE. All of these organisms require small amounts of organic matter for arsenite oxidation and oxidize arsenite to arsenate throughout growth. This contrasts with "A. faecalis" HLE, which oxidizes arsenite in late exponential or stationary phases of growth.

The mechanisms of arsenite oxidation by the organisms in the α-Proteobacteria appear to be different to those in the β-Proteobacteria. The former use soluble periplasmic enzymes whereas the latter (NT-14 and "A. faecalis" HLE) use membrane-bound enzymes. This is also consistent with the fact that those in the α-subgroup are chemolithoautotrophic arsenite oxidizers whereas those in the β-subgroup are heterotrophic. It is possible that the mechanism by which NT-14 oxidizes arsenite is similar to that of "A. faecalis" HLE, which is part of a detoxification mechanism rather than one that can support growth.

Arsenite oxidation by NT-26 is catalyzed by a periplasmic enzyme that consists of two different subunits, with a native molecular weight of 214 kDa, suggesting an $\alpha_2\beta_2$ configuration. The enzyme contains molybdenum and possibly iron as cofactors and the N-terminal sequence of the large subunit is similar to the catalytic subunits of molybdenum-containing enzymes. This is to be compared with the arsenite oxidase of "A. faecalis" HLE, which is a monomeric (85 kDa) oxomolybdoenzyme (see Sec. I). The finding that NT-26 has evolved divergently from "A. faecalis" HLE also supports the finding that the arsenite-oxidizing enzymes are different.

Interestingly, the organisms isolated from Bendigo were different from those isolated from the Northern Territory. The reason for this is unknown, however, the Northern Territory gold mine is located in the Central Australian desert where little organic matter is present. On the other hand, the Bendigo gold mine is located directly under the city of Bendigo, and surface water, which ultimately passes into the gold mine tunnels, contains some organic matter. This difference may explain why strains similar to "A. faecalis" HLE were not isolated from the Northern Territory. It may also explain why more chemolithoautotrophic arsenite oxidizers were found in the Northern Territory compared with only one found in Bendigo.

The chemolithoautotrophic arsenite-oxidizing bacteria described in this chapter may present a novel approach for the oxidation of arsenite to arsenate for the purpose of arsenic bioremediation. At present, chemical methods for the removal of arsenic, for example, by adsorption to iron oxides or clay, favor the adsorption of arsenate, and removal of the most toxic arsenic form, arsenite, is therefore considered more problematic. Other means by which arsenite can be oxidized to arsenate have proven to be time consuming, expensive, and cumbersome, and frequently depend upon acidic conditions. The use of arsenite-oxidizing bacteria may prove to be a much simpler and less expensive approach allowing for the conversion of arsenite to arsenate, which could then be removed by either adsorption to iron oxides (or clays) or by using *C. arsenatis* (arsenate respirer) and *Desulfomicrobium* sp. str. Ben-RB (arsenate/sulfate respirer) (see Chap. 12).

ACKNOWLEDGMENTS

This work was supported by an Australian Research Council Grant (A09925054) and two Central Large La Trobe University grants to JMM. RNV is supported by an Australian Postgraduate Award. We would like to thank S. C. Dawbarn and A. G. Wedd for proofreading of the chapter.

REFERENCES

1. HK Tsai, CM Hsu, BP Rosen. Efflux mechanisms of resistance to cadmium, arsenic and antimony in prokaryotes and eukaryotes. Zool Stud 36:1–16, 1997.
2. HL Ehrlich. Geomicrobiology. 3rd ed. New York: Marcel Dekker, 1996.
3. S Tamaki, WT Frankenberger. Environmental biochemistry of arsenic. Rev Environ Contam Toxicol 124:79–110, 1992.
4. JM Santini, LI Sly, RD Schnagl, JM Macy. A new chemolithoautotrophic arsenite-oxidizing bacterium isolated from a gold mine: Phylogenetic, physiological, and preliminary biochemical studies. Appl Environ Microbiol 66:92–97, 2000.
5. FH Osborne, HL Ehrlich. Oxidation of arsenite by a soil isolate of *Alcaligenes*. J Appl Bacteriol 41:295–305, 1976.
6. HH Green. Description of a bacterium which oxidizes arsenite to arsenate and of one which reduces arsenate to arsenite, isolated from a cattle-dipping tank. S Afr J Sci 14:465–467, 1918.
7. AW Turner. Bacterial oxidation of arsenite. Nature 164:76–77, 1949.
8. AW Turner. Bacterial oxidation of arsenite. I. Description of bacteria isolated from arsenical cattle-dipping fluids. Aust J Biol Sci 7:452–478, 1954.
9. SE Phillips, ML Taylor. Oxidation of arsenite to arsenate by *Alcaligenes faecalis*. Appl Environ Microbiol 32:392–399, 1976.
10. GL Anderson, J Williams, R Hille. The purification and characterization of arsenite

oxidase from *Alcaligenes faecalis*, a molybdenum-containing hydroxylase. J Biol Chem 267:23674–23682, 1992.

11. L McNellis, GL Anderson. Redox-state dependent chemical inactivation of arsenite oxidase. J Inorg Biochem 69:253–257, 1998.

12. AN Ilyaletdinov, SA Abdrashitova. Autotrophic oxidation of arsenic by a culture of *Pseudomonas arsenitoxidans*. Mikrobiologiya 50:197–204, 1981.

13. JM Santini, LI Sly, A Wen, P Durand, D Comrie, JM Macy. New arsenite-oxidizing bacteria isolated from Australian gold-mining environments: Phylogenetic relationship. Geomicrobiology J, in press.

14. C Kisker, H Schindelin, DC Rees. Molybdenum-cofactor-containing enzymes: Structure and mechanism. Annu Rev Biochem 66:233–267, 1997.

15. C Kisker, H Schindelin, D Baas, J Rétey, RU Meckenstock, PMH Kroneck. A structural comparison of molybdenum cofactor-containing enzymes. FEMS Microbiol Rev 22:503–521, 1999.

15

Oxidation of Arsenite by
Alcaligenes faecalis

Gretchen L. Anderson
Indiana University South Bend, South Bend, Indiana

Paul J. Ellis and Peter Kuhn
Stanford Synchrotron Radiation Laboratory, Menlo Park, California

Russ Hille
The Ohio State University, Columbus, Ohio

I. INTRODUCTION

Microbial biotransformation of arsenical compounds results in a biological cycle for arsenic, as originally postulated by Wood (1). The most common valence states of volatile or water-soluble arsenic involved in biotransformation are As^{-3} (e.g., trimethylarsine and dimethylarsine), As^{+1} (e.g., dimethylarsinic acid), As^{+3} (e.g., arsenite and dimethylarsenic acid), and As^{+5} (e.g., arsenate) (Fig. 1). Transformations among these species occur not only in bacteria, but also in plants (2) and animals (3), including humans (4). In addition, biotransformations are not strictly limited to water-soluble species, and relatively insoluble arsenical compounds may be acted upon. Microbial oxidation of orpiment (As_2S_3) (5,6), arsenopyrite (FeAsS) (7), and enargite (Cu_3AsS_4) (7) have been reported, leading to conversion to arsenite.

Most forms of soluble arsenic are toxic to living organisms. Arsenate, a phosphate analog, uncouples respiratory chain phosphorylation due to the facile

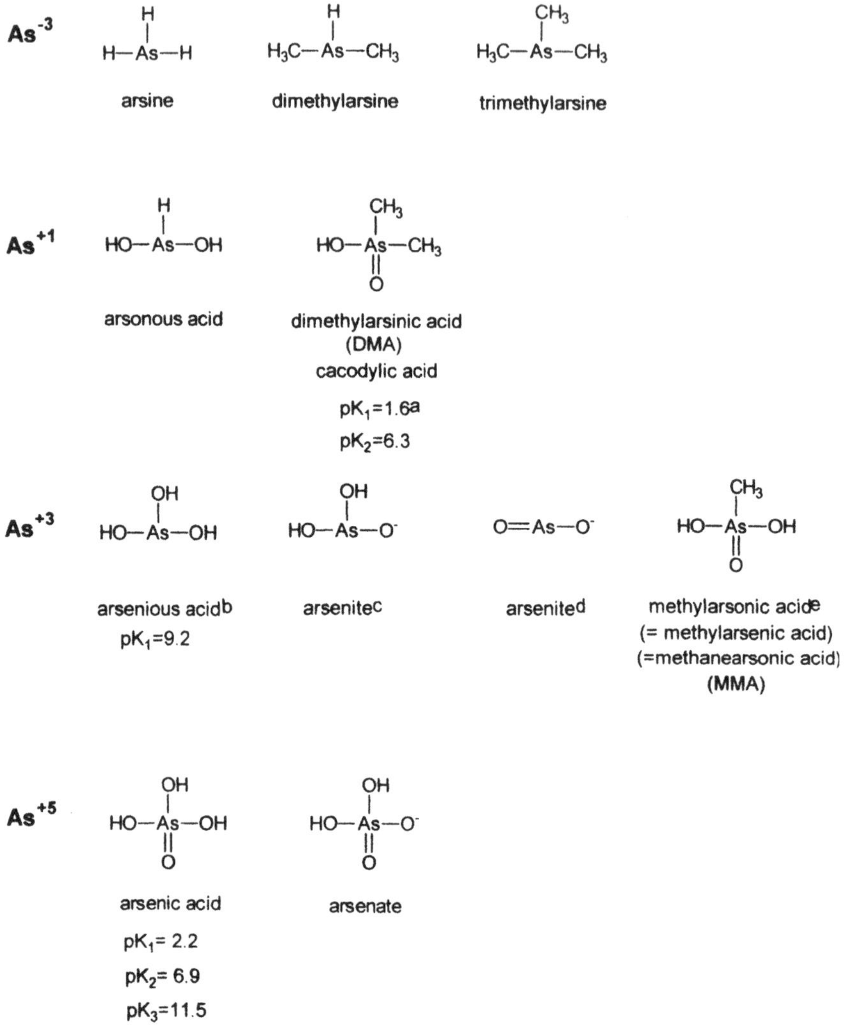

Figure 1 Oxidation states of volatile and soluble arsenic transformed by biological systems. [a]pK_1 represents the equilibrium between the protonated cation and neutral forms of the hydroxyl group of dimethylarsinic acid. pK_2 represents the equilibrium between the neutral form and anionic form (2). [b]Occasionally referred to as arsenous acid (2). HO—As=O has also been referred to as arsenious acid (52). [c]The term arsenite often refers to any of the As(III) oxides. The species will be a function of the solute and the pH. Given that the pKa of arsenious acid is 9.2 (52), the protonated form is most likely the predominant form in aqueous environmental and physiological samples below pH 9. [d]Structure commonly used when discussing arsenite in biological transformations, but is most likely not the predominant form in aqueous solutions. [e]Monosodium methane arsonic acid (MSMA) is the commercially available sodium salt.

nonenzymatic hydrolysis of adenosine diphosphate (ADP)-arsenate and arsenylated catabolites (3,8,9). Arsenite is known to bind to sulfhydryl groups of proteins, and perhaps more importantly can disrupt intracellular redox homeostasis by binding to the dithiol compound glutathione. Arsenite is reported to be 25–60 times more toxic than arsenate and several hundred times more toxic than methylated arsenicals (10). For this reason, microbial oxidation to arsenate represents a detoxification strategy, although optimally some mechanism must also be present (e.g., further chemical modification or active transport out of the cell) to ameliorate the deleterious effects of the arsenate thus formed.

II. ARSENITE OXIDATION AS METABOLIC ENERGY SOURCE

Because the oxidation of arsenite in the presence of oxygen is exergonic, bacteria may be able in principle to derive metabolic energy while detoxifying arsenite. The standard reduction potential, $E_0{}'$, for the two-electron arsenite/arsenate couple of +60 mV (11) suggests that reducing equivalents obtained from the oxidation of arsenite could be transferred to cytochrome c ($E_0{}' = +254$ mV), then to a terminal cytochrome oxidase ($E_0{}' \sim +220$ mV) and eventually to oxygen ($E_0{}' = +816$ mV) (12). Since terminal oxidases translocate protons across the cell membrane to generate a proton gradient (the primary energy storage mechanism of all living organisms), arsenite oxidation may thus be used for the energy requirements of the cell, for example, in the synthesis of adenosine triphosphate. Involvement of such an electron transport chain in arsenite oxidation has been observed in *Pseudomonas arsenoxydans-quinque* (13) and in different isolates of *Alcaligenes faecalis* (14,15). In *P. arsenoxydans-quinque*, mammalian cytochrome c oxidase can be substituted for the terminal oxidase of the organism in the electron transfer chain (13). However, mammalian cytochrome c cannot substitute for its bacterial homologue, presumably because the latter is membrane-bound. The electron transport system of *A. faecalis* also involves both cytochrome c and cytochrome c oxidase, as determined by spectroscopic and inhibitor studies of respiration inhibitors (14). In at least one strain of *A. faecalis*, both azurin and a cytochrome c can be reduced by arsenite oxidoreductase directly (15), with the reduced azurin transferring electrons to the cytochrome c in the absence of arsenite oxidoreductase. This implies that the electron transfer chain is not an obligatory linear pathway. Although several fractions containing cytochrome c were obtained from crude cell extracts of *A. faecalis*, only one of these was capable of accepting electrons from arsenite oxidoreductase (15). The recently obtained x-ray crystallographic structure (16) provides a structural rationale for this observation (see below).

Despite the potential for obtaining energy from arsenite oxidation, *A. faecalis* does not grow chemolithoautotrophically using CO_2 or HCO_3^- as a carbon source and arsenite as energy source (M. Love and G. L. Anderson, unpublished data). Nevertheless, *A. faecalis* cultures survive better in the presence of arsenite than in its absence, suggesting that this bacterium can gain some energy from arsenite oxidation (17). In the stationary growth phase, the rate of arsenite oxidation is approximately 10 times as fast as in the mid-exponential phase (17). Arsenite utilization by *A. faecalis* is not observed until late exponential or early stationary phase (17,18). This has been attributed to the need for a secondary metabolite produced late in the growth phase (18) or to enzyme repression (17). Because of this apparent lag in arsenate accumulation in the media, arsenite oxidation has been suggested to represent a detoxification mechanism rather than a means of energy utilization (17).

The growth of *Sulfolobus acidocaldarius* strain BC in the presence of arsenite also shows a delay in arsenate accumulation until the mid-exponential growth phase (19). However, in this case a low-molecular-weight metabolite, generated when tetrathionate is used as an energy source, apparently reduces arsenate to arsenite approximately 20 times faster than arsenite is oxidized. It has been concluded that *S. acidocaldarius* oxidizes arsenite enzymatically, but the nonenzymatic re-reduction ensures that the predominant form of arsenic in the culture medium is arsenite.

A bacterial isolate designated NT-26 (which may represent a new species in the *Agrobacterium-Rhizobium* sub-branch of α-Proteobacteria), clearly uses arsenite oxidation as an energy source to fix CO_2 (20). Although the growth of this bacterium is not inhibited by the presence of carbon sources in yeast extract, the rate of growth is greater in the presence of arsenite than in its absence. The periplasmic location of arsenite oxidoreductase in this bacterium suggests the involvement of other electron transport proteins, but these have not yet been identified.

Chemolithoautotropic growth on arsenite has also been demonstrated in *Pseudomonas arsenitoxidans* (21), but the growth of this bacterium is considerably slower than of NT-26 (doubling time of 2 days vs. 7.6 hr). In contrast to NT-26, arsenite oxidation does not occur in the presence of organic carbon compounds in *P. arsenitoxidans*, suggesting that this latter organism is an obligate autotroph.

Both NT-26 and *P. arsenitoxidans* are aerobes, and arsenite serves as the source of reducing equivalents to be transferred via an electron transport chain to oxygen to form water. In anaerobic bacteria, arsenate could serve as the terminal electron acceptor in the presence of an electron source of lower reduction potential. This appears to be the case in a strain of anaerobic bacterium, MIT-13, which reduces arsenate when lactate is present as an electron source (22).

III. TOXICITY OF ARSENATE

As indicated above, while less toxic than arsenite, arsenate still compromises cell viability: it is utilized instead of phosphate by adenosine triphosphatase (AT-Pase), and the arsenato-ADP thus obtained spontaneously hydrolyzes, setting up a futile cycle that dissipates the cell's energy stores. For bacteria that oxidize arsenite to arsenate, the generation of arsenate is therefore a potential problem. The appearance of arsenate in the growth medium suggests that, with the possible exception of metabolic reduction of arsenate, the arsenate product is not further modified. Since arsenate is a phosphate analog, it probably enters the cells through the phosphate transport system (3). *Escherichia coli* expresses two major phosphate transport systems that differ in their relative specificities for phosphate and arsenate (23,24). The Pit phosphate transport system has a relatively high K_m for phosphate ($K_m^{phosphate} = 25$ μM) and high K_i for arsenate ($K_i^{arsenate} = 32$ μM) (23), resulting in the facile uptake of both phosphate and arsenate into the cell. In contrast, the Pst phosphate transport system exhibits a much lower K_m and K_i for phosphate and arsenate (0.15 and 6.3 μM, respectively) (23) which provides a 15- to 40-fold preference for phosphate over arsenate (depending on the K_m values used for phosphate) (24). It is currently unknown whether these systems are involved in protecting arsenite-oxidizing cells from arsenate.

Arsenate resistance in some bacteria occurs via an *ars* operon that encodes proteins involved in arsenate reduction to arsenite, and ATP-dependent efflux of arsenite (25–29). The presence of this system has not been demonstrated in arsenite-oxidizing bacteria.

IV. CHARACTERISTICS OF ARSENITE OXIDOREDUCTASE FROM *A. faecalis*

The arsenite oxidoreductase from *A. faecalis* (NCIB 8687) has been purified, characterized (15), and the structure recently determined by x-ray crystallography (16). Immunological precipitation of arsenite oxidoreductase indicates that the enzyme is induced by arsenite but not by arsenate (L. A. Kimpler and G. L. Anderson, unpublished data), suggesting a separate mechanism for cell survival in the presence of arsenate.

A. Spectroscopic Features of Redox-Active Centers

Since detoxification of arsenite occurs via oxidation to arsenate, understanding the mechanism of arsenite oxidoreductase has centered on the three redox-active centers found in the enzyme. These are a molybdenum center, a [3Fe-4S] cluster,

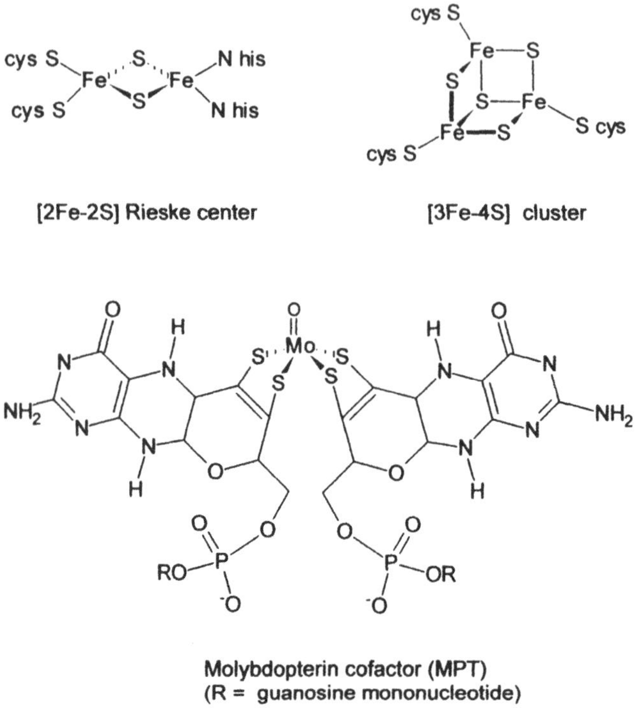

[2Fe-2S] Rieske center   [3Fe-4S] cluster

Molybdopterin cofactor (MPT)
(R = guanosine mononucleotide)

Figure 2 Iron-sulfur clusters and molybdenum cofactor of arsenite oxidoreductase. The molybdopterin cofactor refers to the organic moiety, excluding the molybdenum atom. R = guanine mononucleotide in arsenite oxidoreductase.

and a Rieske-type [2Fe-2S] cluster (15,16) (Fig. 2). Each of these centers contributes to the visible absorption spectrum of the enzyme, which shows a distinct difference spectrum between oxidized and reduced forms of the enzyme (Fig. 3). Reductive titration with arsenite indicates that approximately four reducing equivalents are necessary to fully reduce arsenite oxidoreductase (15). Arsenite [As(III)] oxidation to arsenate [As(V)] most likely occurs by a direct two-electron process (as opposed to sequential one-electron steps) since As(IV) has been observed only rarely and transiently in inorganic systems (30,31). Electron paramagnetic resonance (EPR) spectroscopy, commonly used to study metal centers with an unpaired electron spin, indicates that both the oxidized and reduced forms of the enzyme possess paramagnetic iron-sulfur centers (Fig. 4). The signal seen with oxidized enzyme has g-values of 2.03, 2.01, and 2.00, and arises from the [3Fe-4S] cluster. This signal disappears upon reduction with either arsenite or sodium dithionite to give a second signal with g-values of 2.03, 1.89, and 1.76

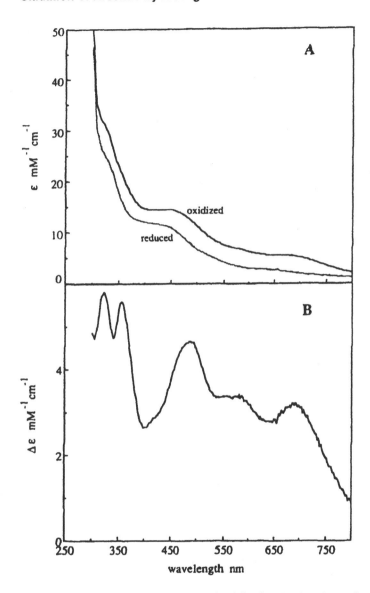

Figure 3 Visible absorption spectra of oxidized and reduced arsenite oxidoreductase. (A) Visible spectrum of oxidized and reduced arsenite oxidoreductase in 50 mM MES, pH 6.0. (B) Difference spectrum of oxidized minus reduced arsenite oxidoreductase. (From Ref. 15.)

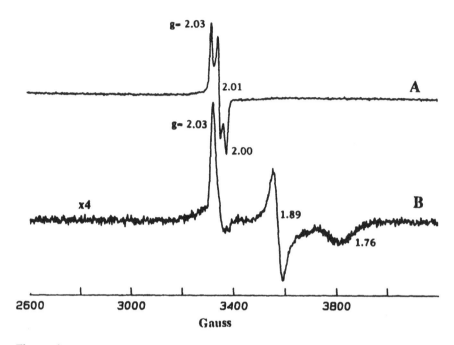

Figure 4 Electron paramagnetic resonance spectra of arsenite oxidoreductase. (A) Oxidized arsenite oxidoreductase. The spectrum is attributed to an oxidized [3Fe-3S] cluster. (B) Arsenite oxidoreductase reduced with a 50-fold excess of dithionite. This spectrum is typical of reduced Rieske-type [2Fe-2S] clusters. EPR parameters: 1 mW power, 9.45 GHz microwave frequency, 12 K temperature, 10 gauss modulation frequency. (From Ref. 15.)

that arises from the Rieske-type [2Fe-2S] center. Both the [3Fe-4S] cluster and the Rieske-type [2Fe-2S] cluster are reduced by one electron equivalent in the presence of arsenite (15).

The [oxidized]-*minus*-[reduced] difference spectrum of arsenite oxidoreductase exhibits a local maximum at ~700 nm that represents the principal spectroscopic signature of the molybdenum center of the enzyme (see Fig. 3). On the basis of this difference spectrum, arsenite oxidoreductase was tentatively assigned to the same family of molybdenum-containing enzymes as DMSO reductase and other bacterial enzymes that catalyze oxygen atom transfer reactions. The recently determined x-ray crystal structure of the enzyme (see below) confirms this assignment. All of these enzymes have an active site in which the molybdenum is coordinated to two equivalents of a unique organic cofactor consisting of a pyranopterin ring with an enedithiolate side chain through which the cofactor binds to the metal (see Fig. 2). (All known eukaryotic enzymes containing a similar cofactor have only a single pyranopterin ring associated with the molybdenum.) This cofactor, usually called "molybdopterin" (MPT) in the literature (despite the fact that the identical cofactor is found in tungsten-containing enzymes), is elaborated as a dinucleotide of guanine, cytosine, adenine, or hypoxanthine in most bacterial enzymes (32). It is the unusual bis(ene-dithiolate) coordination that imparts the unique absorption properties of the center.

Of the biologically relevant oxidation states of molybdenum, Mo(VI), Mo(V), and Mo(IV) (32), only the Mo(V) state has an unpaired electron spin and would be expected to exhibit a signature EPR spectrum (with contributions from hyperfine interactions with the 25% of the naturally occurring isotopes that are ^{95}Mo and ^{97}Mo, with $I = 5/2$ nuclear spin). Despite extensive efforts, a Mo(V) EPR signal has not yet been observed in arsenite oxidoreductase. If the molybdenum center is reduced by two electron equivalents in the course of arsenite oxidation, and both electrons are subsequently transferred to the iron-sulfur clusters of the enzyme, the accumulation of Mo(V) must be either thermodynamically unfavorable (i.e., the reduction potential for the Mo(V)/Mo(IV) couple is higher than that for the Mo(VI)/Mo(V) couple so that little Mo(V) accumulates) or kinetically transient (i.e., the decay of the Mo(V) species is faster than its formation, so that again little Mo(V) accumulates). On the basis of the redox-active cofactors known to be present in the enzyme, the enzyme is able to oxidize two equivalents of arsenite by shuttling electron equivalents among the redox active centers (Scheme 1).

B. Binding of Arsenite

Since it is well established that electron transfer into or out of other molybdenum-containing proteins occurs at the molybdenum center (32–36), this is the most likely site for the binding of arsenite in the case of arsenite oxidoreductase. Furthermore, arsenite is a potent inhibitor of some MPT-containing enzymes (37,38), and for xanthine oxidase has been shown unequivocally to bind within the coordi-

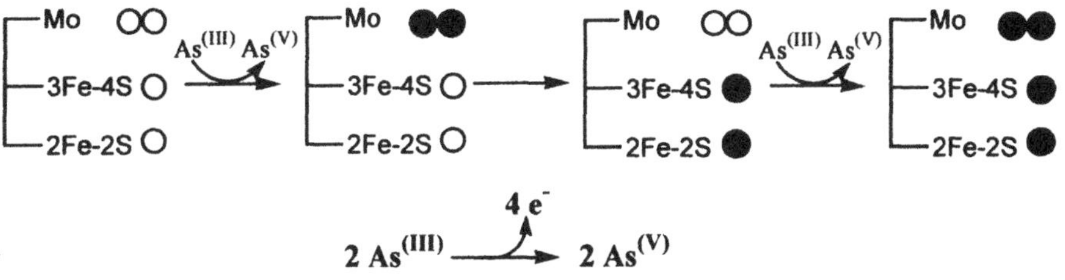

Scheme 1 (Shaded circles represent sites occupied by one electron.)

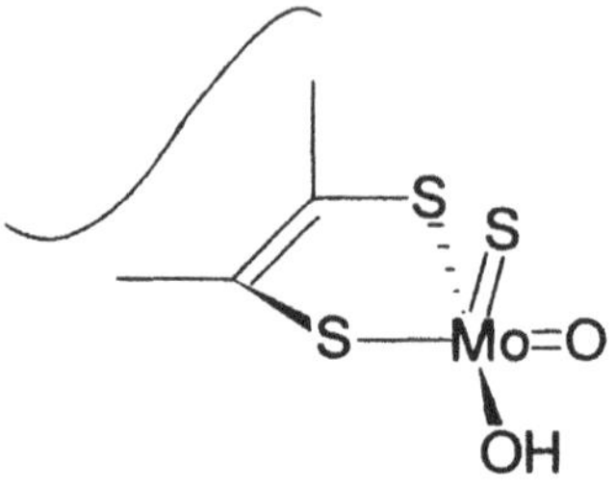

Figure 5 Structure of the molybdenum center in xanthine oxidase. The sulfido ligand to molybdenum is the site at which arsenite binds in xanthine oxidase. This group is absent in arsenite oxidoreductase.

nation sphere of the molybdenum. Specifically, studies of arsenite-inhibited enzyme using x-ray absorption spectroscopy (XAS) suggest a bridging sulfur between arsenic and molybdenum with a Mo(V)-As distance in this Mo-S-As unit of 3.02 Å (39). In addition, the Mo(V) EPR spectrum of arsenite-inhibited xanthine oxidase clearly shows hyperfine structure from the [75]As nucleus (40,41). In binding at the molybdenum center, arsenite blocks the active site and inhibits electron transfer activity in xanthine oxidase via the effect it has on the relative reduction potentials of the redox-active centers (41).

Xanthine oxidase possesses an active site structure differing from that of arsenite oxidoreductase as shown in Figure 5, with a single equivalent of the pterin cofactor and the remainder of the metal coordination sphere taken up by terminal oxo and sulfido groups, as well as a hydroxide. The sulfido sulfur represents the bridging sulfur in the structure of the arsenite-inhibited enzyme, and its removal by reaction with cyanide (which results in release of thiocyanate and replacement of the Mo=S with a second Mo=O group) results in loss of activity and failure to bind arsenite (38,42). These data suggest that arsenite binds to the molybdenum in these MPT-containing enzymes (which contain Mo=S) via a thiolate bridge to give a Mo—S—As moiety. No such Mo=S group is found in arsenite oxidoreductase, as judged by the insensitivity of the enzyme to cyanide (15) or in the x-ray crystallographic structure (16). If arsenite coordinates directly to the molybdenum in arsenite oxidoreductase, it may do so via a Mo—O—As moiety, although there is no evidence for this to date. As discussed further below in the context of the crystal structure of the enzyme, direct coordination of arsenite to molybdenum is not required for efficient oxidation of substrate to arsenate.

C. X-Ray Crystallographic Structure

The x-ray crystal structure of arsenite oxidoreductase from *A. faecalis* has recently been determined (16). The protein was found to be a heterodimer in each of the two crystal forms examined—a $P1$ form with four heterodimers per asymmetric unit (at 1.64 Å resolution) and a $P2_1$ form with two heterodimers per asymmetric unit (at 2.03 Å resolution) (Fig. 6). As the amino acid sequences of the two subunits of the enzyme were not initially known, they were inferred from the electron density. These assignments were subsequently compared with a partial gene sequence corresponding to 70% of the larger subunit (L. Phung and S. Silver, personal communication) and were found to be approximately 90% correct.

The large subunit (consisting of 822 amino acid residues) incorporates the molybdenum center and a [3Fe-4S] cluster that represents the EPR-active site in the oxidized enzyme (see Fig. 6). The molybdenum center lies at the bottom of a funnel-shaped cavity that provides solvent access to the active site. As expected, two equivalents of the pterin cofactor present as the guanosine dinucleotide are coordinated to the metal via the enedithiolate side chains in a manner similar to that seen in other members of this family of molybdenum enzymes, with the two pairs of thiolate sulfurs constituting a four-sided base for the molybdenum coordination sphere and the guanosine dinucleotide portions extending in opposite directions from the metal (see Fig. 6). The molybdenum ion is displaced 0.8 Å from the mean plane defined by the four sulfur atoms. Unlike other enzymes of this group, however, there is no protein ligand to the metal, the fifth ligand being modeled as a Mo=O (with the refined M—O distance at ~1.6 Å) to yield an overall distorted square pyramidal coordination geometry. The amino acid residue corresponding to the cysteine, selenocysteine, or serine coordinating the

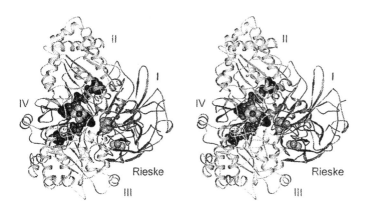

Figure 6 Stereoview ribbon diagram of arsenite oxidoreductase from *Alcaligenes faecalis*. The enzyme is comprised of a large subunit consisting of four domains designated by roman numerals and a small or Rieske subunit. The metal sites in the enzyme are shown as atomic spheres. The molecule is oriented to look down the solvent access channel at the Mo active site.

molybdenum in related proteins is Ala199 in arsenite oxidoreductase, making it quite unambiguous that a protein ligand to the metal is absent in the case of the arsenite-oxidizing enzyme. Arsenite oxidoreductase thus represents a fourth subcategory of the family of molybdenum enzymes that possess two equivalents of the pterin cofactor bound to the metal (the so-called DMSO reductase family), the other three possessing either serine, cysteine, or selenocysteine as a protein ligand to the metal.

The enzyme sample used in the x-ray crystallography was initially oxidized. It is well known, however, that the highly ionizing synchrotron radiation used in the work can reduce the enzyme in the crystal. Indeed, the five-coordinate geometry of the molybdenum center that has been reported (16) suggests that the protein has become reduced in the course of crystallographic data acquisition. Photoreduction of the metal centers is also implied by the observation of a shift of the iron absorption edge toward lower energy upon exposure of the protein crystals to the x-ray beam (16). On the basis of this observation, in conjunction with x-ray absorption spectroscopic studies of both oxidized and reduced arsenite oxidoreductase (George et al., unpublished data), it appears likely that the five-coordinate structure seen crystallographically represents the reduced molybdenum center of the enzyme. The XAS studies are consistent with uptake of hydroxide from solvent upon reoxidation of the molybdenum, giving a six-coordinate $Mo(VI)O(OH)(mpt)_2$ species as shown in Figure 7.

The solvent access channel to the active site of arsenite oxidoreductase has a highly polar surface, being formed almost entirely from serine, aspartate, asparagine, glutamate, glutamine, lysine, histidine, arginine, and tyrosine residues (Fig. 8). Manually fitting an $As(OH)_3$ molecule into the base of the channel close to the apical oxygen of the molybdenum center suggested that four of these residues, His195, Glu203, Arg419, and His423 constitute the substrate binding site of the enzyme, with the enzyme-substrate complex stabilized by hydrogen bonds (Fig. 9). The involvement of His195 and His423 is notable and it is likely that at least one of these is a histidine residue whose chemical modification results in reduction of activity in the enzyme (43). Binding of arsenite positions it well

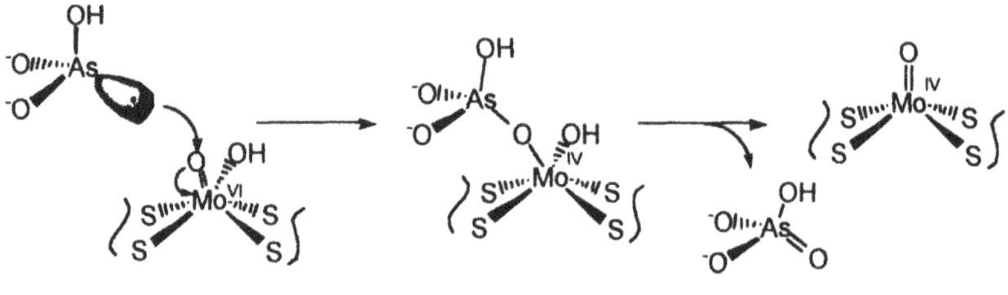

Figure 7   Proposed reaction mechanism of arsenite oxidation in arsenite oxidoreductase.

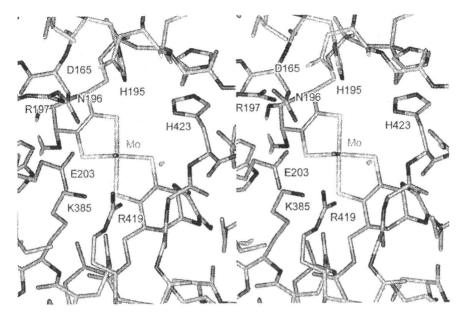

Figure 8 Stereoview of the solvent access channel near the active site. The solvent access channel has a highly polar surface. This view also shows the distorted square pyramidal geometry of the (presumably reduced) Mo center.

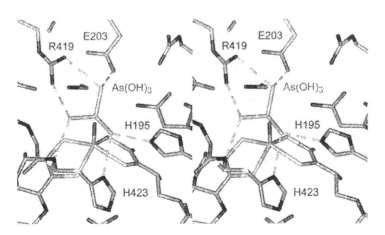

Figure 9 Proposed binding interactions of arsenite in the active site of arsenite oxidoreductase. The stereoview shows possible hydrogen bonding with Arg419, Glu203, His195, and His423 to optimally position arsenite for interaction with the Mo center.

for reaction with the molybdenum center in a simple oxygen atom transfer process. By analogy to the chemistry seen in the reaction of a variety of model complexes with phosphines and related compounds (44–46), the reaction could proceed via nucleophilic attack of the arsenite lone pair on the Mo(VI)=O group (bearing in mind that arsenite will have hydrated in aqueous solution to the form shown in the figure). This would create the As—O bond of product, and at the same time lead to the formal two-electron reduction of the molybdenum to Mo(IV) since all four electrons in the Mo=O bond belong to the oxygen in a formal valence count. This represents obligatory two-electron chemistry, which possibly contributes to the failure to observe a Mo(V) EPR signal in the course of the reaction. According to this mechanism, the reaction is completed by dissociation of bound product, reoxidation of the metal by intramolecular electron transfer to the iron-sulfur centers with concomitant binding and deprotonation of hydroxide to regenerate the Mo(VI)=O group.

The overall protein fold of the large subunit of arsenite oxidoreductase is laid out into four domains, designated I–IV from the N-terminus (see Fig. 6). One of the pterin cofactors (that designated "P" by convention on the basis of its interaction within the polypeptide) lies between domains II and IV, and the other (designated "Q") between III and IV; domains II and III possess homologous dinucleotide-binding folds and are related to one another within the protein structure by a pseudo two-fold axis of symmetry. The principal folding motif for each domain is that of an $\alpha\beta\alpha$ helix-sheet-helix sandwich, with that for domain II having an $\alpha_5\beta_7\alpha_7$ fold and domain III an $\alpha_6\beta_5\alpha_9$ fold. The [3Fe-4S] cluster is found in domain I of the large subunit, which possesses three separate β sheets and six helices. Domain I exhibits significant structural homology to other iron-sulfur-possessing members of this family, including a Cys-X_2-Cys-X_3-Cys motif that coordinates the iron-sulfur cluster. Both formate dehydrogenase from *E. coli* (47) and the dissimilatory nitrate reductase of *Desulfovibrio desulfuricans* (48) have a fourth cysteine ligand and the iron-sulfur cluster is present as a [4Fe-4S] center in these systems. This cysteine is replaced by Ser99 in arsenite oxidoreductase. Domain I provides the principal interaction with the small subunit of the heterodimer. The overall sequence identity between the large subunit of arsenite oxidoreductase and formate dehydrogenase on the one hand and the dissimilatory nitrate reductase on the other is 23 and 20%, respectively (16), and the disposition of the molybdenum center and [3Fe-4S] cluster with respect to one another are very similar in all three enzymes.

The small subunit of arsenite oxidoreductase, consisting of 134 amino acid residues, can be divided into two domains, each consisting principally of β-sheet (see Fig. 6). The overall fold is similar to the Rieske-containing subunits of cytochrome b_6f (49) and bc_1 (50,51), and the Rieske-containing domain of phthalate-1,2-dioxygenase (52). The Cys60-X-His62-X_{15}-Cys78-X_2-His81 sequence binding the Rieske-type [2Fe-2S] cluster conforms to the consensus Cys-X-His-X_{15-}

$_{17}$-Cys-X$_2$-His motif previously observed for Rieske proteins. The two loops possessing the iron-binding ligands in arsenite oxidoreductase are held together by a disulfide bond between Cys65 and Cys80. The iron coordinated by His62 and His81 lies nearer the surface of the subunit, with the iron coordinated by the two cysteine ligands lying more buried in the subunit. Both histidine residues lie at the surface of the subunit, with His62 buried within the subunit-subunit interface and His81 exposed to the solvent.

The overall layout of the redox-active centers clearly implies a linear sequence of electron transfer from the molybdenum center to the [3Fe-4S] cluster in the large subunit, then on to the Rieske center of the small subunit. The [3Fe-4S] cluster is approximately 14 Å from the molybdenum atom, with the "Q" pterin cofactor intervening. Several alternate electron-transfer pathways involving covalent- and hydrogen-bonding interactions presumably mediate electron transfer between the pterin cofactor and the iron-sulfur cluster. The two iron-sulfur clusters are separated by a distance of 12 Å. The shortest pathway for electron transfer passes through Ser99 of the large subunit (the residue whose mutation from cysteine is responsible for formation of a [3Fe-4S] rather than a [4Fe-4S] cluster in the protein) and His62 of the small subunit (which lies at the interface between the two subunits). Electron transfer out of arsenite oxidoreductase to its physiological oxidants, either copper-containing azurin or a c-type cytochrome (15), is presumably via His81, which is the sole solvent-exposed ligand to the Rieske center. The immediate vicinity of this residue is a large concave surface that seems well suited for binding small globular proteins such as azurin or cytochrome c. The residues forming this surface are an obvious target for site-directed mutagenesis to study the interaction between arsenite oxidoreductase and these small electron-carrier proteins.

D. Steady State Kinetics

The steady state kinetics of arsenite oxidoreductase from *A. faecalis* indicate a so-called double displacement (or "ping-pong") mechanism (15) in which the enzyme cycles between oxidized and reduced forms in its reaction with arsenite and azurin (or cytochrome c). This overall kinetic scheme is common in redox-active proteins. Arsenite must bind, the oxygen atom transfer chemistry take place, and arsenate dissociate before the subsequent reaction of a second molecule of substrate. Since arsenate is not an inhibitor of arsenite oxidoreductase (43), product dissociation must be effectively irreversible. The turnover number (k_{cat}) of 27 sec^{-1} and K_m for arsenite of 8 µM are reasonable parameters for the detoxification of arsenite, especially since *A. faecalis* is able to survive in at least 80 mM (1%) sodium arsenite. The considerable catalytic power of the enzyme is reflected by the kinetic parameter k_{cat}/K_m of 3.4×10^6 M^{-1}sec^{-1}, which is fairly close to the diffusion-controlled maximum of 10^8–10^{10} M^{-1}sec^{-1} for proteins in aqueous solution. The K_m for azurin is significantly larger than that for arsenite, at 68 μM, and is a relatively high value for a (presumably physiological) protein-protein interaction. A relatively high concentration of azurin in the periplasm may help compensate for this, but it may also be that azurin is not the preferred physiological partner. The specific cytochrome c that also acts as an electron acceptor for arsenite oxidoreductase has not yet been isolated and characterized.

The activity of arsenite oxidoreductase from A. faecalis is affected by essential histidines (43). Approximately three histidine residues in the oxidized enzyme are readily accessible to chemical modification by diethylpyrocarbonate, and at least one of these modulates the activity of the oxidized enzyme. However, if arsenite oxidoreductase is first reduced by either dithionite (a low potential generic reductant) or by arsenite, approximately three histidines can be modified, without affecting arsenite oxidoreductase activity. The reductive half reaction of arsenite oxidoreductase may therefore be dependent on histidine residue(s) either for the process of electron transfer or for the correct conformation of the oxidized protein. As indicated above, His195 and His423 form part of the binding site and one of these may be the residue whose modification in oxidized enzyme results in loss of activity.

V. SUMMARY

Arsenite is relatively abundant in the environment and represents a significant toxic threat to micro-organisms. Its oxidation to arsenate significantly reduces the toxicity of arsenic and in some cases may also be used as an energy source for the organism. This oxidation constitutes an important step in the environmental arsenic cycle. An arsenite oxidoreductase from the soil pseudomonad Alcaligenes faecalis has been purified and characterized, and found to effectively catalyze the oxidation of arsenite to arsenate. The crystal structure of the enzyme has recently been elucidated. The enzyme is found to be a member of the DMSO reductase family of molybdenum-containing enzymes, and possesses a [3Fe-4S] and a [2Fe-2S] Rieske center in addition to the active site molybdenum. The structure provides a plausible chemical rationale for the reaction mechanism of the enzyme, which appears to be closely related to the oxygen atom transfer chemistry seen in small inorganic complexes of molybdenum that serve as models for these enzymes.

REFERENCES

1. JM Wood. Biological cycles for toxic elements in the environment. Science 183: 1049–1052 1974.

2. FC Knowles, AA Benson. The biochemistry of arsenic. Trends Biol Sci May:178–180 1983.

3. KA Coddington. Review of arsenicals in biology. Toxicol Environ Chem 11:281–290 1986.

4. P Mahieu, JP Buchet, HA Roels, R Lauwerys. The metabolism of arsenic in humans acutely intoxicated by As_2O_3. Its significance for the duration of BAL therapy. Clin Toxicol 18:1067–1075 1981.

5. HL Ehrlich. Bacterial action on orpiment. Econ Geol 58:991–994 1963.

6. MP Silverman, DG Lundgren. Studies on the chemoautotrophic iron bacterium Ferrobacillus ferrooxidans: An improved medium and a harvesting procedure for securing high cell yields. J Bacteriol 77:642–647 1959.

7. HL Ehrlich. Bacterial oxidation of arsenopyrite and enargite. Econ Geol 59:1306–1312 1964.

8. DR Sanadi, DM Gibson, P Ayengar, L Quellet. Evidence for a new intermediate in the phosphorylation coupled to α-ketoglutarate oxidation. Biochim Biophys Acta 13: 146–148, 1954.

9. RK Crane, F Lipmann. The effect of arsenate on aerobic phosphorylation. J Biol Chem 201:235–243, 1953.

10. GMP Morrison, GE Batley, TM Florence. Metal Speciation and toxicity. Chem Br 25:791, 1981.

11. AJ Bard, R Parsons, J Jordon. Standard Potentials of Aqueous Solutions. New York: Marcel Dekker, 1985.

12. RH Garret, CM Grisham. Biochemistry. 2d ed. Fort Worth, TX: Saunders, 1999.

13. JW Legge. Bacterial oxidation of arsenite IV. Some properties of the bacterial cytochromes. Aust J Biol Sci 7:496–503, 1954.

14. FH Osborne, HL Ehrlich. Oxidation of arsenite by a soil isolate of Alcaligenes. J Appl Bacteriol 41:295–305, 1976.

15. GL Anderson, J Williams, R Hille. The purification and characterization of arsenite oxidase from Alcaligenes faecalis, a molybdenum-containing hydroxylase. J Biol Chem 267:23674–23682, 1992.

16. PJ Ellis, T Conrads, R Hille, P Kuhn. Crystal structure of the 100 kDa arsenite oxidase from Alcaligenes faecalis in two crystal forms at 1.64 Å and 2.03 Å. Accepted for publication, Structure, 2001.

17. HL Ehrlich. Inorganic energy sources for chemolithotrophic and mixotrophic bacteria. Geomicrobiol J 1:65–83, 1978.

18. SE Phillips, ML Taylor. Oxidation of arsenite to arsenate by Alcaligenes faecalis. Appl Environ Microbiol 32:392–399, 1976.

19. HM Sehlin, EB Lindstrom. Oxidation and reduction of arsenic by Sulfolobus acidocalarius strain BC. FEMS Microbiol Lett 93:87–92, 1992.

20. JM Santini, Ll Sly, RD Schnagl, JM Macy. A new chemolithoautotrophic arsenite-oxidizing bacterium isolated from a gold mine: Phylogenetic, physiological, and preliminary biochemical studies. Appl Environ Microbiol 66:92–97, 2000.

21. AN Ilyaletdinov, SA Abdrashitova. Autotrophic oxidation of arsenic by a culture of Pseudomnas arsenitoxidans. Mikrobiologiya 50:197–204, 1981.

22. D Ahmann, AL Roberts, LR Krumholz, FMM Morel. Microbe grows by reducing arsenic. Nature 371:750, 1994.

23. H Rosenberg, RG Gerdes, K Chegwidden. Two systems for the uptake of phosphate in Escherichia coli. J Bacteriol 131:505, 1977.

24. GR Willsky, MH Malamy. Characterization of two genetically separable inorganic phosphate transport systems in Escherichia coli. J Bacteriol 144:356, 1980.

25. C Rensing, M Ghosh, BP Rosen. Families of soft-metal-ion-transporting ATPases. J Bacteriol 181:5891–5897, 1999.

26. S Silver, LT Phung. Bacterial heavy metal resistance: new surprises. Annu Rev Microbiol 50:753–789, 1996.

27. S Silver. Bacterial resistances to toxic metals: A review. Gene 179:9–19, 1996.

28. C Cervantes, G Ji, JL Ramirez, S Silver. Resistance to arsenic compounds in microorganisms. FEMS Microbiol Rev 15:355–367, 1994.

29. S Silver, M Walderhaug. Gene regulation of plasmid and chromosome determined inorganic ion transport in bacteria. Microbiol Rev 56:195–228, 1992.

30. KM Kadish, C Erben, Z Ou, VA Adamian, S Will, E Vogel. Corroles with group 15 metal ions. Synthesis and characterization of octaethylcorroles containing As, Sb, and Bi ions in +3, +4, and +5 oxidation states. Inorg Chem 39:3312–3319, 2000.

31. S Nishida, M Kimura. Kinetic studies of the oxidation reaction of arsenic (III) to arsenic (V) by peroxodisulfate ion in aqueous alkaline media. J Chem Soc Dalton Trans 2:357–360, 1989.

32. R Hille. The mononuclear molybdenum enzymes. Chem Rev 96:2757–2816 1996.

33. NA Turner, RC Bray, GP Diakun. Information from E.X.A.F.S. spectroscopy on the structures of different forms of molybdenum in xanthine oxidase and the catalytic mechanism of the enzyme. Biochem J 260:563, 1989.

34. RC Bray, MT Lamy, S Gutteridge, T Wilkinson. Evidence from electron-paramagnetic-resonance spectroscopy for a complex of sulphite ions with the molybdenum center of sulphite oxidase. Biochem J 201:241, 1982.

35. Y Kubo, N Ogura, H Nakagawa. Limited proteolysis of the nitrate reductase from spinach leaves. J Biol Chem 263:19684, 1988.

36. R Hille, GN George, MK Eidsness, SP Cramer. EXAFS analysis of xanthine oxidase complexes with alloxanthine, violapterin, and 6-pteridylaldehyde. Inorg Chem 28:4018, 1989.

37. G Barry, G Bunbury, EL Kennaway. The effect of arsenic upon some oxidation-reduction reactions. Biochem J 22:1102–1111, 1928.

38. MP Coughlan, KV Rajagapalan, P Handler. The role of molybdenum in xanthine oxidase and related enzymes. J Biol Chem 244:2658–2663, 1969.

39. SP Cramer, R Hille. Arsenite-inhibited xanthine oxidase; Determination of the Mo—S—As geometry by EXAFS. J Am Chem Soc 107:8164–8169, 1985.

40. GN George, RC Bray. Reaction of arsenite ions with the molybdenum center of milk xanthine oxidase. Biochemistry, 22:1013–1021, 1983.

41. R Hille, RC Stewart, JA Fee, V Massey. The interaction of arsenite with xanthine oxidase. J Biol Chem 258:4849–4856, 1983.

42. D Edmondson, V Massey, G Palmer, LM Beacham III, GB Elion. Resolution of active and inactive xanthine oxidase by affinity chromatography. J Biol Chem 247:1597–1604, 1972.

43. L McNellis, GL Anderson. Redox-state dependent chemical inactivation of arsenite oxidase. J Inorg Biochem 69:253–257, 1998.
44. JM Berg, RH Holm. Synthetic approaches to the mononuclear active sites of molybdoenzymes: catalytic oxygen atom transfer reactions by oxomolybdenum (IV, VI) complexes with saturation kinetics and without molybdenum (V) dimer formation. J Am Chem Soc 106:3035–3036, 1984.
45. JM Berg, RH Holm. A model for the active sites of oxo-transfer molybdoenzymes: synthesis, structure and properties. J Am Chem Soc 107:917–925, 1985.
46. JP Caradonna, PR Reddy, RH Holm. Kinetics, mechanisms, and catalysis of oxygen atom transfer reactions of S-oxide and pyridine N-oxide substrates with molybdenum (IV, VI) complexes: Relevance to molybdoenzymes. J Am Chem Soc 110:2139–2144, 1988.
47. JC Boyington, VN Gladyshev, SV Khangulov, TC Stadtman, PD Sun. Crystal structure of formate dehydrogenase H: Catalysis involving Mo, molybdopterin, selenocysteine, and an Fe_4S_4 cluster. Science 275:1305–1308, 1997.
48. JM Dias, ME Than, A Humm, R Huber, GP Bourenkov, HD Bartunik, S Bursakov, J Calvete, J Caldeira, C Carneiro, JJG Moura, I Moura, MR Romao. Crystal structure of the first dissimilatory nitrate reductase at 1.9 Å solved by MAD methods. Structure 7:65–79, 1999.
49. CJ Carrell, H Zhang, WA Cramer, JL Smith. Biological identity and diversity in photosynthesis and respiration: structure of the lumen-side domain of the chloroplast Rieske protein. Structure 5:1613–1625, 1997.
50. S Iwata, M Saynovits, TA Link, H Michel. Structure of the water soluble fragment of the 'Rieske' iron-sulfur protein of the bovine heart mitochondrial cytochrome bc₁ complex determined by MAD phasing at 1.5 Å resolution. Structure 4:567–579, 1996.
51. S Iwata, JW Lee, K Okada, JK Lee, M Iwata, B Rasmussen, TA Link, S Ramaswamy, BK Jap. Complete structure of the 11-subunit bovine mitochondrial cytochrome bc₁ complex. Science 281:64–71, 1998.
52. B Kauppi, K Lee, E Carredano, RE Parales, DT Gibson, H Eklund, S Ramaswamy. Structure of an aromatic-ring-hydroxylating dioxygenase—naphthalene 1,2-dioxygenase. Structure 6:571–586, 1998.
53. NE Korte, Q Fernando. A review of arsenic (III) in groundwater. Crit Rev Environ Control 21:1–39, 1991

16
Volatilization of Arsenic

William T. Frankenberger, Jr.
University of California, Riverside, California

Muhammad Arshad
University of Agriculture, Faisalabad, Pakistan

I. ARSENIC AND THE ENVIRONMENT

Arsenic can occur in many different species in soil and water environments. The most often encountered arsenic forms are highly toxic arsenious acid [As(III)] and arsenic acid [As(V)]. Methylated species, monomethyl arsonic acid (MMAA) and dimethyl arsenic acid (DMAA), which are less-toxic forms, dominate in biomass, but have also been detected in soil (1). The methylated arsenic forms are believed to be part of a detoxification mechanism in living organisms. Other quaternary organic arsenic compounds, such as arsenobetaine and arsenocholine, are nontoxic. In air, arsenic can exist as gaseous arsines, which are extremely toxic compounds and are formed mainly in anoxic, reducing environments. Arsenate can replace phosphate in energy transfer phosphorylation reactions and arsenite has a high affinity for thiol groups of proteins, inactivating many enzymes.

Arsenic may accumulate to phytotoxic levels in soil from continuous arsenical usage and this depends on the rates of transfer and transformation processes (2). Once these arsenicals reach the soil either directly or as crop residues, they can be transformed by microorganisms to inorganic arsenic and/or volatile arsines [arsine, AsH_3; methylarsine, $CH_3(AsH_2)$; dimethylarsine, $(CH_3)_2AsH$; and trimethylarsine, $(CH_3)_3As$]. Production of these arsines results in transfer of arsenic from soil to the atmosphere. It is estimated that as much as 2.10×10^7 kg of arsenic is lost to the atmosphere through volatilization from land surfaces annually (3). The continental vapor flux is about 8 times that of the continental dust flux, indicating that the biogenic contribution may play a significant role in cycling of arsenic. The ubiquity of arsenic in the environment, its biological toxicity, and its redistribution are factors evoking public concern. Tamaki and Frankenberger (4) have thoroughly reviewed the biochemistry of arsenic transformations in the environment.

II. BIOTRANSFORMATION OF ARSENIC
IN THE ENVIRONMENT

The biological transformations (methylation, demethylation, oxidation, and reduction) of arsenic were recognized in the mid-nineteenth century (5–7). Wood (8) pointed out that, in a reduced environment such as flooded soils, arsenate is reduced to arsine and then methylated to form methylarsenic acid forms. These arsenic compounds may further be reduced to methylarsines that volatilize to the atmosphere (7,8).

Numerous bacteria, fungi, yeasts, and algae are able to transform arsenic compounds. Among the transforming processes mediated by microorganisms are oxidation (9,10), reduction (11), demethylation (12,13), and methylation reactions (12–16). Biological transformations of arsenic in soil are illustrated in Figure 1 (17).

A. Mineralization and Immobilization

Organic forms of arsenic may be mineralized into inorganic forms, e.g., through demethylation processes, while microorganisms and higher plants can assimilate inorganic arsenic and convert it into organic arsenic compounds (immobilization). The inorganic forms of arsenic can be biomethylated by certain microbes to gaseous arsines or to MMAA and DMAA, while other microbes can demethylate organic forms to inorganic arsenic species (18). Gao and Burau (19) observed that arsenate was the main metabolite from degradation of MMAA and sodium cacodylate (CA). The rate of MMAA mineralization was slower than that of CA under the same conditions. The amount of CA mineralized was linearly related to the concentration of CA added to the soil, indicating that the rate was first order. Mineralization of CA increased as soil moisture increased from 50 to 550 g H_2O kg^{-1} soil, and the process was strongly stimulated when soil moisture was >350 g H_2O kg^{-1} soil (−0.03 MPa). The overall percentage of CA and MMAA mineralized was up to 87% after 70 days.

Degradation of organic arsenicals has been studied in soil and culture media. Methylated organic arsenicals including trimethylarsine oxide (TMAO), CA, and MMAA are converted to less-methylated compounds in culture media by sediment microorganisms (20). Microorganisms isolated from soil degraded disodium methane arsonate (DSMA) to arsenate (21). Woolson and Kearney (22) showed that CA was degraded to arsenate in soil under aerobic conditions but

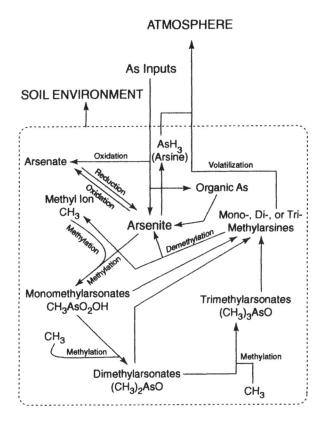

Figure 1   Biological transformations of arsenic in soils. (From Ref. 17.)

not under anaerobic conditions, and the persistence of CA was a function of soil type. Degradation of methanearsonate was shown to be associated with soil organic matter oxidation. In a loamy soil, degradation of methanearsonate increased with increasing soil organic matter content and also from the addition of plant material (23). Akkari et al. (24) studied the degradation of methanearsonate in soils at a low range of concentrations, 0–5 mg As kg^{-1} soil. It was found that degradation followed first-order kinetics. Andreae (25) suggested that biological demethylation is the dominant process for the generation of inorganic arsenic from organic arsenicals.

B. Oxidation/Reduction

Arsenite [As(III)] can be oxidized to arsenate [As(V)] or vice versa via reduction under a given set of soil and environmental conditions. In general, As(V) compounds predominate in aerobic soils, whereas As(III) compounds predominate in slightly reduced conditions. *Bacillus* and *Pseudomonas* spp. have been isolated that can oxidize arsenite to the less-toxic arsenate. In addition, a strain of *Alcaligenes faecalis* obtained from raw sewage was capable of oxidizing arsenite (10). Oxidation of arsenite by heterotrophic bacteria plays an important role in detoxifying the environment, catalyzing as much as 78–96% of the arsenite to arsenate (26). Oxidized forms of arsenic [As(V)] can be transformed to As(III) under reduced conditions, which can be further transformed to arsine gas (AsH$_3$) under highly reduced conditions. These reduction reactions can indirectly or directly be mediated by soil microorganisms (17,27). Microbial volatilization of arsenate occurs via its reduction into arsenite. Soil microorganisms are capable of converting arsenate and arsenite to several reduced forms, largely methylated arsines, which are volatile (17). In addition, methylphenylarsinic acid and dimethylphenylarsine oxide are also reduced by *Candida humicola* to dimethylphenylarsine. Both *Pseudomonas* and *Alcaligenes* reduce soluble arsenate (AsO$_4^{3-}$) to arsine (AsH$_3$), which is volatile (28). Challenger (27) reported the reduction of arsenic oxide, sodium arsenate, DSMA, and CA by *Scopulariopsis* and *Aspergillus* fungal species. At the present time, seven diverse species of Eubacteria and two species of Crenoarchaea have been isolated capable of respiratory reduction of arsenate to arsenite (see Chap. 11). In addition to biological reduction, chemical reduction of arsenate also occurs in the soil environment. Detailed information on the biochemistry of arsenic oxidation/reduction has already been reviewed in Chapters 8 and 10–15.

C. Volatilization

Arsenic volatilization is a dissipation mechanism to remove arsenic from soil and water systems. Many soil and water microorganisms are capable of mediating arsenic volatilization, largely in the form of methylated arsines. Under highly reduced conditions, arsenate can be transformed to trimethylarsine as shown in Figure 2.

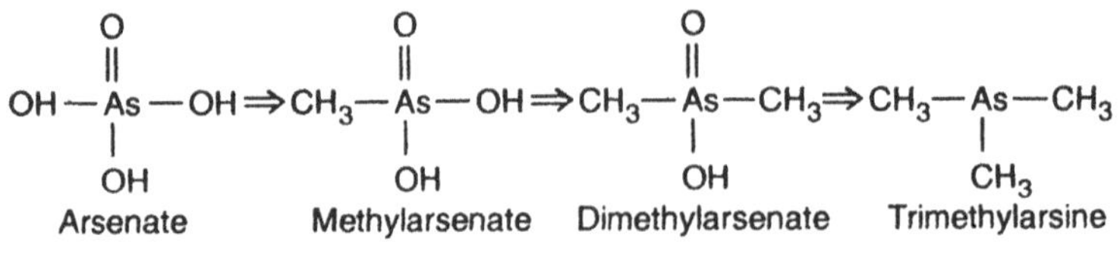

Figure 2   Biomethylation of arsenic. (From Ref. 28.)

III. BIOCHEMISTRY OF ARSENIC VOLATILIZATION

A. Microbial Volatilization of Arsenic from Soil

The formation of arsines, a volatile form of arsenic, is known to occur via microbial activity (27). Soil microbes produce volatile arsenicals by a reductive methylation pathway from inorganic and methylated forms of arsenic (29). Methylation in which a methyl ion is added to an arsenite ion is strictly a biological process (8,30). Studies on arsenic evolution from arsenicals in contaminated environments have shown wide variation in rates of release. High rates of arsine evolution in soils have been reported (22,31,32), suggesting that arsenical usage may not lead to arsenic accumulation in soils due to the high volatilization rates. However, much slower rates of arsine evolution were reported from a calcareous California soil amended with As(V), MMAA, and CA (2). It is not certain that the loss of arsenic as volatile arsines is a significant arsenic removal pathway from all soils. Turpeinen et al. (33) studied the effects of microbial activity on the mobilization and speciation of arsenic in soil and in microcosm experiments under laboratory conditions. They found that microbes enhanced mobilization of arsenic from soil by 19–24%. However, formation of dissolved methylated arsenic species by microbes was low (<0.1%) during 5 days of incubation. Turpeinen et al. (33) concluded that even though methylation may function as a detoxification method, it was of minor importance in the soil they tested. According to Sandberg and Allen (34), as much as 35% of the arsenic species in soil may be eventually volatilized as arsine, dimethylarsine (DMA), and trimethylarsine (TMA).

1. Microflora Involved

Bacteria, fungi, and algae have been reported to be involved in the biotransformations of arsenic in soils. A diverse group of soil microorganisms can generate different or similar biochemical products of arsenic. Even the same phenotype of soil microorganisms may produce different arsenic species under variable soil conditions. Soils amended with inorganic and methylated arsenic herbicides often produce DMA and TMA (22,31,35,36). The organisms responsible for arsenic volatilization come from diverse environments, suggesting that a number of species have the capacity to produce alkylarsines. For example, mixed communities of soil microorganisms produced DMA and TMA, which were trapped in bell jars over soil and lawns that were previously treated with methylarsenicals (37). The methylation of arylarsonic acids is important because of their wide use as food supplements for swine, turkeys, and poultry. The first evidence of bacterial methylation of arsenic by the anaerobic *Methanobacterium* was provided by McBride and Wolfe (38). Later, several strains of soil bacteria were isolated that oxidized arsenite to arsenate and were also involved in the methylation and alkylation of arsenic in soils (39–41). A summary of studies reporting microbial biomethylation of arsenic is given in Table 1.

2. Volatilization Mediated by Bacteria

Bacterial methylation of inorganic arsenic has been studied extensively using methanogenic bacteria. The production of volatile arsenic can be accomplished either aerobically or anaerobically. Anaerobically, bacteria such as *Methanobacterium, Pseudomonas,* and *Alcaligenes* have been reported as active organisms in arsenic volatilization. The volatilization of arsenic by bacteria under aerobic conditions have been demonstrated with *Staphylococcus aureus* and *Escherichia coli* (42). Methanogenic bacteria are a morphologically diverse group capable of producing methane as their principal metabolic end product. They are present in large numbers in anaerobic ecosystems, such as sewage sludge, freshwater sediments, and composts, where organic matter is decomposing (43). It has been shown that at least one species of *Methanobacterium* is capable of methylating inorganic arsenic to produce volatile DMA. Arsenate, arsenite, and MMAA can serve as substrates in DMA formation. Inorganic arsenic methylation is coupled to the CH_4 biosynthetic pathway and may be a widely occurring mechanism for arsenic removal/detoxification. *Methanobacterium* strain MOH cell-free extracts, when incubated under anaerobic conditions with [^{74}As]Na_2AsO_4, a methyl donor (methylcobalamin, CH_3CoB_{12}), H_2, and ATP, produced a volatile [^{74}As]dimethylarsine (38). The pathway involves the reduction of arsenate to arsenite with subsequent methylation involving a low-molecular-weight cofactor, coenzyme M (CoM). Methanearsonic acid added to cell-free extracts was not reduced to methylarsine, but required an additional methylation step before reduction. However, DMAA was reduced to DMA even in the absence of a methyl donor (38).

Whole cells of methanogenic bacteria under anaerobic conditions also produce DMA as a biomethylation end product of arsenic, but not heat-treated cells, indicating that this is a biological reaction. Furthermore, samples collected from a number of different anaerobic ecosystems (anaerobic sewage digestor sludge and rumen from cattle) that produced methane also transformed arsenate into DMA. Under anaerobic conditions, biomethylation of arsenic proceeds only to DMA, which is stable in the absence of oxygen, but is rapidly oxidized under aerobic conditions. Dimethylarsine in anaerobic environments can react with disulfide bonds present on particulates, thus reducing the concentration of soluble arsenic. Resting cell suspensions of *Pseudomonas* sp. and *Alcaligenes* sp. incubated with either arsenite or arsenate under anaerobic conditions produced arsine but no other intermediates (29). *Aeromonas* sp. methylated DMAA to TMAO (35).

Honschopp et al. (43) isolated an arsenic-resistant and arsenic-methylating bacterium belonging to the *Flavobacterium-Cytophaga* group from soil with an

Table 1 Biomethylation of Arsenic by Microorganisms

Organism	Description and effectiveness	Ref.
Penicillium brevicaule (*Scopulariopsis brevicaulis*)	Produced TMA in the presence of MMAA or DMAA.	27
Methanogenic bacteria (*Methanobacterium* strain MOH)	Methylates arsenic, arsenite and MMAA to DMAA.	38
Candida humicola, Gliocladium roseum and *Penicillium* sp.	Converted MMAA and DMAA to TMA. *C. humicola* used As(III) and As(V) as substrates to produce TMA.	14
Flavobacterium sp.	Methylated dimethylarsinic acid to trimethylarsinic oxide.	58
Candida humicola	Methylated a wood preserving fungicide, chromated copper arsenate, into a volatile As species.	45
Alcaligenes sp., *Pseudomonas* sp.	Methylated As(III) or As(V) into arsine under aerobic conditions.	29
Penicillium sp.	Produced TMA from MMAA and DMAA. The addition of carbohydrates and sugar acids to the minimal medium suppressed trimethylarsine production. The amino acids phenylalanine, isoleucine, and glutamine promoted TMA production with an enhancement ranging from 10.2- to 11.6-fold over control without amino acid enhancement.	46
Scopulariopsis bevicaulis and *Phaeolus schweinitzii*	Both fungi coverted As_2O_3 into volatile form of arsenic with *P. schweinitzii* being the more efficient. The formation of volatile As was increased with increasing concentration of As(III) in the growth medium. The volatile arsenic was tentatively identified as TMA.	47
Penicillium sp.	MMAA was used as a substrate which was transformed into TMA by the fungus. As(III) stimulated TMA production (3.4-fold), while As(V) inhibited (14–78%) TMA production with increasing concentration, most likely due to fungal substrate toxicity.	57

Table 2 Formation of Trimethylarsine (TMA)
by *Flavobacterium-Cytophaga* sp.

Arsenic species in the media	Concentration of arsenic (ppm)	Incubation time (hr)	Concentration of TMA in the gas phase (ppb)[a]
As(III)	50	24	35
As(III)	50	48	75
As(III)	100	24	11
As(III)	100	48	61
As(V)	50	24	<10
As(V)	50	48	<10
As(V)	100	24	<10
As(V)	100	48	20

[a] Values are averages of eight experimental samples.
Source: Ref. 43.

arsenic content of 1.5 ppm. The growth of the bacterium was enhanced in the presence of arsenic compounds in concentrations up to 200 ppm in the cultural medium with a stronger effect of As(V) than of As(III) compounds. Trimethylarsine (Me_3As), a product of the methylation of both NaH_2AsO_3 and NaH_2AsO_4, was formed and detected by mass spectrometry (Table 2). The intracellular accumulation of arsenic in the methylating strain was compared with two nonmethylating strains from the same soil. Results showed that the *Flavobacterium-Cytophaga* sp. accumulated significantly less arsenic due to its biomethylation ability than the strains of nonmethylating bacteria.

3. Volatilization Mediated by Fungi

The involvement of soil fungi in the volatilization of arsenicals was first reported by Challenger (27). It is now well established that fungi are capable of transforming inorganic and organic arsenic compounds into volatile methylarsines (4). The importance of fungal metabolism of arsenic dates back to the early 1800s when a number of poisoning incidents in Germany and England were caused by trimethylarsine gas (27). The victims lived in musty rooms with a characteristic garlic-like odor. Molds growing on wallpaper decorated with arsenical pigments (Scheel's green and Schweinfürter green) produced the toxic trimethylarsine gas. Since then, several species of fungi have been identified that are able to volatilize arsenic (5). The volatilized arsenic dissipates from the cells, reducing arsenic exposure to the fungus. Three different fungal species, *Candida humicola, Gliocladium roseum*, and *Penicillium* sp. were shown to be capable of converting

MMAA and DMAA to TMA (14). In addition, *C. humicola* used arsenate and arsenite as substrates to produce TMA. Cell-free homogenates of *C. humicola* transformed arsenate into arsenite, MMAA, and DMAA (12). *Candida humicola* is capable of methylating benzenearsonic acid to produce volatile dimethylphenylarsine (44). In addition, methylphenylarsinic acid and dimethylphenylarsine oxide are also reduced by *C. humicola* to dimethylphenylarsine. The adaptiveness of *C. humicola* in methylating arsenic is evident by the fact that dilute solutions of the highly effective wood preserving fungicide, chromated copper arsenate (CCA), are depleted of arsenic through volatilization (45).

Huysmans and Frankenberger (46) isolated a *Penicillium* sp. from agricultural evaporation pond water capable of producing TMA from MMAA and DMAA. Pearce et al. (47) investigated the ability of fungi to produce volatile arsenic compounds in pure cultures using two strains each of *Scopulariopsis brevicaulis* (an inhabitant of PVC cot mattress covers), and *Phaeolus schweinitzii* (a wood decay fungus). Volatile arsenic compounds were detected from all cultures grown on arsenic-supplemented media. Arsenic concentrations in test papers exposed to the volatile products of fungi growing on medium containing As(III) were between 1 and 2 orders of magnitude above the assay background. More arsenic was trapped from cultures containing the higher concentrations of As(III) and both strains of *P. schweinitzii* volatilized more arsenic than either of the strains of *S. brevicaulis*. Since alkylated derivatives have been detected in studies of biological volatilization of arsenic (42), it is likely that the volatile arsenic trapped in this study was TMA (47).

B. Algae

Arsenic is metabolized into various methylated forms by freshwater algae. Arsenite is methylated by at least four freshwater species of green algae, including *Ankistrodesmus* sp., *Chlorella* sp., *Selenastrum* sp., and *Scenedesmus* sp. (48). All four species methylated arsenite when present in media at 5000 µg L^{-1}, approximately the same level of As(III) used to control aquatic plants in lakes (49). The levels of recovered methylated arsenic species were quite high on a per g dry weight basis. Each of these organisms transformed As(III) to MMAA and DMAA and all, except *Scenedesmus*, produced detectable levels of TMAO. Unlike fungi, volatile methylarsines were not produced (48), but instead, limnetic (freshwater) algae like marine algae synthesize lipid-soluble arsenic compounds.

IV. FACTORS AFFECTING ARSENIC VOLATILIZATION

Although microorganisms are known to be involved in arsenic volatilization, many environmental factors, such as pH, redox potential, presence of other elements, and organic matter content, can influence the abundance of different arsenic forms in the environment (31). Because microorganisms play a key role in controlling the speciation and cycling of arsenic in soil, it is important to have a better understanding of the major factors that influence microbial activity to the biogeochemistry of arsenic (42).

A. Arsenic Speciation

Studies on arsine evolution from soil have shown that volatilization is affected by the soil environment and by arsenic speciation (22,31,50). When environmental conditions change, the speciation and mobility of arsenic may also change. Evolution of arsines is much higher from organic arsenicals than from inorganic arsenicals (31,32). Woolson (31) found that in soil treated with 10 mg As kg^{-1} as [^{74}As]sodium arsenate, [^{14}C]methanearsonate, and [^{14}C]cacodylic acid and continuously flushed the closed system with air, 18.0, 12.5, and 1.0% of the As was volatilized from CA, methanearsonate, and arsenate, respectively, in 16 days. When flushed with N_2, 7.8, 0.8, and 1.8% were lost. The main volatile forms were identified as DMA and TMA. When arsenicals were applied to the surface of an established lawn in Florida, the most rapid evolution was from CA, a slightly slower rate was found from MMAA, but a very much slower rate was observed with arsenite (32). Arsine evolution rate from soil followed the order: CA > MMAA > sodium arsenite [As(III)] = sodium arsenate [As(V)]. The arsenic volatilization rate increased linearly with CA concentration in the range of 0–100 mg As kg^{-1} soil (19).

B. Environmental Factors

In the presence of vegetation and increasing soil organic matter, microbial activity is enhanced, which may have a high impact on the cycling of elements. A study of gaseous evolution of arsenic from soil treated with ^{74}As-labeled DSMA at 100 mg kg^{-1} showed that the greatest losses of arsenic (11%) was from a soil that contained 11% organic matter and was maintained under reduced (wet) soil conditions for 60 days (50). Cellulose addition enhanced arsine evolution from a silty clay soil (19). Huysmans and Frankenberger (46) found that evolution of TMA from MMAA by a *Penicillium* sp. was suppressed by the addition of carbohydrates and sugar acids to the minimal medium, while addition of amino acids (phenylalanine, isoleucine, and glutamine) promoted TMA production with an enhancement ranging from 10.2- to 11.6-fold over the control without amino acid enhancement. Thomas and Rhue (51) reported that glucose addition had no effect on arsenic volatilization under low oxic conditions, whereas arsenic volatilization was stimulated by glucose addition under high oxic conditions.

Soil temperature affects microbiological activity in soils, and therefore can modify biotransformations of arsenic in soils (46). We found that the *Penicillium* sp. mediated transformation of MMAA into TMA was optimum at 20°C. Similarly, more arsines were produced in a silty clay loam at 25°C than at 5°C (19). Methylation of arsenic is pH dependent with the highest rates occurring at pH 3.5–5.5, suggesting that arsenic mobilization from the sediments to the overlying water phase is enhanced by acidification (35).

Redox conditions may affect biological transformations of arsenic in soils (30,52). Chiu et al. (53) reported that lowering of soil redox potential increased the ratio of As(III) and promoted arsenic methylation. Methylation of arsenic compounds by yeast and bacteria under oxic conditions plays a significant role, whereas methanogenic bacteria are important under anoxic conditions in releasing volatile arsenic from the soil to the atmosphere (39,41). Woolson and Kearney (22) studied the loss of CA from soils under aerobic and anaerobic conditions. Over 24 weeks, about 35% was evolved as a volatile organoarsenical compound under aerobic incubation, but 61% was evolved under anaerobic conditions, possibly as DMA. A *Fusarium* sp. that was isolated from the soil was placed in arsenic media under high and low oxic conditions (51). It was observed that low levels of oxygen stimulated the release of volatile arsenic species (Fig. 3). Loss of arsenic from waterlogged reduced soils has long been attributed to volatilization of As arsines (6,50,54).

Hassler et al. (36) found that arsenic volatilization in retorted oil shales would occur only when a nutrient source was provided and that total soluble arsenic levels decreased with time if no nutrient source were made available.

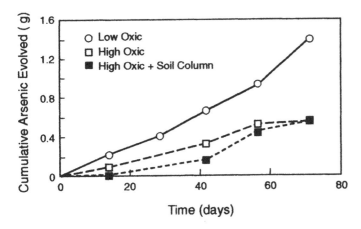

Figure 3   Volatilization of arsenic by *Fusarium* sp. under various oxic conditions with and without a soil column scrubber. (From Ref. 51.)

However, Onken and Adriano (55) demonstrated that arsenic added as sodium arsenate and sodium arsenite became less available in soil with time, rendering no loss via volatilization up to 68 days. The conversion (curing) of arsenic to more insoluble forms occurred in saturated soils where the pe/pH status was in the arsenite stability field and in subsaturated soils where the pe/pH status was in the arsenate form. The rate at which arsenic was converted to more insoluble (and possibly less bioavailable) forms was as rapid in saturated soils as in unsaturated soils.

The presence of heavy metals and other elements may inhibit or enhance microbiological transformations of arsenic in soil systems. It was observed that presence of phosphate and selenate causes inhibition of methylated evolution of arsenic (5,46,56). Frankenberger (57) studied the effect of 21 trace elements for their activation or inhibition on methylated arsine production by a *Penicillium* sp. from MMAA. Metals and metalloids at an elemental concentration of 0, 0.1, 1, 10, 100, and 1000 µM were tested for their influence on arsenic volatilization by the *Penicillium* sp. The effect of trace elements varied considerably depending on the speciation and concentration. At the lower elemental concentrations (0.1 and 1 µM), the metals and metalloids that stimulated arsenic volatilization were

Table 3 Effect of Trace Elements on Arsenic Volatilization (ng ml⁻¹)

Trace element	Oxidation state	Arsenic volatilized at different concentrations (µM) of added trace elements					
		0	0.1	1.0	10	100	1000
Ag	I	168	158	160	5.1	1.3	<1.0
Ba	II	168	153	145	123	138	<1.0
Cu		165	174	302	41	7.1	<1.0
Hg		168	315	640	<1.0	<1.0	<10
Mn		167	163	146	115	88	<1.0
Ni		166	155	90	76	<1.0	<1.0
Sn		165	125	101	81	68	<1.0
Zn		168	206	230	168	91	<1.0
Al	III	167	237	217	256	253	<1.0
B		164	164	110	138	29	<1.0
Fe		169	248	423	418	390	<1.0
Te	IV	167	170	149	48	<1.0	<1.0
V		165	72	68	77	81	<1.0
Cr	VI	168	161	153	84	55	<1.0
Mo		166	157	148	113	95	<1.0
Te		166	131	138	133	57	<1.0

Source: Ref. 57.

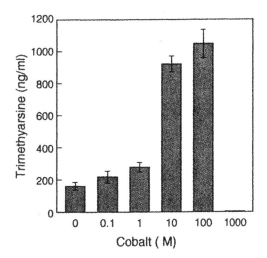

Figure 4 Influence of cobalt added to a culture medium in the presence of *Penicillium* sp. on arsenic volatilization. Standard error of mean was calculated based on five replicates incubated for 7 days (From Ref. 57.)

ranked as follows: Hg stimulated a 3.80-fold enhancement followed by Fe (2.50-fold), Cu (1.83-fold), Co (1.70-fold), Se(VI) (1.51-fold), Al (1.42-fold), Zn (1.37-fold), Se(IV) (1.27-fold), and As(III) (1.19-fold) (Table 3). The elements that inhibited arsenic volatilization were ranked in order of the most inhibitory element: V (59%), followed by Ni (46%). The stimulatory effect of Co^{2+} on TMA formation is shown in Figure 4 (57).

Gao and Burau (19) systematically examined the effects of soil moisture on arsenic transformations in samples of a Sacramento silty clay soil. The optimum soil moisture level for arsine evolution was between 250 and 350 g H_2O kg^{-1} soil (-0.3–0.03 MPa).

V. VOLATILIZATION OF ARSENIC AS A BIOREMEDIATION STRATEGY

Both oxidation and methylation are microbial transformations involved in the redistribution and global cycling of arsenic. Oxidation involves the conversion of toxic arsenite to less toxic arsenate. Bacterial methylation of inorganic arsenic under anaerobic conditions may be a mechanism of arsenic detoxification. Fungi also transform inorganic and organic arsenic compounds into volatile methylarsines. However, unlike methylated selenium which is nontoxic, the volatile arsine

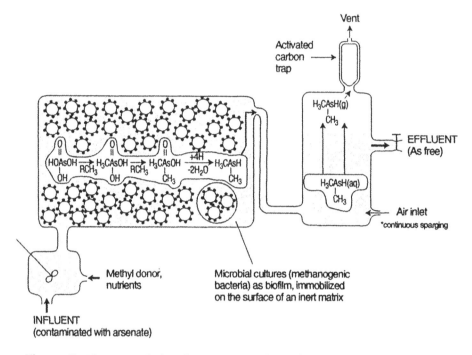

Figure 5 Bioreactor design for treatment of arsenic-contaminated water. (From Ref. 59.)

gases (mono-, di-, and trimethylarsine) are relatively more toxic and need to be disposed of properly. Under certain conditions where trapping of the gaseous methylated arsenic is manageable, enhanced volatilization of methylated arsenic could be used to clean up contaminated matrices (Fig. 5).

IV. CONCLUDING REMARKS

It is well established that microbiota plays a very significant role in the various transformations of arsenic, including mineralization/immobilization, oxidation/reduction, and methylation/demethylation. Some of these biotransformations lead to less toxic forms of arsenic that can be used in the detoxification of the arsenic-contaminated environments. Biomethylation of arsenic results in formation of mono-, di-, and trimethylarsines which are volatile; however, these gaseous arsenic forms are also toxic. In developing a bioremediation strategy to clean the environment of arsenic, special attention must be paid to the toxic nature of the microbially transformed arsenic forms.

REFERENCES

1. JP Buchet, R Lauwerys. Evaluation of exposure to inorganic arsenic in man. Analytical techniques for heavy metals in biological fluids. Amsterdam: Elsevier, 1983, pp 75–89.
2. RG Burau. Kinetics of transformations of arsenicals in soils under oxidative conditions. Final Report to Western Region Pesticide Impact Assessment Program, Department of Land, Air and Water Resources, University of California, Davis, CA, 1981.
3. FT Mackenzie, RJ Lantzy, V Paterson. Global trace metal cycles and predictions. Math Geol 11:99–142, 1979.
4. S Tamaki, and WT Frankenberger Jr. Environmental biochemistry of arsenic. Rev Environ Contam Tox 124:79–110, 1992.
5. DP Cox. Microbiological methylation of arsenic. In: EA Woolson, ed. Arsenical Pesticides. ACS Symp. Ser. 7. Washington, DC: American Chemical Society, 1975, pp 81–96.
6. EA Epps, MB Sturgis. Arsenic compounds toxic to rice. Soil Sci Soc Am Proc 4: 215–218, 1939.
7. M Fleischer. Cycling and control of metals. Proceedings of Environmental Resources Conference, National Environmental Resources Center, Cincinnati, 1973.
8. JM Wood. Biological cycles for toxic elements in the environment. Science 183: 1049–1052, 1974.
9. FH Osborne, HL Ehrlich. Oxidation of arsenite by a soil isolate of *Alcaligenes*. J Appl Bacteriol 41:295–305, 1976.
10. SE Phillips, ML Taylor. Oxidation of arsenite to arsenate by *Alcaligenes faecalis*. Appl Environ Microbiol 32:392–399, 1976.
11. D Ahmann, AL Roberts, LR Krumholz, and FMM Morel. Microbe grows by reducing arsenic. Nature 371:750, 1994.
12. WR Cullen, BC McBride, and AW Pickett. The transformation of arsenicals by *Candida humicola*. Can J Microbiol 25:1201–1205, 1979.
13. M Shariatpanahi, AC Anderson, AA Abdelghani, AJ Englande. Microbial metabolism of an organic arsenical herbicide. In: TA Oxley and S Barry, eds. Biodeterioration. 5. New York: Wiley, 1983, pp 268–277.
14. DP Cox, M Alexander. Production of trimethylarsine gas from various arsenic compounds by three sewage fungi. Bull Environ Contam Toxicol 9:84–88, 1973.
15. RW Cullen, BC McBride, and M Reimer. Induction of the aerobic methylation of arsenic by *Candida humicola*. Bull Environ Contam Toxicol 21:157–161, 1979.
16. M Shariatpanahi, AC Anderson, AA Abdelghani, AJ Englande, J Hughes, RF Wilkinson. Biotransformation of the pesticide sodium arsenate. J Environ Sci Health Part B. 16:35–47, 1981.

17. M Sadiq. Arsenic chemistry in soils: An overview of thermodynamic predictions and field observations. Water Air Soil Pollut 93:117–136, 1997.
18. Y Sohrin, M Masakazu, M Kawashima, M Hojo, H Hasegawa. Arsenic biogeochemistry affected by eutrophication in Lake Biwa, Japan. Environ Sci Technol 31:2712–2720, 1997.
19. S Gao, and RG Burau. Environmental factors affecting rates of arsine evolution from and mineralization of arsenicals in soil. J Environ Qual 26:753–763, 1997.
20. K Hanaoka, S Hasegawa, N Kawabe, S Tagawa, and T Kaise. Aerobic and anaerobic degradation of several arsenicals by sedimentary microorganisms. Appl Organomet Chem 4:239–243, 1990.
21. DW Von Endt, Kearney, Kaufman. Degradation of monosodium methanearsonic acid by soil microorganisms. J Agric Food Chem 16:17–20, 1968.
22. EA Woolson, PC Kearney. Persistence and reactions of [^{14}C] cacodylic acid in soils. Environ Sci Technol 7:47–50, 1973.
23. R Dickens, AE Hiltbold. Movement and persistence of methanearsonate in soil. Weeds 15:299–304, 1967.
24. KH Akkari, RE Frans, TL Lavy. Factors affecting degradation of MSMA in soil. Weed Sci 34:781–787, 1986.
25. MO Andreae. Arsenic speciation in seawater and interstitial waters: The influence of biological-chemical interactions on the chemistry of a trace element. Limnol Oceanogr 24:440–452, 1979.
26. No Wakao, H Koyatsu, Y Komai, H Shimokawara, Y Sakurai, H Shiota. Microbial oxidation of arsenite and occurrence of arsenite-oxidizing bacteria in acid mine water from a sulfur-pyrite mine. Geomicrobiology J 6:11–24, 1988.
27. F Challenger. Biological methylation. Chem Rev 36:315–361, 1945.
28. M Coyne. Introduction to Soil Microbiology. New York: Delmar, 1999.
29. CN Cheng, DD Focht. Production of arsine and methylarsines in soil and in culture. Appl Environ Microbiol 38:494–498, 1979.
30. AW Turner. Bacterial oxidation of arsenite. Nature 164:76–77, 1949.
31. EA Woolson. Generation of alkylarsines from soil. Weed Sci 25:412–416, 1979.
32. RS Braman. Arsenic in the environment. In: EA Woolson, ed. Arsenical Pesticides. ACS Symp. Ser. 7. Washington, DC: American Chemical Society, 1975, pp 108–123.
33. R Turpeinen, M Pantsar-Kallio, M Haggblom, T Kairesalo. Influence of microbes on the mobilization, toxicity and biomethylation of arsenic in soil. Sci Total Environ 236:173–180, 1999.
34. G Sandberg, IK Allen. A proposed arsenic cycle in an agronomic ecosystem. In:EA Woolson, ed. Arsenical Pesticides. ACS Symp. Ser. 7. Washington, DC: American Chemical Society, 1975, pp 124–147.
35. MD Baker, WE Inniss, CI Mayfield, PTS Wong, YK Chau. Effect of pH on the methylation of mercury and arsenic by sediment microorganisms. Environ Technol Lett 4:89–100, 1983.
36. RA Hassler, DA Klein, RR Meglen. Microbial contributions to soluble and volatile arsenic dynamics in retorted oil shale. J Environ Qual 13:466–470, 1984.
37. RS Braman, CC Foreback. Methylated forms of arsenic in the environment. Science 182:1247–1249, 1973.

38. BC McBride, RS Wolfe. Biosynthesis of dimethylarsine by *Methanobacterium*. Biochem 10:4312–4317, 1971.
39. RW Boyle, IR Jonasson. The geochemistry of arsenic and its use as an indicator element in geochemical prospecting. J Geochem Explor 2:251–296, 1973.
40. M Mandl, P Matulova, H Docekalova. Migration of arsenic (III) during bacterial oxidation of arsenopyrite in chalcopyrite concentrate by *Thiobacillus ferrooxidans*. Appl Microbiol Biotechnol 38:429–431, 1992.
41. ED Weinberg. Microorganisms and Minerals. New York: Marcel Dekker, 1977, p 492.
42. WR Cullen, KJ Reimer. Arsenic speciation in the environment. Chem Rev 89:713–764, 1989.
43. BC McBride, H Merilees, WR Cullen, W Pickett. Anaerobic and aerobic alkylation of arsenic. In: FE Brickman and JM Bellama, eds. Organometals and Organometalloids Occurrence and Fate in the Environment. AS Symp. Ser. 82. Washington, DC: American Chemical Society, 1978, pp 94–115.
43. S Honschopp, N Brunken, A Nehrkorn, HJ Breunig. Isolation and characterization of a new arsenic methylating bacterium from soil. Microbiol Res 151:37–41, 1996.
44. WR Cullen, AE Erdman, BC McBride, AW Pickett. The identification of dimethylphenylarsine as a microbial metabolite using a simple method of chemofocussing. J Microbiol Method 1:297–303, 1983.
45. WR Cullen, BC McBride, AW Pickett, J Regalinski. The wood preservative chromated copper arsenate is a substrate for trimethylarsine biosynthesis. Appl Environ Microbiol 47:443–444, 1984.
46. KD Huysmans, WT Frankenberger Jr. Evolution of trimethylarsine by a *Penicillium* sp. isolated from agricultural evaporation pond water. Sci Total Environ 105:13–28, 1991.
47. RB Pearce, ME Callow, LE Macaskie. Fungal volatilization of arsenic and antimony and the sudden infant death syndrome. FEMS Microbiol Lett 158:261–265, 1998.
48. MD Baker, PTS Wong, YK Chau, CI Mayfield, WE Inniss. Methylation of arsenic by freshwater green algae. Can J Fish Aquat Sci 40:1254–1257, 1983.
49. RD Hood and Associates. Cacodylic acid: Agricultural uses, biological effects, and environmental fate. Washington, DC: U.S. Government Printing Office, 1985, p 164.
50. MB Akins, RJ Lewis. Chemical distribution and gaseous evolution of arsenic-74 added to soils as DSMA-74 arsenic. Soil Sci Soc Am J 40:655–658, 1976.
51. JE Thomas, RD Rhue. Volatilization of arsenic in contaminated cattle dipping vat soil. Bull Environ Contam Toxicol 59:882–887, 1997.
52. LE Devel, AR Swoboda. Arsenic solubility in a reduced environment. Soil Sci Soc Am Proc 36:276–278, 1972.
53. C-H Chiu, G-C Li, C-C Young, R-Y Chiou. A study on the transformation of arsenite in soils after poultry compost and poultry manure application. J Chinese Agric Chem Soc 36:380–393, 1998.
54. SL McGeehan, DV Naylor. Sorption and redox transformation of arsenite and arsenate in two flooded soils. Soil Sci Soc Am J 58:337–342, 1994.
55. BM Onken, DC Adriano. Arsenic availability in soil with time under saturated and subsaturated conditions. Soil Sci Soc Am J 61:746–752, 1997.

56. DP Cox, M Alexander. Effect of phosphate and other anions on trimethylarsine formation by *Candida humicola*. Appl Microbiol 25:408–413, 1973.
57. WT Frankenberger Jr. Effects of trace elements on arsenic volatilization. Soil Biol Biochem 30:269–274, 1998.
58. YK Chau, PTS Wong. Occurrence of biological methylation of elements in the environment. ACS Symp. Ser. 82. Washington, DC: American Chemical Society, 1978, pp 39–53.
59. WT Frankenberger Jr, ME Losi. Applications of bioremediation in the cleanup of heavy metals and metalloids. In: Bioremediation: Science and Applications. Soil Sci Soc Am Special Pub 43, 1995, pp 173–210.

Index

T - #0030 - 111024 - C0 - 229/152/23 - PB - 9780367396565 - Gloss Lamination